Studies in Economic Theory

Editors

Charalambos D. Aliprantis
Purdue University
Department of Economics
West Lafayette, IN 47907-1310
USA

Nicholas C. Yannelis
Department of Economics
University of Illinois
Champaign, IL 61820
USA

Springer-Verlag Berlin Heidelberg GmbH

Titles in the Series

M. Ali Khan and Nicholas C. Yannelis (Eds.)
Equilibrium Theory in Infinite Dimensional Spaces

Charalambos D. Aliprantis, Kim C. Border and
Wilhelmus A. J. Luxemburg (Eds.)
Positive Operators, Riesz Spaces, and Economics

Donald G. Saari
Geometry of Voting

Charalambos D. Aliprantis and Kim C. Border
Infinite Dimensional Analysis

Jean-Pierre Aubin
Dynamic Economic Theory

Mordecai Kurz (Ed.)
Endogenous Economic Fluctuations

Jean-François Laslier
Tournament Solutions and Majority Voting

Ahmet Alkan, Charalambos D. Aliprantis and
Nicholas C. Yannelis (Eds.)
Theory and Applications

James C. Moore
Mathematical Methods for Economic Theory 1

James C. Moore
Mathematical Methods for Economic Theory 2

Mukul Majumdar, Tapan Mitra and Kazuo Nishimura
Optimization and Chaos

Katri K. Sieberg
Criminal Dilemmas

Monique Florenzano and Cuong Le Van
Finite Dimensional Convexity and Optimization

Karl Vind
Independence, Additivity, Uncertainty

Timothy Cason · Charles Noussair (Eds.)

Advances in Experimental Markets

Springer

Prof. Timothy Cason
Associate Prof. Charles Noussair
Purdue University
Department of Economics
1310 Krannert School of Management
West Lafayette, IN 47907-1310
USA

First published in: Economic Theory Vol. 16 Nr. 3 (2000)

ISBN 978-3-642-62657-9

Library of Congress Cataloging-in-Publication Data
Die Deutsche Bibliothek – CIP-Einheitsaufnahme

ISBN 978-3-642-62657-9 ISBN 978-3-642-56448-2 (eBook)
DOI 10.1007/978-3-642-56448-2

http://www.springer.de

Originally published by Springer-Verlag Berlin Heidelberg New York in 2001
Softcover reprint of the hardcover 1st edition 2001

Hardcover Design: Erich Kirchner, Heidelberg

SPIN 10839257 42/2202-5 4 3 2 1 0 – Printed on acid-free paper

Contents

The experimental study of market behavior

Timothy Cason and Charles Noussair

Department of Economics, Krannert School of Management, Purdue University, West Lafayette, IN 47907-1310, USA (e-mail cason@mgmt.purdue.edu and noussair@mgmt.purdue.edu).

Over the past two decades, experimental methods have entered the mainstream as one of the empirical methodologies of economic science. Early experimental work focused on issues in individual decision-making and industrial organization, but in recent years economists have applied laboratory methods to study topics in macroeconomics, international economics, information economics, finance, and other fields. Applied economists have also conducted experiments to gather data expressly for use in policy debates. The broadening of the domain of experimental economics has created several clearly identifiable branches of the field. The focus of this volume is one of these branches, the study of the behavior of markets. The papers in this volume represent several current directions of research on experimental markets.

The study of experimental markets is a topic unique to economics, distinguishing it from more fundamental questions of rationality and strategic interaction that have been studied experimentally by psychologists and political scientists as well as economists. For excellent surveys of research on experimental markets, see Holt (1995), Plott (1989), Sunder (1995), and Davis and Holt (1993, chapters 1, 3, 4 and 7). It is one of the oldest strands of experimental economics, dating to Chamberlain (1948) and Smith (1962).[1] These initial experiments focused on the market-clearing assumption that lies at the core of the fundamental model of microeconomics, the theory of supply and demand. They explored the conditions under which market prices would stabilize and the price and quantity exchanged would be at the level predicted by competitive theory. In Chamberlain's experiment, trading was decentralized, and the results were not

We would like to thank Roko Aliprantis for his advice and encouragement, and the dozens of anonymous referees who assisted us in reviewing manuscripts.

[1] Roth (1995a) describes the early history of other important strands of experimental research that have origins in psychologists' individual choice experiments, and also discusses early tests of game-theoretic solution concepts.

supportive of competitive theory. Smith's (1962) study identified market rules, called continuous double auction rules, under which observed prices and quantities traded corresponded to the competitive equilibria of single market economies. This finding was important because it showed that the optimality properties of competitive equilibria could be achieved with appropriate market rules even with a small number of traders, and it illustrated the great influence that the trading institution could have on outcomes.

Continuous double auction markets

A continuous double auction market operates using the following rules: At any time, any potential buyer or seller can submit an order specifying an offer price and a maximum quantity he is offering to buy or sell at that price. The order is then publicly displayed. At any time, any agent can accept an offer submitted by an agent on the other side of the market, and upon acceptance a trade is concluded at the offered price. Using robot traders that employ very simple decision rules, Gode and Sunder (1993) demonstrated that continuous double auction trading rules allow the market to realize a very high percentage of the gains from trade even with minimal rationality on the part of human traders. Essentially, most of the gains from trade are realized in a single market even if traders randomly generate offers with only the constraint that they do not trade at a loss.

There has been far less work, however, on the operation of multiple simultaneous double auction markets in which different goods are traded in the different markets. A few papers have explored the behavior of input and output markets that operate simultaneously (e.g., Goodfellow and Plott, 1990, and Noussair et al., 1995). In the case of the latter work, multiple outputs trade simultaneously in different markets. These markets are linked by a production technology that transforms the good that trades in one market to a good that trades in the other. These studies show convergence to the market equilibrium, but convergence is slower than when a single market is operating. The structure of these multiple-market economies is considerably simplified by the fact that the market demand for output is separable, in the sense that the marginal utility of each unit of a good consumed is independent of the level of consumption of other goods.

The first paper in this volume, by Williams, Smith, Ledyard and Gjerstad, studies the behavior of two simultaneous double auction markets, in which the demand for the two goods trading in the market is interdependent. Buyers' demands for the two commodities arise through constant elasticity of substitution earnings functions. Despite the complexity introduced by the demand interdependence, the competitive equilibrium model yields a rather accurate description of the market outcomes. As in earlier research with single commodities, strategic behavior in these double auction markets does not prevent the markets from clearing at near the most efficient allocation. The authors thus demonstrate that the market clearing result for continuous double auction markets extends to the multiple-good case with demand interdependencies.

Posted offer markets

An alternative market institution to the continuous double auction is the posted offer market. Posted offer rules are of interest because they are simple and in common use, particularly in modern retail markets. In a posted offer market, each seller submits a price-quantity pair to the market in each period. This offer indicates the price at which they are willing to sell, and the maximum quantity they are offering to sell at that price. Buyers then purchase from the seller(s) of their choice, presumably from the sellers who offer the lowest prices. Sauermann and Selten (1959) and Fouraker and Siegel (1963) were the first to study this type of market. Their work focused on developing experimental methods to test oligopoly models with small numbers of traders. Fouraker and Siegel studied games in which two or three participants simultaneously made price decisions, in both incomplete and complete information conditions. In the incomplete information condition subjects observed only their own profit, while in the complete information condition they also observed their rivals' profits. Outcomes in most incomplete information sessions corresponded to the (competitive) Bertrand equilibrium, but were much more variable in the complete information treatment.[2]

Although these early experiments demonstrated that competitive outcomes could arise when sellers post prices, one of the important early discoveries in research on experimental markets was that outcomes, particularly during the convergence process, differed between the posted offer and double auction trading institutions. In double auction markets, prices during the convergence process tend to favor the side earning smaller rents in the competitive equilibrium (Smith and Williams, 1982). For example, if equilibrium consumer surplus exceeds producer surplus, prices tend to converge from above. The opposite is true if producer surplus is greater than consumer surplus. In posted offer markets, however, prices tend to converge from above regardless of the relative equilibrium consumer and producer surplus, although this institutional effect appears to interact with other design conditions (Ketcham, Smith and Williams, 1984; Davis and Williams, 1986).

In posted offer markets prices usually converge to competitive equilibria with a small number of agents on each side of the market (typically four buyers and four sellers are sufficient to ensure competitive outcomes). But if there are very few traders in the market, if traders are given the opportunity to communicate, or if demand and supply have a structure that gives individual traders the ability to influence market prices, prices can deviate from the competitive level. These deviations can typically be traced to two root causes (1) strategic behavior, consisting of underrevelation of demand or supply with the intent of shifting prices in one's favor, and (2) other-regarding ("fairness") preferences. Strategic behavior seems more pronounced in posted offer than in double auction markets, since prices are almost always more competitive in the double auction.

[2] Fouraker and Siegel (1963) also conducted quantity-setting oligopoly experiments, and in their incomplete information sessions outcomes were closer to the Cournot equilibrium than to collusive or competitive levels.

For example, it is possible to construct market environments with a small number of traders in which (large capacity) individual traders have market power. Market power here refers to the ability to profitably and unilaterally shift prices away from the competitive equilibrium. Holt, Langan and Villamil (1986) studied a double auction market in an environment with seller market power, and they observed supercompetitive prices in about half of their sessions. Davis and Williams (1991) reported experiments using this same market power environment but with trade organized through posted offer rules. Posted offer sellers were significantly more successful in raising prices above the competitive level than double auction sellers. The opportunity for strategic behavior thus had a larger impact on outcomes in the posted offer institution.

When sellers have an opportunity to communicate and conspire to fix prices above the competitive level, they are also more successful in manipulating outcomes in their favor in the posted offer than in the double auction institution. Isaac and Plott (1981) allowed sellers to discuss anything they wished between double auction trading periods, except that they could not mention side payments, threats, or quantitative details of their payoff schedules and costs. In all experiments the participants clearly recognized their potential gains from cooperation and reached specific price fixing agreements. Nevertheless, they were typically able to raise transaction prices above the competitive level by only a very modest amount. Isaac, Ramey and Williams (1984) compared sellers' ability to raise prices through explicit conspiracies in double auction and posted offer markets. They found that posted offer conspirators were able to charge higher transaction prices relative to double auction conspirators. The discrete timing and enforced commitment to a single posted price per period for each seller seems to help them implement agreements. The double auction, in contrast, allows a continuous temptation to defect, since any seller can post a new price or accept a buyer offer at any time.

The next two papers in this volume also consider the extent and impact of strategic behavior in posted offer markets. Most previous research using posted offer trading has studied only seller behavior, with buyer behavior controlled using demand-revealing "robots." This has permitted increased experimenter control because it eliminates strategic uncertainty on the buyers' side of the market, and the available evidence indicates that it provides a harmless simplification in environments with many buyers. The paper by Ruffle indicates, however, that significant buyer demand withholding does arise in posted offer markets, particularly when the division of surplus in equilibrium strongly favors sellers or when there are few buyers. This strategic buyer behavior exerts a downward influence on transaction prices.

The paper by Davis and Wilson studies strategic behavior in posted offer markets on the part of sellers rather than buyers. The focus is on sellers' ability to exercise market power following the introduction of production cost savings arising from a reallocation of capacity across sellers. The authors show that cost savings on marginal units can reduce market power and generate more competitive outcomes, while cost savings on inframarginal units can exacerbate market power problems and lead to supercompetitive prices. Their study is particularly important from a policy perspective, as it illustrates for antitrust authorities the interrelationships between market power and firm-specific cost savings. In particular, it shows that the ability and incentive to influence prices can depend on quite subtle differences in individuals' cost functions.

Buyers and sellers also sometimes exercise strategic behavior in order to more "fairly" distribute the exchange surplus among traders. For example, Smith (1981) observed buyers resisting high price offers made by monopoly sellers in a double auction market. Monopoly sellers were more successful in extracting monopoly rents in an otherwise identical market that was conducted under posted offer rules. One possible reason for this difference between institutions is the possibility of price discrimination in the double auction. After the monopolist has sold to high valuation buyers, he has an incentive to sell at lower prices to marginal buyers. However, if more than one period is played, all buyers become aware of the price marginal buyers are paying, and hold out for that price. Over time, the seller's monopoly power erodes. By contrast, the posted offer market requires the monopolist to precommit to a single price for the entire market period. With a single price, the relationship between the price charged and the resulting profit becomes clear after a few periods, and the seller is often able to identify and charge the monopoly price. Moreover, buyers often do not observe low prices, and they cannot withhold demand in anticipation of price concessions within a trading period.

Smith (1976) reports an experiment with an extremely inequitable division of gains from exchange in equilibrium, which dramatically illustrates the potential role of information regarding "unfair" payoffs on market outcomes. In this environment, all sellers have the same cost value, and all buyers have the same resale value. For example, four sellers have a total of 16 units of capacity, with each unit costing \$ 3.10, while four buyers have a total of 11 units they can redeem, with each resale value equal to \$ 4.20. Since there is an excess supply of 5 units at every price between \$ 3.10 and \$ 4.20, and no trader has an individual capacity greater than 4 units, the competitive price is at \$ 3.10. All exchange surplus at this equilibrium price is realized by buyers, however. When all values and costs are private information, Holt, Langan and Villamil (1986) and Smith and Williams (1990) find that the prices converge to the competitive equilibrium under double auction trading rules. Cason and Williams (1990) obtain a similar result under posted offer trading rules, although convergence to equilibrium is slower and the equilibrium is less stable.

Smith (1976) reports results from a double auction market experiment under similar supply and demand conditions, but with complete payoff information. The instructions publicly disclose the seller costs and the buyer resale values. Prices in this complete information treatment failed to reach the competitive equilibrium level of \$ 3.10, although they were closer to \$ 3.10 than \$ 4.20. Complete payoff information apparently allows traders to identify and resist unfair distributions of exchange surplus (also see Smith, 1994, for a discussion of this issue).

The paper by Abrams, Sefton and Yavas in this volume demonstrates that traders also resist equilibria with highly asymmetric divisions of surplus that arise due to incomplete market information. They modify the posted offer institution by introducing buyer search costs, in a posted offer market with complete payoff information. Buyer values are uniform and constant, as are seller costs. Buyers must incur monetary costs to search. When buyers observe only one seller's price per search, the search costs lead to a unique theoretical equilibrium with all transactions at the buyer valuations. This is sometimes called the "Diamond paradox" (Diamond, 1971). But when buyers observe two sellers' prices per search, in equilibrium sellers compete aggressively as in a standard Bertrand model and price at cost. Because of the uniform and constant values and costs, these two theoretical equilibria correspond to conditions with the same maximum asymmetry in exchange surplus described above.

Traders, however, resist these extreme-surplus equilibria. Prices are higher in the "Diamond" treatment than in the "Bertrand" treatment as predicted, but only rarely does one side of the market obtain more than 90 percent of the surplus. The impact of search costs in the Diamond treatment is much smaller than predicted, with mean prices rarely rising above 75 percent of the equilibrium level. The prices observed in Abrams, Sefton and Yavas' experiment suggest that the preferences for fairness observed in experiments with complete payoff information can exert an influence on market prices even with large numbers of traders. Consequently, the data support only the equilibrium comparative static, but not the point, predictions. Cason and Friedman (1999) obtain a similar result in their Diamond and Bertrand treatments with human buyers, but when sellers know that buyers are computerized "robots" the prices correspond closely to the "unfair" Diamond/Monopoly price equilibrium.

In contrast to the complete payoff information condition studied by Abrams et al., the paper by Rustichini and Villamil in this volume follows the more standard procedure in market experiments of keeping valuation information private. But unlike most earlier experiments, which typically feature stationary repetition of the same buyer values and seller costs for a sequence of trading periods, Rustichini and Villamil introduce period-to-period variation in buyer values. In addition to differential information, their environment features repeated (intertemporal) pricing with a random termination date and a Markov Process generating substantial serial correlation in the buyer's private value. The authors compare the results of 32 sessions with the equilibria in stationary Markov strategies derived in Rustichini and Villamil (1996). They find that the posted offer institution allows sellers to acquire information about the buyer values, although substantial inefficiencies persist relative to a full information benchmark. Prior experience in both the buyer and seller roles in the experiment appears to significantly influence behavior.

Asset markets

The papers in the first half of this volume concern experimental markets for goods with a life of one period, which form the domain of classical supply and demand theory. However, there is a rich experimental literature that focuses on multi-period asset markets (see Sunder, 1995 or Duxbury, 1995 for surveys). The sixth paper in this volume, by Smith, Van Boening, and Wellford, concerns the behavior of laboratory markets for long-lived assets. These markets, when populated by inexperienced subjects, are characterized by the robust phenomenon of price bubbles and market crashes, first documented by Smith, Suchanek and Williams (1988).

Follow-up work has shown that the existence of bubbles is not dependent on the specific experimental design of Smith et al. (1988). Van Boening et al. (1993) showed that bubbles also form in call markets (a call market follows rules in which all offers are submitted to the market simultaneously and the uniform market price is determined by the intersection of the submitted demand and supply curves). Fisher and Kelly (1997) showed that when two assets traded at the same time bubbles would form in both and crashes would occur at the same time in both markets. Lei et al. (2001) observed that bubble will occur in asset markets even when there are markets other than asset markets operating at the same time. Porter and Smith (1995) showed that the uncertainty in the dividend payment was also not necessary to create a bubble. King et al. (1993) showed that the bubble would also occur when short selling is permitted, margin buying is allowed, agents had equal endowment, a transaction fee is present, price changes each period are limited, and when professional traders participate as subjects. Bubbles also occur if the fundamental value is constant over time (Noussair et al., 2001).

In their paper, Smith, Van Boening and Wellford further explore possible causes of the bubble and crash phenomenon. By comparing markets for assets that pay dividends frequently with assets that pay off dividends only once, the authors find that frequent dividend payments increase the likelihood and magnitude of bubbles. Bubbles are least likely to form when a single dividend is paid at the end of the trading horizon. The authors interpret this result to suggest that the concentration of payments at this single point in time helps to create common expectations of future prices.

Auction markets

Our discussion thus far has been largely confined to two-sided markets with multiple sellers and buyers. A simpler market situation occurs in a one-sided auction, such as when one seller interacts with several buyers. This setting isolates all competition to one side of the market, and is easier to model theoretically than two-sided competition. The most basic case is when one seller wishes to sell one item to a group of buyers. The development of game-theoretic models of auctions for this case (e.g., Vickrey, 1961; Milgrom and Weber, 1982) has led to a rich experimental literature (see the survey by Kagel, 1995). Most of this literature has been dedicated to testing game-theoretic models of bidding in auctions.

These models require assumptions on the preferences and beliefs of players. Vickrey (1961) introduced the independent private values framework as an environment under which to model and compare different auction processes. In the independent private values framework, bidders draw valuations for the object available for purchase independently from a distribution that is common knowledge, and bidders' own valuations are their private information. Under these assumptions, it can be shown that any auction, in which there is a Bayesian equilibrium where the bidder with the highest valuation always receives the item and where any bidder with the lowest possible valuation receives an expected payoff of zero, generates the same revenue in the equilibrium (Myerson, 1981).

Coppinger et al. (1980), Cox et al. (1982), Kagel and Levin (1993), and other authors have studied whether the revenue equivalence result is supported in comparisons between four auctions modeled by Vickrey (1961). The four auctions are the English auction, the Dutch auction, the first-price sealed-bid auction and the second-price sealed-bid auction. The English auction is an open ascending price auction, whereas the Dutch auction is an open descending price auction. In a first price sealed-bid auction, all bidders submit their bids simultaneously. The high bidder receives the item for sale and pays the amount of his bid. The second price sealed bid auction follows the same rules as the first-price auction except that the winning bidder pays an amount equal to the second highest bid. The English auction and the second price sealed bid auctions have dominant strategy equilibria. In the English auction, each player has a dominant strategy to participate in the bidding until the price reaches his valuation and then drop out of the auction. In the second price sealed bid auction each player has a dominant strategy to bid an amount equal to his valuation. The Dutch auction and the first-price sealed bid auction lack dominant strategies, but in the independent private values environment they have Bayesian equilibria in which the expected value of the highest bid is equal to the expected value of the second highest valuation.

Experimental research has found that in the English auction, bidders tend to use the dominant strategy after very few iterations. The dominant strategy therefore seems transparent in the English auction. In contrast, most bidders in second price sealed bid auctions do not bid an amount equal to their valuations, but use a variety of other dominated strategies. In the first price sealed bid auction, bidders with high valuations tend to bid higher than the strategy ascribed to their type by the Bayesian equilibrium, and this leads the auction to generate more revenue than in equilibrium, as well as more revenue on average than the English or the Dutch auctions generate.

There have been several strands of research that have studied these and other auctions under different assumptions on preferences and information of bidders. Kagel et al. (1987) study bidding behavior in auctions in which valuations of the bidders are affiliated, and several authors (see for example Kagel and Levin, 1986; Lind and Plott, 1991; and Holt and Sherman, 1994) have studied common value auctions in which an adverse selection problem can cause a "winner's curse." The true value of the good for sale is common to all agents, but unknown at the time of bidding. Before the auction takes place, each agent receives a noisy but unbiased signal of the true value. The adverse selection problem arises because a bidder is more likely to win the auction the more his signal exceeds the true value. Typically, bidders in the auction fail to adjust bids downward sufficiently to compensate for this winner's curse, and the winning bidder often pays an amount greater than the value of the good.

The seventh paper in this symposium, by Asker, studies the behavior of the first price sealed bid auction in the independent private values environment, when bidders can default on the transaction after the auction occurs. The game proceeds as a standard first price sealed bid auction in that bidders submit sealed bids for the object, and the high bidder receives the object and pays the amount he bid. There is an exogenous probability, however, that the item turns out after purchase to have low value. In some treatments, default by the winning bidder is allowed. In equilibrium, allowing such default raises the auctioneer's revenue. The experimental data support this prediction.

Coordination through markets

The central function of markets is to coordinate the allocation of resources, and simple one-sided auctions and posted offer markets are capable of fulfilling this function in some environments. However, more complex market institutions are needed in other settings. The coordination function of markets is a focus of the next two papers of the volume. Multiple equilibria often exist in games, and experiments can provide a behavioral criterion for equilibrium selection. See Ochs (1995) for a survey of experiments on coordination.

Experiments have uncovered some behavioral patterns in equilibrium selection that are unanticipated by standard game theory. For example, in many games with multiple Pareto rankable equilibria, agents coordinate on equilibria other than the Pareto-dominant one (see, for example, Cooper et al., 1990, or Van Huyck et al., 1990). Equilibrium selection is influenced by out-of-equilibrium payoffs, even the payoffs resulting from players using strictly dominated strategies.

The paper by Plott and Williamson demonstrates how markets can help players to coordinate on a specific equilibrium. Subjects trade the right to play a 2×2 Battle-of-the-Sexes game in a double auction market that precedes the game. The games trade at prices that allow the players to infer the strategies that will be played, allowing coordination on one of the two pure strategy equilibria. As in previous studies, the particular equilibrium upon which agents in a given session coordinate is highly dependent on the history of play (Van Huyck, Battalio and Beil, 1991).

The paper by Rapoport, Seale and Winter also presents evidence on coordination, in an entry game with two potential markets. Subjects in relatively large

(n=20) groups must simultaneously decide which of two independent markets to enter, if any, immediately after the capacity of the markets is publicly announced. The capacities vary randomly across periods and are sometimes equal in the two markets. Higher capacities permit a greater number of profitable entrants. There exist many asymmetric pure-strategy equilibria, as well as one symmetric mixed-strategy equilibrium. The market entry decisions exhibit a remarkable level of coordination, which seems to arise through decision rules that feature "cutoff" values for the entry decisions based on announced capacity. The experimental sessions run for 100 periods, allowing the authors to fit a reinforcement-based adaptive learning model for individual subjects.

Learning

The learning model estimated by Rapoport et al. provides an example of the application of one of a new generation of behavioral economic models (see Roth and Erev, 1998; Cheung and Friedman, 1997; McKelvey and Palfrey, 1995; and Camerer and Ho, 1999). In contrast to earlier models inspired by pure logic, these models draw their basic principles from observed behavior in experiments. A self-imposed criterion for measuring these new models' performance is the level of accuracy with which they account for the laboratory data.

Two distinct approaches to learning dominate the literature. The belief-based approach assumes that players keep track of past strategy choices of the players they are matched with, and thereby form beliefs about what others will do in the future (Brown, 1951; Cheung and Friedman, 1997; Fudenberg and Levine, 1995 and 1998). They tend to choose a best response to those beliefs, perhaps with some decision error (e.g., Capra, Goeree, Gomez and Holt, 1999). The other dominant approach is choice reinforcement. It assumes that players previous strategies are "reinforced" by their previous payoffs, but that they do not have beliefs about the possible choices by other players, and how others' choices affect their payoffs (Bush and Mosteller, 1955; Roth and Erev, 1998).

The paper by Anderson and Camerer in this volume is an important extension of earlier work by Camerer and Ho (1999), which introduced the Experience-Weighted Attraction (EWA) model. The model predicts the propensity of agents to choose actions in a normal form game of complete information as they gain more experience playing the game. The EWA model provides a synthesis of reinforcement learning and belief learning, in that each of these two dominant approaches are special cases of the general EWA model. The Anderson and Camerer paper generalizes EWA to signaling games, making two significant extensions of the original model: to dynamic games and to games of incomplete information. The model is then tested on experimental data on signaling games, and found to fit the data well.

Policy experiments

Researchers have also conducted market experiments to inform government policy, such as regulatory, antitrust or tax policy. To be informative, such experiments must provide some insight into incentives and markets that exist in the naturally occurring economy. They are held to higher standards of "parallelism" than other types of experiments (Plott, 1987). Because of their strength in providing clear institutional comparisons (because they can hold all environmental factors constant), experimental methods have frequently been used to evaluate the impact of trading institutions and rules on market performance.

For example, in an early policy experiment, Hong and Plott (1982) evaluated the impact of advance posting of tariff rates, a measure that was being considered at the time by the U.S. Interstate Commerce Commission for barge traffic. To evaluate this policy proposal, Hong and Plott conducted experimental sessions with bilateral contract negotiation (which was the current trading institution) and sessions with posted prices (the new institution proposed by the competing railroad industry). To increase parallelism, the authors carefully implemented supply and demand conditions similar to a specific portion of the barge industry on the Mississippi River. In the posted price markets, observed prices were higher and gains from exchange were lower than under bilateral negotiation, contrary to the railroad industry lobbyists' claims. The proposal was never implemented.

Grether and Plott (1984) report another example of this type of policy experiment that helped support a case brought by the U.S. Federal Trade Commission (FTC) against manufacturers of lead-based gasoline additives. They examined a variety of potentially anti-competitive market practices, such as delivered pricing, most-favored nation clauses, and advance notice of price increases, that were challenged in court by the FTC. The experiments were designed to match many of the structural features of this industry, and they were decisive in showing that the practices could have a substantial anti-competitive effect.

Other policy topics that have been examined with market experiments include the design and evaluation of new auction institutions for allocating pollution emission rights (Cason, 1995), radio spectrum licenses (Ledyard, Porter and Rangel, 1997), airport landing slots (Rassenti, Smith and Bulfin, 1982), space station resources (Banks, Ledyard and Porter, 1989) and electric power (Noussair and Porter, 1992).

The above studies have considered policy questions at a microeconomic level, studying market rules for the allocation of specific goods. Experimental economics has only recently begun to be employed to study policy questions in macroeconomics and public finance. The final paper in this volume, by Kerschbamer and Kirchsteiger, uses experimental methods to study a fundamental concept of public finance, the Liability Side Equivalence Principle. In competitive markets, the principle asserts that the share of a sales tax borne by buyers and sellers should not depend on which side of the market is actually required to pay the tax. Kerschbamer and Kirchsteiger examine this proposition within the context of simple bargaining game, the ultimatum game. In this game, which was first studied by Güth et al. (1982), there is a known amount of surplus to divide. One player, called the proposer, offers a division of the surplus. The second player, the responder, can then accept or reject the proposal. An acceptance leads to implementation of the proposal, and a rejection leads to zero payoff to both players. In the subgame perfect equilibrium of the game, responders accept any offer yielding positive surplus, and proposers offer responders the minimum feasible positive amount. Previous experimental work has shown a tendency towards far more equal division of surplus than predicted, as well as rejection of offers that yield responders relatively little surplus. The Kerschbamer and Kirchsteiger paper shows that the distributional effect of a tax does depend upon which party has the obligation to pay it. They find that the legal obligation to pay the tax results in the liable party in fact bearing most of the cost, contrary to the Liability Side Equivalence Principle.

References

Banks J, Ledyard J, Porter D (1989) Allocating uncertain and unresponsive resources, an experimental approach. RAND Journal of Economics 20: 1–25
Brown G (1951) Iterative solution of games by fictitious play. In: Activity Analysis of Production and Allocation. New York, John Wiley & Sons
Bush R, Mosteller F (1955) Stochastic models for learning. New York, John Wiley & Sons
Camerer C, Ho T (1999) Experience-weighted attraction learning in normal form games. Econometrica 67(3): 827–783
Capra C M, Goeree J, Gomez R, Holt C (1999) Anomalous behavior in a traveler's dilemma? American Economic Review 89: 678–690
Cason T (1995) An experimental investigation of the seller incentives in EPA's emission trading auction. American Economic Review 85: 905–922
Cason T, Williams A (1990) Competitive equilibrium convergence in a posted-offer market with extreme earnings inequities. Journal of Economic Behavior and Organization 14: 331–352
Cason T, Friedman D (1999) Customer search and market power: Some laboratory evidence. In: Baye M (ed) Advances in applied microeconomics. Greenwich, CT, JAI Press
Chamberlain E (1948) An experimental imperfect market. Journal of Political Economy 56: 95–108
Cheung Y-W, Friedman D (1997) Individual learning in games: some experimental results. Games and Economic Behavior 19: 46–76
Cooper R, DeJong D, Forsythe R, Ross T (1990) Selection criteria in coordination games: some experimental results. American Economic Review 80: 218–233
Coppinger V, Smith V, Titus J (1980) Incentives and behavior in English, Dutch and sealed–bid auctions. Economic Inquiry 43: 1–22
Cox J, Roberson B, Smith V (1982) Theory and behavior of single-object auctions. In: Smith VL (ed) Research in experimental economics. Greenwich, CT, JAI Press
Davis D, Holt C (1993) Experimental economics. Princeton University Press, Princeton NJ
Davis D, Williams A (1986) The effects of rent asymmetries in posted offer markets. Journal of Economic Behavior and Organization 7: 303–316
Davis D, Williams A (1991) The Hayek hypothesis in experimental auctions: Institutional Effects and Market Power. Economic Inquiry 29: 261–274
Diamond P (1971) A model of price adjustment. Journal of Economic Theory 3: 156–168
Duxbury D (1995) Experimental asset markets within finance. Journal of Economic Surveys 9(4): 331–371
Fisher E, Kelly F (2001) Experimental foreign exchange markets. Pacific Economic Review 5: 365–387
Forsythe R, Palfrey T, Plott C (1982) Asset valuation in an experimental market. Econometrica 50: 537–568

Friedman D, Harrison G, Salmon J (1984) The informational efficiency of experimental asset markets. Journal of Political Economy 92: 349–408

Fouraker L, Siegel S (1963) Bargaining behavior. McGraw–Hill, New York

Fudenberg D, Levine DK (1995) Consistency and cautious fictitious play. Journal of Economic Dynamics and Control 19: 1065–1090

Fudenberg D, Levine DK (1998) Theory of learning in games. Cambridge, MIT Press

Gode D, Sunder S (1993) Allocative efficiency of markets with zero intelligence (ZI) traders: market as a partial substitute for individual rationality. Journal of Political Economy 101: 119–137

Goodfellow J, Plott C (1990) An experimental examination of the simultaneous determination of input and output prices. Southern Economic Journal 56: 969–983

Grether D, Plott C (1984) The effects of market practices in oligopolistic markets. An experimental examination of the Ethyl case. Economic Inquiry 22: 479–507

Güth W, Schmittberger R, Schwarze B (1982) An experimental analysis of ultimatum bargaining. Journal of Economic Behavior and Organization 3: 367–388

Holt C (1995) Industrial organization: a survey of laboratory research. In: Kagel J, Roth A (eds) The handbook of experimental economics. Princeton University Press, Princeton NJ

Holt C, Sherman R (1994) The loser's curse. American Economic Review 84: 642–652

Holt C, Langan L, Villamil A (1986) Market power in oral double auctions. Economic Inquiry 24: 107–123

Hong J, Plott C (1982) Rate filing policies for inland water transportation. An experimental approach. Bell Journal of Economics 13: 1–19

Isaac RM, Plott C (1981) The opportunity for conspiracy in restraint of trade. Journal of Economic Behavior and Organization 2: 1–30

Isaac RM, Ramey V, Williams A (1984) The effects of market organization on conspiracies in restraint of trade. Journal of Economic Behavior and Organization 5: 191–22

Kagel J (1995) Auctions. A survey of experimental research. In: Kagel J, Roth A (eds) The handbook of experimental economics. Princeton University Press, Princeton NJ

Kagel J, Levin D (1986) The winner's curse and public information in common value auctions. American Economic Review 76: 894–920

Kagel J, Levin D (1993) Independent private value auctions: bidder behavior in first-, second- and third-price auctions with varying numbers of bidders. Economic Journal 103: 868–879

Kagel J, Harstad R, Levin D (1987) Information impact and allocation rules in auctions with affiliated private values. A laboratory study. Econometrica 55 1275–1304

Ketcham J, Smith V, Williams A (1984) A comparison of posted-offer and double-auction pricing institutions. Review of Economic Studies 51: 595–614

King R, Smith V, Williams A, Van Boening M (1993) The robustness of bubbles and crashes in experimental stock markets. In: Prigogine I, Day R, Chen P (eds) Nonlinear dynamics and evolutionary economics. Oxford University Press

Ledyard J, Porter D, Rangel A (1997) Experiments testing multi-object allocation mechanisms. Journal of Economics and Management Strategy 6: 639–675

Lei V, Noussair C, Plott C (2001) Non-speculative bubbles in experimental asset markets: lack of common knowledge of rationality vs. actual irrationality. Econometrica (forthcoming)

Lind B, Plott C (1991) The winner's curse: experiments with buyers and sellers. American Economic Review 81: 335–346

McKelvey R, Palfrey T (1995) Quantal response equilibria in normal form games. Games and Economic Behavior 7: 6–38

Milgrom P, Weber R (1982) A theory of auctions and competitive bidding. Econometrica 50(5): 1089–1122

Myerson R (1981) Optimal auction design. Mathematics of Operations Research 6: 58–73

Noussair C, Porter D (1992) Allocating priority with auctions: An experimental analysis. Journal of Economic Behavior and Organization 19: 169–195

Noussair C, Plott C, Riezman R (1995) An experimental investigation of the patterns of international trade. American Economic Review 85: 462–491

Noussair C, Robin S, Ruffieux B (2001) Price bubbles in laboratory asset markets with constant fundamental values. Experimental Economics (forthcoming)

Ochs J (1995) Coordination problems. The handbook of experimental economics. Princeton University Press, Princeton NJ

Plott C (1987) Dimensions of parallelism: some policy applications of experimental methods. In: Roth A (ed) Laboratory experimentation in economics: six points of view. Cambridge, Cambridge University Press
Plott C (1989) An updated review of industrial organization. applications of experimental methods. In: Schmalensee R, Willig R (eds) Handbook of industrial organization, vol. 2. North Holland, Amsterdam
Plott C, Sunder S (1988) Rational expectations and the aggregation of diverse information in laboratory security markets. Econometrica 56: 1085–1118
Porter D, Smith V (1995) Futures contracting and dividend uncertainty in experimental asset markets. Journal of Business 68(4): 509–541
Rassenti S, Smith V, Bulfin R (1982) A combinatorial auction for airport time slot allocation. Bell Journal of Economics 13: 402–417
Roth A (1995a) Introduction to experimental economics. In: Kagel J, Roth A (eds) The handbook of experimental economics. Princeton University Press, Princeton NJ
Roth A (1995b) Bargaining experiments: In: Kagel J, Roth A (eds) The handbook of experimental economics. Princeton University Press, Princeton NJ
Roth A, Erev I (1998) Predicting how people play games: reinforcement learning in experimental games with unique, mixed strategy equilibria. American Economic Review 88: 848–881
Rustichini A, Villamil A (1996) Intertemporal pricing in markets with differential information. Economic Theory 8: 211–227
Sauermann H, Selten R (1959) Ein Oligopolexperiment. Zeitschift für die Gesamte Staatswissenschaft 115: 427–471
Smith V (1962) An experimental study of competitive market behavior. Journal of Political Economy 111–137
Smith V (1976) Experimental economics. Induced value theory. American Economic Review Papers and Proceedings 66: 274–279
Smith V (1981) An empirical study of decentralized institutions of monopoly restraint. In: Horwich G, Quirk J (eds) Essays in contemporary fields of economics in honor of E.T. Weiler (1914–1979). West Lafayette, Purdue University Press
Smith V (1982) Microeconomic systems as an experimental science. American Economic Review 72: 923–955
Smith V (1994) Economics in the laboratory. Journal of Economic Perspectives 8: 113–131
Smith V, Suchanek G, Williams A (1988) Bubbles, crashes, and endogenous expectations in experimental spot asset markets. Econometrica 1119–1151
Smith V, Williams A (1982) The effect of rent asymmetries in experimental double auction markets. Journal of Economic Behavior and Organization 99–116
Smith V, Williams A (1990) The boundaries of competitive price theory: convergence, expectations and transactions costs. In: Green L, Kagel J (eds) Advances in Behavioral Economics. New York, Ablex Publishing
Sunder S (1995) Experimental asset markets: a survey. In: Kagel J, Roth A (eds) The handbook of experimental economics. Princeton University Press, Princeton NJ
Van Boening M, Williams A, LaMaster S (1993) Price bubbles and crashes in experimental call markets. Economics Letters 41: 179–185
Van Huyck J, Battalio R, Beil R (1990) Tacit coordination games, strategic uncertainty, and coordination failure: American Economic Review 80: 234–248
Van Huyck J, Battalio R, Beil R (1991) Strategic uncertainty, equilibrium selection, and coordination failure in average opinion games. Quarterly Journal of Economics 106: 885–910
Vickrey W (1961) Counterspeculation, auctions and competitive sealed tenders. Journal of Finance 16: 8–37

Concurrent trading in two experimental markets with demand interdependence

Arlington W. Williams[1], **Vernon L. Smith**[2], **John O. Ledyard**[3], **and Steven Gjerstad**[4]

[1] Department of Economics, Wylie Hall 105, Indiana University, Bloomington, IN 47405, USA (e-mail: williama@indiana.edu)
[2] Economics Science Laboratory, McClelland Hall 116, University of Arizona, Tuscon, AZ 85721, USA (e-mail: smith@econlab.arizona.edu)
[3] Division of Humanities and Social Sciences, 228 – 77, California Institute of Technology, Pasadena, CA 91125, USA (e-mail: jledyard@hss.caltech.edu)
[4] T.J. Watson Research Center, IBM Corporation, Route 134, Kitchawan Road, Yorktown Heights, NY 10598, USA (e-mail: gjerstad@us.ibm.com)

Summary. We report results from fifteen computerized double auctions with concurrent trading of two commodities. In contrast to prior experimental markets, buyers' demands are induced via CES earnings functions defined over the two traded goods, with a fiat money expenditure constraint. Sellers receive independent marginal cost arrays for each commodity. Parameters for buyers' earnings functions and sellers' costs are set to yield a stable, competitive equilibrium. In spite of the complexity introduced by the demand interdependence, the competitive model is a good predictor of market outcomes, although prices tend to be above (below) the competitive prediction in the low-price (high-price) market.

Keywords and Phrases: Induced utility, General equilibrium, Double auction.

JEL Classification Numbers: C92, D44, D51, D83.

1 Introduction

One of the most widely known and well documented results from laboratory experimentation with private good allocation mechanisms is that double auction (DA) trading of a single commodity under conditions of a static competitive equilibrium (CE) will generate actual price and quantity observations that are near the CE.[1] This CE convergence property is quite robust with respect to the number of market participants, the specific supply and demand configurations used (for example, see Smith and Williams, 1990). DA market performance has rarely been surpassed by alternative institutions to which it has been systematically compared.[2]

This paper reports fifteen experimental sessions, conducted between June 1982 and December 1983, which examine the predictive power of the competitive model in a two-market trading environment with demand interdependence. The demand implementation method corresponds to an integer-discrete version of the standard two-good indifference curve exposition of consumer theory presented in classrooms as the theoretical foundation for the demand functions utilized in competitive market analysis. Trading decisions are made in real time in the presence of considerable price uncertainty as the market adjusts toward a behavioral equilibrium. Several initialization parameters are systematically varied across experimental replications in order to explore the sensitivity of market outcomes to variations in market structure within the basic two-commodity trading framework described below.

Consider a laboratory market with m buyers and n sellers trading two pure private goods, A and B, whose fiat money denominated prices P_A and P_B are individually negotiated for each unit traded. Agents are assumed to have monotonic increasing subjective utility functions U_i for U.S. currency. Each buyer $i \in \{1, 2, \ldots, m\}$ is given (1) an experimenter induced quasi-concave earnings function $\pi_i(A_i, B_i)$ which associates a specific cash payment with integer commodity bundles[3] and (2) an endowment of T_i "tokens" that can be used to purchase units of either good. One can think of T_i as implicitly being generated by a labor endowment, L_i, supplied to sellers at a token-wage of w such that $T_i = w \cdot L_i$. Buyer i can continue to purchase commodity units as long as the price is less than or equal to the number of "tokens remaining" defined as T_i minus expenditures on goods A and B.

A utility maximizing buyer would be concerned with maximizing $U_i[\pi_i(A_i, B_i)]$ subject to the budget constraint implied above. Assuming P_A and P_B con-

[1] This conclusion is for classical environments, but exceptions were reported by Smith (1965) where 11 buyers each have one unit valued at \$4.60 and 13 sellers each have one unit costing \$3.10. With constant excess supply of only 2 units, convergence was incomplete and erratic. A similar result has been demonstrated by Holt, Langan, and Villamil (1986) and by Davis and Williams (1991), in the context of a "market power" design where five buyers and five sellers each have five units, and two sellers need only withhold one unit each to convert one unit of excess supply to one unit of excess demand. In non-classical environments with externalities or indivisibilities, the double auction fails to yield efficient competitive equilibria. For externalities, see Plott (1983); for an example of indivisibilities in which firms have avoidable fixed costs, see Van Boening and Wilcox (1996) and Durham et al. (1996). With indivisibilities, the market failure occurs because of the tendency of the double auction to converge to one price. Smith and Williams (1992) provide a nontechnical comparison of computerized DA, posted offer, and call market institutions with a nonstationary CE.

[2] Modest exceptions include one of four different continuous feedback uniform-price auctions (McCabe, Rassenti, and Smith, 1991), and the double-dutch auction (McCabe, Rassenti, and Smith, 1992).

[3] See Smith (1982) for a more formal and comprehensive discussion of induced valuation techniques in experimental environments.

stant, the solution set of the above constrained maximization problem yields m individual demand functions for each commodity and thus market demand functions $D_A(P_A, P_B)$ and $D_B(P_B, P_A)$.

Each seller has an additively separable induced token-cost function $C_i(A_i, B_i)$ presented in the form of increasing, independent marginal cost arrays for integer increments of each good. Sellers earn the difference between price and marginal cost for each unit traded plus a 0.05 token commission to cover subjective transaction costs. Sellers' token profits are automatically converted into U.S. dollars on a one to one basis. The sellers' cost functions can be thought of as implicitly derived from independent labor-product transformation functions and a token-wage rate of w paid to buyers for each labor input unit supplied.

Assuming P_A and P_B constant, the marginal cost arrays which correspond to n individual supply arrays thus determine the aggregate market supply functions $S_A(P_A)$ and $S_B(P_B)$. The individual demand and cost parameters are such that a unique Walrasian equilibrium exists in each market, with equilibrium prices (P_A^*, P_B^*) and quantities (Q_A^*, Q_B^*) determined by the equations $D_A(P_A^*, P_B^*) = S_A(P_A^*) = Q_A^*$ and $D_B(P_B^*, P_A^*) = S_B(P_B^*) = Q_B^*$, where P_A and P_B are measured in terms of tokens (or labor), the third commodity.[4] The general equilibrium closure conditions are thus satisfied at (P_A^*, P_B^*): (1) all units of goods A and B produced by sellers are consumed by buyers, and (2) all token (labor) endowments to buyers are paid out to sellers.

Since our method of inducing (interdependent) utility on two commodities is a substantial change relative to the use of explicitly stated marginal valuations, it is an open question as to whether, or in what sense, the results of one-commodity experiments will extend to this new environment.[5] In one-commodity markets, buyer i's maximum willingness-to-pay for each successive unit is well-defined by the "resale values" (limit prices) assigned to i for successive units that might be purchased. This follows if we assume merely that buyers are nonsatiated in money since buyer i is paid in cash the difference between the assigned value of each unit purchased and the unit's purchase price (realized consumer surplus) in the experimental market. In the two-commodity market just described, maximum willingness-to-pay for each successive unit of A is defined by the uncertain opportunity costs of spending tokens in market B and vice versa. This opportunity cost demand, relative to limit price demand, may be so weak that

[4] This economy has a unique stable competitive equilibrium. For an experimental design of exchange economies with multiple equilibria, see Gjerstad (1996).

[5] Forsythe, Palfrey, and Plott (1984) use an earnings function that depends on individual holdings of an asset at the end of two distinct trading periods in a double-auction market. While their study does deviate from the traditional (unit independent) induced valuation structure, the specific focus of their work is very different from ours. Our interdependent reward structure is somewhat akin to Smith's (1980) experiments with a public good allocation mechanism using a Cobb-Douglas reward function defined over final holdings of a private and a public good. Lian and Plott (1998) employ a quadratic utility function over two commodities, and fiat currency that carries over across periods. The utility inducing technique in their experiment is similar, but the currency carry-over creates the possibility of equilibria in the multi-period economy that are not equilibria of the single period economy.

the convergence properties of one-commodity markets fail to extend to this new environment.

This paper is organized as follows: Section 2 explains the two-commodity double-auction trading mechanism. Section 3 explains the experimental design and initialization of market parameters. Section 4 reports the results of the experimental markets and is divided into three subsections, one focusing on market price and quantity convergence to the CE, one focusing on individual buyer quantity decisions, and the last is a brief summary of our experimental results.

2 Two commodity double auction trading

The market software employed in this study is a revised version of the one-commodity DA mechanism developed by Smith and Williams (Williams, 1980; Smith and Williams, 1983). Figure 1 (2) shows the basic screen display for a buyer (seller) during a market session.

Bargaining and exchange in both the market for good A and the market for good B occurs simultaneously. Buyers and sellers indicate the market in which they desire to be currently active by touching the rectangular area on their display screen labeled "MARKET A" or "MARKET B". The area corresponding to the market in which the subject is currently active is shaded as a reminder of this choice. Market participants are able to switch markets at any time during a trading period.

Buyers (sellers) enter bids to buy (offers to sell) one commodity unit by typing their entry and then touching the rectangular area on their display screen labeled "ENTER BID" ("ENTER OFFER") at which time the entry is made public unless it violates a DA procedural rule. Any buyer (seller) is free to accept any seller's offer (buyer's bid) in the market in which he/she is currently active by touching a display screen area labeled "ACCEPT OFFER" ("ACCEPT BID"). The acceptor must then touch an area labeled "CONFIRM CONTRACT" at which time a binding contract is formed and the information is logged in both the maker's and taker's private record sheets. Buyers are not allowed to enter bids or accept offers that exceed their "tokens remaining". Sellers are not allowed to enter offers or accept bids that are below the marginal cost of the unit being sold. Price quotes which violate these rules automatically generate a descriptive error message and are subsequently ignored by the auction system.

Price quotes must progress so as to reduce the bid-ask spread. Only the highest bid to buy and the lowest offer to sell in each market are displayed to the participants and are open to acceptance. Any quotation which does not provide better terms to the other side of the market is placed in a queue which ranks bids from highest to lowest (offers from lowest to highest). After a contract occurs, the highest queued bid and the lowest queued offer are automatically entered as the new bid-ask spread. The price quotes contained in each queue are unknown to market participants, however, the maker of a queued price quote is given continuously updated information on the position of the quote in the queue.

PROFIT TABLE for BUYER 2 ← B →

Units	0	1	2	3	4	5	6	7	8
0	0.00	0.41	0.83	1.24	1.66	2.07	2.49	2.90	3.32
1	0.16	0.70	1.16	1.61	2.04	2.48	2.91	3.34	3.77
2	0.31	0.91	1.40	1.86	2.32	2.77	3.21	3.65	4.09
3	0.47	1.11	1.62	2.10	2.57	3.03	3.48	3.93	4.37
4	0.63	1.30	1.83	2.32	2.80	3.27	3.73	4.19	4.64
5	0.78	1.49	2.03	2.54	3.02	3.50	3.97	4.43	4.89
6	0.94	1.67	2.23	2.74	3.24	3.72	4.20	4.67	5.13
7	1.09	1.85	2.42	2.94	3.45	3.94	4.42	4.90	5.37
8	1.25	2.03	2.61	3.14	3.66	4.16	4.64	5.12	5.60

↑ A ↓

Period 11 Purchases			
	A	2.62	2.62
	B	5.38	5.38

Token Endowment = 16.00, Spent = 16.00, Remaining = 0.00

MARKET A
SELLER 3 OFFERS $2.71
BUYER 3 BIDS $2.65

MARKET B
SELLER 5 OFFERS $5.45
BUYER 2 BIDS $5.40

ENTER BID ≫$ MARKET A

ACCEPT OFFER

CONFIRM CONTRACT

A Contracts: 2.68,2.65,2.68,2.68,2.67,2.50,2.63,2.62,2.63
Last 8 B Contracts: 5.37,5.38,5.37,5.37,5.38,5.38,5.40,5.40
Vote to end: 10 of 12 (you voted yes), SECONDS REMAINING: 58

Figure 1. Buyer's screen display in two-commodity trading experiment

Queued entries may be withdrawn at any time by pressing a key labeled -EDIT-. If queued, participants must exit the queue before being allowed to switch markets. Note that the "rank-queue" in each market is simply an electronic limit order file or "specialist's book". Smith and Williams (1983) found that this element of the DA trading mechanism significantly improved (relative to three alternatives) the rate of convergence to the CE in single commodity DA markets.

It is important to stress that bids, offers, and subsequent contracts in both markets are the only public price information made available to market participants through the DA mechanism. The bid-ask spread and a list of recent contract prices for both markets is presented to each subject regardless of the market in which they are active.

Trading takes place during a sequence of market periods each lasting a maximum of 360 seconds. Buyers' token endowments are replenished prior to the beginning of each trading period and buyers begin each period at the $(A_i, B_i) = (0, 0)$ commodity bundle. Unspent tokens can not be carried over from one trading period to another. Participants can bypass the 360 second stopping rule by unanimously voting to end the trading period. Registering a vote to end a period does not affect the individual's ability to actively participate in the markets. The bid-ask spread, contract prices, the number of seconds remaining,

RECORD SHEET for SELLER 4 TRADING PERIOD 11

Commodity A	Units→ 1	2	3	4	5
Selling Price	2.68	2.67			
Production Cost	2.20	2.50	2.80	3.10	3.40
Profit + $.05 commission	0.53	0.22			

Commodity B	Units→ 1	2	3	4	5
Selling Price	5.37	5.40			
Production Cost	4.90	5.40	5.50	6.00	6.10
Profit + $.05 commission	0.52	0.05			

MARKET A
SELLER 3 OFFERS $2.71
BUYER 3 BIDS $2.65

MARKET B
SELLER 5 OFFERS $5.45
BUYER 2 BIDS $5.40

ENTER OFFER >$ MARKET A

ACCEPT BID

CONFIRM CONTRACT

A Contracts: 2.68,2.65,2.68,2.68,2.67,2.50,2.63,2.62,2.63
Last 8 B Contracts: 5.37,5.38,5.37,5.37,5.38,5.38,5.40,5.40
Vote to end:10 of 12 (you voted yes), SECONDS REMAINING: 57

Figure 2. Seller's screen display in two-commodity trading experiment

and the current vote to end the period are presented as shown at the bottom of Figures 1 and 2. Screen displays are updated every one or two seconds. At the end of each trading period, participants are able to review all contract prices and their personal record sheets from any completed trading period.

3 Experimental design

In all experiments reported below, the profit function of buyer i has a constant elasticity of substitution (CES) specification $\pi_i(A, B) = \delta_i \cdot (\alpha_i A^{\rho_i} + \beta_i B^{\rho_i})^{1/\rho_i}$. The ρ_i parameter determines the buyer's elasticity of commodity substitution, $\frac{1}{1-\rho_i}$, and hence the curvature of the buyer's indifference contours. The function reduces to Cobb-Douglas as $\rho \to 0$, has linear indifference contours for $\rho = 1$, and tends to Leontief preferences as $\rho \to -\infty$. The CES parameter values $\{(\delta_i, \alpha_i, \beta_i, \rho_i)\}_{i=1}^m$ and the token endowments $\{T_i\}_{i=1}^m$ are initialized by first specifying the desired CE price (P_A^*, P_B^*) in each market, then specifying (for each buyer) integer CE quantities of both commodities, the profit associated with the CE bundle, and ρ_i. Using the first-order conditions for utility maximization, the computer then calculates and stores each buyer's token endowment T_i, α_i, and $\beta_i = (1 - \alpha_i)$. The δ_i parameter is then set so as to yield the desired profit at the CE bundle. Each seller's marginal cost arrays are then initialized by the experimenter to generate market supply arrays that are consistent with the CE prices and aggregate quantities specified in the demand initialization procedure.

Table 1. Buyer parameters in 6 buyer and 4 buyer base markets

6 Buyer Market

	Parameter				
Buyer	ρ	δ	α	β	T
1	0.25	0.554	0.174	0.826	23.50
2	0.75	0.580	0.268	0.732	23.50
3	0.25	0.700	0.325	0.675	20.00
4	0.75	0.700	0.325	0.675	20.00
5	0.25	0.759	0.523	0.477	16.50
6	0.75	0.818	0.388	0.612	16.50

4 Buyer Market

	Parameter				
Buyer	ρ	δ	α	β	T
1	0.25	0.606	0.223	0.777	33.50
2	0.75	0.781	0.364	0.636	26.50
3	0.25	0.759	0.447	0.553	26.50
4	0.75	0.620	0.288	0.712	33.50

Table 2. Sellers' unit marginal costs for goods *A* and *B*

	Units of *A*					Units of *B*				
Seller	1	2	3	4	5	1	2	3	4	5
1	2.95	3.05	3.55	3.65	4.15	6.35	6.65	6.95	7.25	7.55
2	2.75	3.25	3.35	3.85	3.95	6.25	6.75	6.85	7.35	7.45
3	2.75	3.25	3.35	3.85	3.95	6.45	6.55	7.05	7.15	7.65
4	2.85	3.15	3.45	3.75	4.05	6.25	6.75	6.85	7.35	7.45
5	2.95	3.05	3.55	3.65	4.15	6.35	6.65	6.95	7.25	7.55
6	2.85	3.15	3.45	3.75	4.05	6.45	6.55	7.05	7.15	7.65

The specific buyer parameters used in our base market designs are given in Table 1. Sellers' unit marginal cost schedules for the base market design are given in Table 2. The markets have either four or six buyers and six sellers who, at a CE, exchange a total of twelve units in each market. Figure 3 displays buyer profit tables generated by the parameter values shown in Table 1 for a six-buyer market. Figure 4 displays buyer profit tables for a four-buyer market. The highlighted quantities and payoffs in these tables are the equilibrium quantities and payoffs for each buyer. At the CE, the aggregate earnings of buyers is $8.40 per trading period in both the four-buyer and six-buyer designs while sellers earn $3.00 per period in each market, net of commissions ($7.20 total including commissions).

In one-commodity market experiments it is quite common to shift the individual induced marginal valuation and cost arrays (and thus the CE) by an arbitrary, uniform additive constant in order to disguise the CE price when replicating experiments using the same subject population. The technique employed to accomplish this in a two-commodity environment with interdependent demand is not quite as simple as in the one-commodity case. In order to implement a rescaling that is (potentially) behaviorally benign, the ratio of market prices must be held constant. However, this means that the CE price gap between markets will be altered. When replicating the basic market designs shown in Table 1, we

PROFIT TABLE for BUYER 1 ← B →

A \ Units	0	1	2	3	4	5	6	7	8
0	0.00	0.26	0.51	0.77	1.03	1.29	1.54	1.80	2.06
1	0.00	0.55	0.99	1.40	1.80	2.18	2.56	2.94	3.30
2	0.00	0.63	1.11	1.55	1.98	2.39	2.80	3.20	3.59
3	0.00	0.69	1.19	1.66	2.11	2.54	2.97	3.39	3.80
4	0.00	0.73	1.26	1.75	2.22	2.67	3.10	3.54	3.96
5	0.00	0.77	1.32	1.82	2.31	2.77	3.22	3.66	4.10
6	0.00	0.81	1.37	1.89	2.38	2.86	3.32	3.78	4.22
7	0.00	0.84	1.42	1.95	2.46	2.94	3.41	3.88	4.33
8	0.00	0.87	1.46	2.01	2.52	3.02	3.50	3.97	4.43

PROFIT TABLE for BUYER 2 ← B →

A \ Units	0	1	2	3	4	5	6	7	8
0	0.00	0.38	0.77	1.15	1.53	1.91	2.30	2.68	3.06
1	0.10	0.58	0.99	1.40	1.80	2.20	2.59	2.99	3.38
2	0.20	0.73	1.16	1.58	1.99	2.40	2.80	3.20	3.60
3	0.30	0.86	1.31	1.74	2.16	2.57	2.98	3.39	3.80
4	0.40	0.99	1.45	1.89	2.32	2.74	3.16	3.57	3.98
5	0.50	1.11	1.59	2.04	2.47	2.90	3.32	3.74	4.15
6	0.60	1.23	1.72	2.18	2.62	3.05	3.48	3.90	4.32
7	0.70	1.35	1.85	2.31	2.76	3.20	3.63	4.06	4.48
8	0.80	1.47	1.97	2.45	2.90	3.35	3.78	4.21	4.64

PROFIT TABLE for BUYER 3 ← B →

A \ Units	0	1	2	3	4	5	6	7	8
0	0.00	0.15	0.29	0.44	0.58	0.73	0.87	1.02	1.16
1	0.01	0.70	1.13	1.52	1.88	2.22	2.55	2.87	3.18
2	0.02	0.89	1.40	1.85	2.26	2.66	3.03	3.40	3.75
3	0.02	1.04	1.60	2.10	2.56	2.99	3.40	3.79	4.18
4	0.03	1.16	1.78	2.31	2.80	3.26	3.70	4.12	4.53
5	0.04	1.27	1.93	2.50	3.01	3.50	3.96	4.41	4.84
6	0.05	1.37	2.07	2.67	3.21	3.72	4.20	4.66	5.11
7	0.05	1.47	2.20	2.82	3.39	3.92	4.42	4.90	5.36
8	0.06	1.56	2.32	2.97	3.55	4.10	4.62	5.12	5.60

PROFIT TABLE for BUYER 4 ← B →

A \ Units	0	1	2	3	4	5	6	7	8
0	0.00	0.41	0.83	1.24	1.66	2.07	2.49	2.90	3.32
1	0.16	0.70	1.16	1.61	2.04	2.48	2.91	3.34	3.77
2	0.31	0.91	1.40	1.86	2.32	2.77	3.21	3.65	4.09
3	0.47	1.11	1.62	2.10	2.57	3.03	3.48	3.93	4.37
4	0.63	1.30	1.83	2.32	2.80	3.27	3.73	4.19	4.64
5	0.78	1.49	2.03	2.54	3.02	3.50	3.97	4.43	4.89
6	0.94	1.67	2.23	2.74	3.24	3.72	4.20	4.67	5.13
7	1.09	1.85	2.42	2.94	3.45	3.94	4.42	4.90	5.37
8	1.25	2.03	2.61	3.14	3.66	4.16	4.64	5.12	5.60

PROFIT TABLE for BUYER 5 ← B →

A \ Units	0	1	2	3	4	5	6	7	8
0	0.00	0.04	0.08	0.12	0.16	0.20	0.24	0.27	0.31
1	0.06	0.76	1.07	1.33	1.56	1.77	1.97	2.16	2.34
2	0.11	1.11	1.52	1.85	2.14	2.41	2.66	2.90	3.12
3	0.17	1.40	1.89	2.28	2.62	2.93	3.22	3.49	3.75
4	0.23	1.66	2.21	2.65	3.04	3.38	3.70	4.00	4.29
5	0.28	1.91	2.52	3.00	3.42	3.79	4.14	4.47	4.78
6	0.34	2.14	2.80	3.32	3.77	4.18	4.55	4.90	5.24
7	0.40	2.36	3.07	3.63	4.11	4.54	4.94	5.31	5.67
8	0.46	2.57	3.33	3.92	4.43	4.89	5.31	5.70	6.07

PROFIT TABLE for BUYER 6 ← B →

A \ Units	0	1	2	3	4	5	6	7	8
0	0.00	0.43	0.85	1.28	1.70	2.13	2.55	2.98	3.40
1	0.23	0.82	1.30	1.77	2.23	2.68	3.13	3.57	4.02
2	0.46	1.12	1.64	2.13	2.60	3.07	3.54	4.00	4.45
3	0.69	1.40	1.94	2.45	2.95	3.43	3.91	4.38	4.84
4	0.93	1.67	2.24	2.76	3.27	3.77	4.25	4.73	5.21
5	1.16	1.94	2.52	3.06	3.58	4.09	4.59	5.08	5.56
6	1.39	2.20	2.80	3.36	3.89	4.40	4.91	5.40	5.90
7	1.62	2.46	3.07	3.64	4.18	4.71	5.22	5.73	6.22
8	1.85	2.71	3.34	3.92	4.47	5.01	5.53	6.04	6.54

Figure 3. Buyers' profit tables: six-buyer markets

PROFIT TABLE for BUYER 1 ← B →

A \ Units	0	1	2	3	4	5	6	7	8
0	0.00	0.22	0.44	0.66	0.89	1.11	1.33	1.55	1.77
1	0.00	0.61	1.05	1.46	1.85	2.23	2.60	2.97	3.32
2	0.00	0.72	1.21	1.67	2.10	2.52	2.92	3.32	3.70
3	0.00	0.80	1.33	1.82	2.28	2.72	3.15	3.57	3.98
4	0.01	0.86	1.43	1.94	2.43	2.89	3.34	3.77	4.20
5	0.01	0.92	1.52	2.05	2.55	3.03	3.50	3.95	4.39
6	0.01	0.97	1.59	2.15	2.66	3.16	3.64	4.10	4.56
7	0.01	1.02	1.66	2.23	2.77	3.28	3.77	4.24	4.71
8	0.01	1.07	1.73	2.31	2.86	3.38	3.89	4.37	4.85

PROFIT TABLE for BUYER 2 ← B →

A \ Units	0	1	2	3	4	5	6	7	8
0	0.00	0.43	0.85	1.28	1.71	2.14	2.56	2.99	3.42
1	0.20	0.78	1.26	1.73	2.19	2.64	3.09	3.53	3.98
2	0.41	1.05	1.56	2.05	2.53	2.99	3.46	3.91	4.37
3	0.61	1.30	1.84	2.34	2.83	3.31	3.79	4.26	4.72
4	0.81	1.54	2.10	2.62	3.13	3.62	4.10	4.58	5.05
5	1.02	1.78	2.35	2.89	3.40	3.91	4.40	4.89	5.37
6	1.22	2.01	2.60	3.15	3.68	4.19	4.69	5.18	5.67
7	1.42	2.24	2.85	3.41	3.94	4.46	4.97	5.47	5.96
8	1.63	2.47	3.09	3.66	4.20	4.73	5.24	5.75	6.25

PROFIT TABLE for BUYER 3 ← B →

A \ Units	0	1	2	3	4	5	6	7	8
0	0.00	0.07	0.14	0.21	0.28	0.35	0.42	0.50	0.57
1	0.03	0.76	1.13	1.44	1.73	2.00	2.25	2.49	2.73
2	0.06	1.05	1.52	1.91	2.26	2.58	2.89	3.18	3.46
3	0.09	1.29	1.83	2.28	2.67	3.04	3.39	3.72	4.03
4	0.12	1.50	2.10	2.60	3.03	3.44	3.82	4.17	4.52
5	0.15	1.69	2.35	2.88	3.36	3.79	4.20	4.58	4.95
6	0.18	1.87	2.58	3.15	3.66	4.12	4.55	4.96	5.35
7	0.21	2.04	2.79	3.40	3.94	4.42	4.88	5.31	5.72
8	0.24	2.20	3.00	3.64	4.20	4.71	5.19	5.64	6.07

PROFIT TABLE for BUYER 4 ← B →

A \ Units	0	1	2	3	4	5	6	7	8
0	0.00	0.39	0.79	1.18	1.58	1.97	2.36	2.76	3.15
1	0.12	0.62	1.05	1.47	1.88	2.29	2.70	3.11	3.51
2	0.24	0.79	1.24	1.67	2.10	2.52	2.94	3.35	3.77
3	0.35	0.94	1.41	1.86	2.30	2.73	3.15	3.57	3.99
4	0.47	1.09	1.57	2.03	2.48	2.92	3.35	3.78	4.20
5	0.59	1.23	1.73	2.20	2.65	3.10	3.54	3.97	4.40
6	0.71	1.37	1.88	2.36	2.82	3.27	3.72	4.16	4.59
7	0.83	1.51	2.03	2.52	2.99	3.44	3.89	4.34	4.78
8	0.94	1.65	2.18	2.67	3.15	3.61	4.07	4.51	4.96

Figure 4. Buyers' profit tables: four-buyer markets

varied: (1) the designation of A or B as the market with the higher (lower) CE price, and (2) the absolute size of the CE price gap ($|P_A^* - P_B^*| = 2.80$, 3.50, or 4.20), holding the ratio of CE prices constant at 2.077. The latter was accomplished by multiplying the token endowments and CE prices (P_A^*, P_B^*) shown in Table 1 by either 0.8 or 1.2 and shifting the supply arrays by an additive constant such that $S_A(P_A^*) = S_B(P_B^*) = 12$. The effects of this rescaling on price dynamics are, a priori, unclear but our working hypothesis is that dynamics will be unaffected by the CE price gap differential. In Section 4 we report some evidence regarding the empirical validity of this hypothesis.

Table 3 displays the value of P_A^* and P_B^*, the number of buyers and sellers, and the final trading period for each experiment. The final period was set so that each experimental session lasted between 2 and 2.5 hours. All subjects were experienced with the basic trading mechanism in the sense that each had

Table 3. Experiment classification

Experiment	Number of buyers	Number of sellers	P_A^*	P_B^*	Final Period
4pda06	6	6	3.25	6.75	11
4pda07	6	6	6.75	3.25	11
4pda10	6	6	2.60	5.40	13
4pda11	6	6	8.10	3.90	13
4pda14	6	6	5.40	2.60	11
4pda16	6	6	3.90	8.10	10
4pda12	4	6	8.10	3.90	12
4pda17	4	6	2.60	5.40	10
4pda19	4	6	3.25	6.75	10
4pda08x	6	6	2.60	5.40	11
4pda18x	6	6	6.75	3.25	11
4pda20x	6	6	8.10	3.90	11
4pda09x	4	6	8.10	3.90	12
4pda13x	4	6	2.60	5.40	11
4pda15x	4	6	3.25	6.75	13

participated in at least one single-commodity computerized DA experiment.[6] Experiments with an "x" suffix used subjects who had also participated in a previous two-commodity experiment (level 2 experience). Buyers in level 2 experiments were also buyers in their first two-commodity experiment. In the other nine experiments subjects had no previous experience with the two-commodity environment (level 1 experience). All subjects were volunteers drawn from the student populations at Indiana University and the University of Arizona. Most of the experiments were run "multisite" with subjects participating simultaneously from labs at both locations.

After arriving at the experiment site, participants were each paid $3 for keeping their appointment and were then randomly assigned to individual computer terminals. The double-auction program then (1) assigns each terminal to the buyer or seller condition, (2) presents the instructions at an individually controlled pace, (3) waits for everyone to finish the instructions, and then (4) executes the experiment and stores the resulting data on disk for later recall and analysis.

4 Experimental results

4.1 Market convergence to CE

Figure 5 displays 9 sets of time series for each of the experience level 1 markets. The graph shows, for each of these 9 markets, the average price in each period in the high-price and in the low-price market. The mean contract prices for the

[6] One pilot experiment using completely inexperienced subjects was run. It was clear that the two-commodity environment was too complex for subjects with no previous exposure to computerized DA markets.

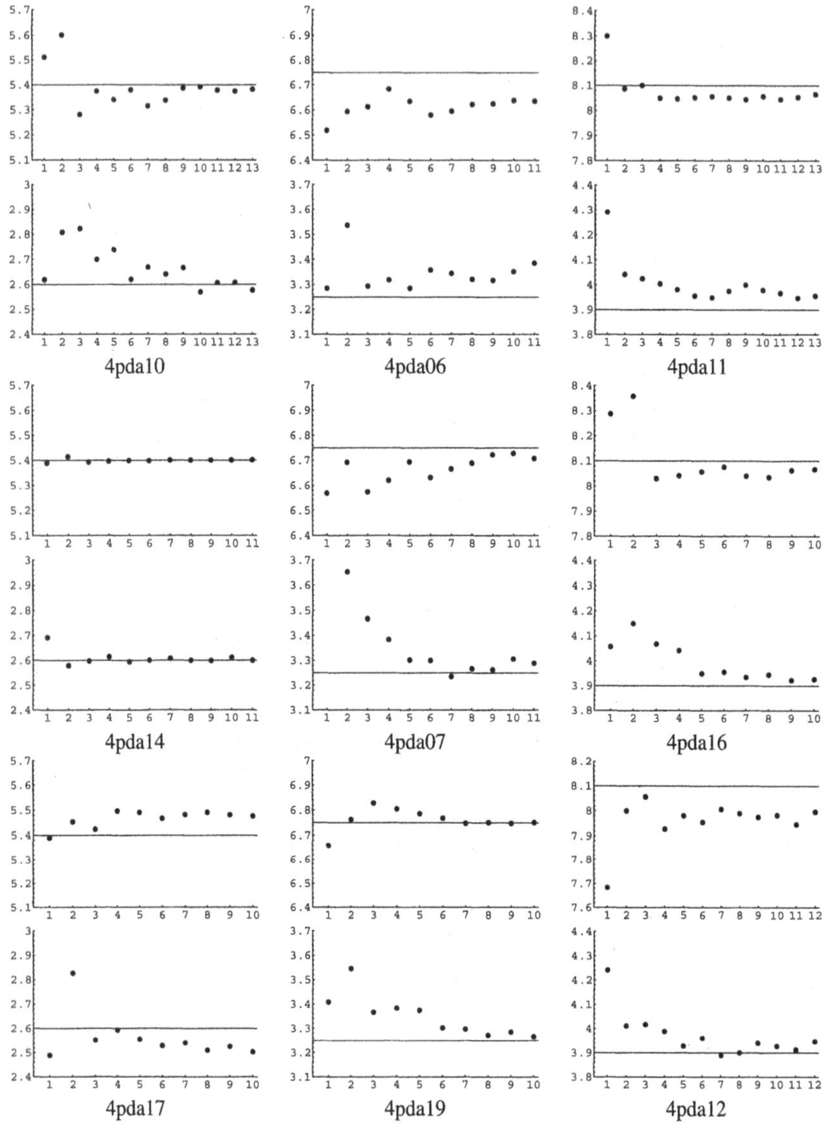

Figure 5. Mean contract prices by period: experience level 1 markets

experience level 2 markets are shown in Figure 6.[7] The time series of individual contract prices for three selected experiments are presented in Figures 7–9. In

[7] Ten (of more than 3500 total) price observations are deleted as outliers. These contract prices represent documented human errors by buyers who made a contract in the low-price market when they thought they were contracting in the high-price market. The erroneous contracts occurred in: 4pda07, period 3; 4pda09x, periods 2, 4, 9; 4pda10, period 3; 4pda11, period 8 (twice by the same buyer); 4pda15x, period 13; 4pda17, period 9; 4pda20, period 3. The markets showed no sustained price reaction to the errant contracts. In all cases the price series returned to their previous range immediately; it appeared that everyone was aware that the contract simply represented a human error and thus did not significantly affect anyone's price or profit expectations. It should also be noted that similar errors were impossible in the high-price market since sellers were not able to enter or accept

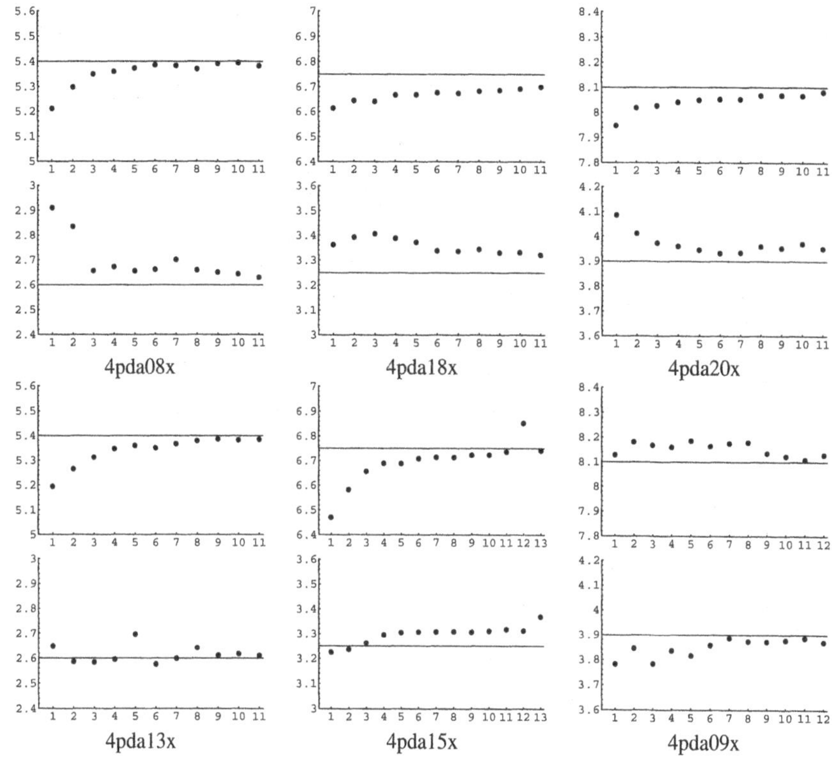

Figure 6. Mean contract prices by period: experience level 2 markets

these figures, each dot represents a transaction price. Transactions in successive periods are separated by a vertical line, and the number of transactions in each period is indicated below the vertical lines separating periods. It is clear from these charts that the CE has considerable predictive power although CE convergence is not as rapid nor as pervasive as in single-commodity DA markets with a stationary CE and similarly experienced traders. Given the task complexity inherent in our two-commodity environment, this result is not terribly surprising.

The price data presented in Figures 5–9 are summarized in Figure 10 by the 95% confidence bands for the time sequence of population period-mean prices. Figure 10 shows that prices tend to stabilize slightly below the CE in the high-price markets and slightly above the CE in the low-price markets. (Two experiments, 4pda09 and 4pda17, clearly run counter to this general result.) After about five trading periods, the confidence bands are quite tight and nearly stationary. This behavioral price equilibrium is further characterized in Figure 11 which displays histograms of individual contract prices in the low-price and in the high-price markets pooled across all replications of periods 6 - 10. Although prices tend to be "close" to the CE in both sample distributions, t-tests indicate that we should reject the null hypothesis that the population mean is equal to the

price quotes below marginal cost (just as buyers were not able to enter or accept quotes that violated their budget constraints).

CE prediction ($t = -11.3$ for the high-price market and $t = 5.7$ for the low-price market).[8]

In addition to the generally small, although sustained, price deviations from the CE, Figures 7–9 reveal a tendency for market volume to be slightly higher in the low-price markets than in the high price markets (even though the CE volume is 12 in both). This is also reflected in Figure 11 by the difference in the sample sizes for the high-price and low-price frequency polygons. Over periods 6–10, trading volume in low-price markets is 98.4% of the CE prediction (889 of 900) while in high-price markets trading volume is 90% of the CE prediction (810 of 900). Further, the mean volume in the low-price markets is larger than in the high-price markets in all trading periods.

The marginal (linear) effect of the market design variations on the speed of price converge to the CE is summarized by OLS estimation of the following equation:

$$r(t) = a + b\,t + c\,X + d\,H + e\,L + f\,F + g\,B + u_t.$$

The dependent variable, $r(t)$, is the root-mean-square-error of contract prices from the CE price prediction in period t:

$$r(t)^2 = \frac{1}{Q(t)} \cdot \sum_{i=1}^{Q(t)} [P_i(t) - CE]^2,$$

where $P_i(t)$ and $Q(t)$ are the ith contract price and the number of contracts in trading period t, respectively. Since $r(t)$ depends on both the mean and variance of prices, $r(t) = 0$ only for a zero variance price series at the CE. This occurred only twice in the 170 total trading periods used in this study. (Zero variance price sequences at the equilibrium occured in market 4pda14, periods 8 and 11, commodity B; see Figure 8.) The independent variables are defined as follows:

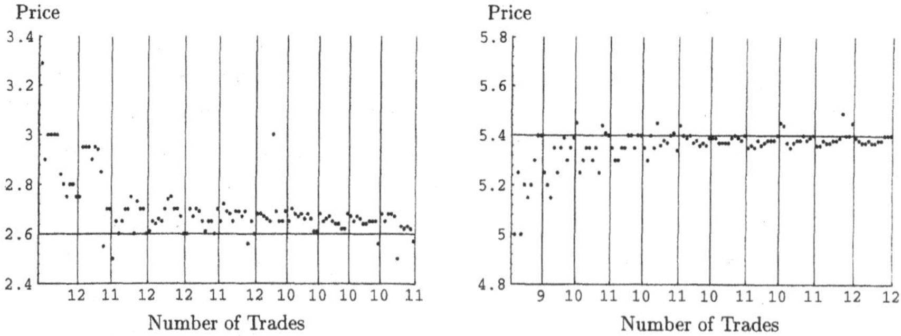

Figure 7. Prices in markets A (left) and B (right) for experiment 4pda08

[8] These sample distributions use all price observations (except as noted in footnote 5) rather than the fifteen period-mean prices. While this grand pooling captures intraperiod price variation it also is likely to violate the independence assumption needed to apply the t-distribution in the hypothesis test. This t-test should thus be viewed as describing an average price deviation from CE price rather than as a formal statistical result.

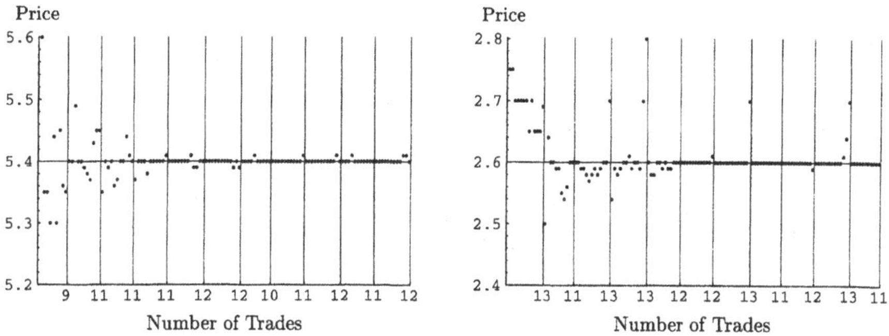

Figure 8. Prices in markets A (left) and B (right) for experiment 4pda14

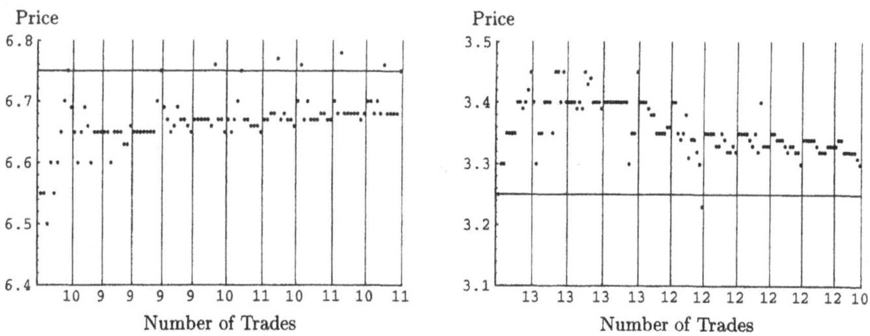

Figure 9. Prices in markets A (left) and B (right) for experiment 4pda18

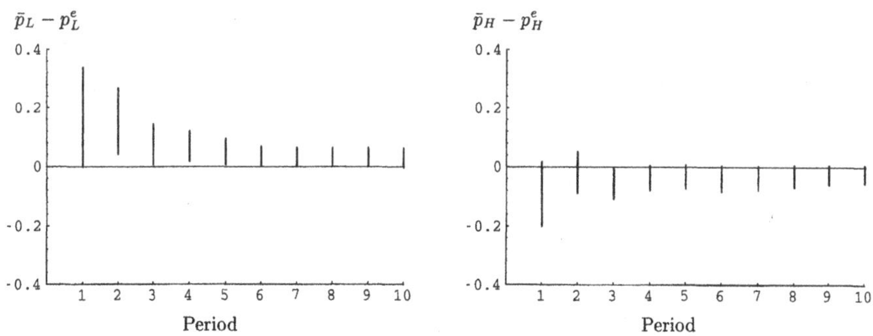

Figure 10. Confidence intervals for period mean prices in low-price markets (left) and high-price markets (right)

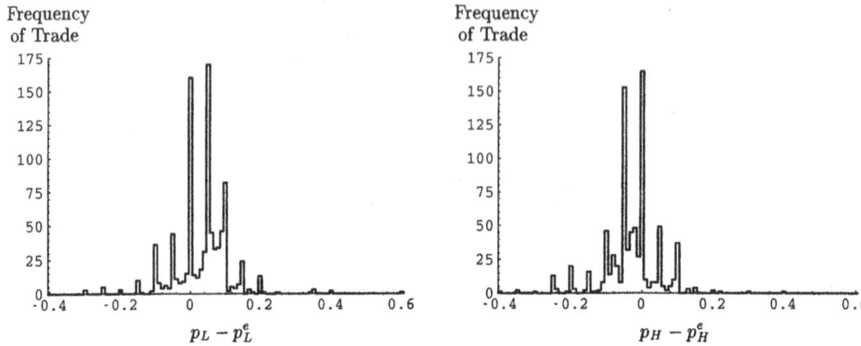

Figure 11. Histograms of deviation of transaction price from theoretical equilibrium price

$$t \;=\; \text{trading period time index (1-13)}$$

$$X = \begin{cases} 1, & \text{if subject experience level 2 applies} \\ 0, & \text{otherwise} \end{cases}$$

$$H = \begin{cases} 1, & \text{if the "high" CE price gap of 4.20 applies} \\ 0, & \text{otherwise} \end{cases}$$

$$L = \begin{cases} 1, & \text{if the "low" CE price gap of 2.80 applies} \\ 0, & \text{otherwise} \end{cases}$$

$$F = \begin{cases} 1, & \text{if the four-buyer market design applies} \\ 0, & \text{otherwise} \end{cases}$$

$$B = \begin{cases} 1, & \text{if market B is the high-price market} \\ 0, & \text{otherwise.} \end{cases}$$

Our a priori expectation was that only the coefficients of t, X and the constant term would be significantly different from zero (and negatively signed). In the OLS regression estimates that follow, t-values are shown in parentheses and "*" indicates significance at the 95% level of confidence.

High-price Market: $N = 170$, $R^2 = 0.329$, $s = 0.062$

$$\begin{aligned} r(t) \;=\; & 0.182 - 0.012\,t - 0.031\,X + 0.003\,H - 0.037\,L + 0.002\,F + 0.018\,B \\ & (12.5^*)\;(-8.1^*)\;\;(-3.1^*)\quad(.19)\quad(-3.1^*)\quad(.02)\quad(1.6) \end{aligned}$$

Low-price Markets: $N = 170$, $R^2 = 0.2671$, $s = 0.103$

$$\begin{aligned} r(t) \;=\; & 0.272 - 0.016\,t - 0.052\,X - 0.025\,H - 0.049\,L - 0.022\,F + 0.009\,B \\ & (11.2^*)\;(-6.9^*)\;\;(-3.1^*)\quad(-1.2)\quad(-2.4^*)\quad(-1.2)\quad(0.46) \end{aligned}$$

The subject experience (X) and time trend (t) variables are significant as expected. All other independent variables are not significant except for L. In both the high-price and low-price markets the low CE price difference tends to speed convergence to the CE relative to the other two CE price differentials. However, the high price gap (H) clearly does not retard CE convergence. We have no explanation for this empirical result other than "subject group effects" that would fade with additional market replications. To further examine the effects of the design treatments on CE convergence, we regress $r(t)$ on the X, H, L, F, and B dummy variables for the separate ($n = 15$) cross-sectional poolings corresponding to $t = 1$, 5, and 10. In all six (3 high-price, 3 low-price) OLS estimates, none of the dummy variable coefficients are significant at the 90% confidence level. We thus conclude that none of the design treatments have a consistent, robust effect on market behavior.

4.2 Individual buyer behavior

The buyer parameter sets used in our experimental designs (Table 1) were initialized so that two buyers were assigned CE commodity bundles consisting of high-price and low-price units as follows: $(A_i, B_i) = (1, 3)$, $(2, 2)$, or $(3, 1)$ in the six-buyer design and $(A_i, B_i) = (2, 4)$ or $(4, 2)$ in the four-buyer design. The buyers paired at a particular CE bundle had different elasticities of substitution, either -1.33 ($\rho = 0.25$) or -4 ($\rho = 0.75$); the latter implying increased commodity substitutability and flatter indifference contours. In terms of the buyer profit tables shown in Figures 3 and 4, a higher r value translates into a reduction in the monetary opportunity cost associated with deviations from the CE bundle. For example, in the six-buyer design with $P_A^* = 3.25$ and $P_B^* = 6.75$, a buyer with $\rho = 0.75$ and a CE commodity bundle at $(A_i, B_i) = (1, 3)$ could choose to purchase only two units of the high-price commodity and spend the remaining tokens on three low-price units ending up at $(A_i, B_i) = (3, 2)$. This would result in a \$0.09 reduction in profit relative to that obtained at the CE bundle (\$1.31 vs. \$1.40). In contrast, the buyer with $\rho = 0.25$ and the same CE bundle would give up \$0.21 (\$1.19 vs. \$1.40) making the same move from $(A_i, B_i) = (1, 3)$ to $(3, 2)$. Based on the differential in the opportunity cost associated with deviations from the CE bundle, one might predict that deviations from the CE bundle will occur more frequently for buyers with $\rho = 0.75$ than for similar buyers with $\rho = 0.25$.

Table 4 presents, for each buyer classification, the percentage of end-of-period commodity bundles held by buyers that correspond to the CE prediction. The individual column labels in Table 4 specify the number of low-price and high-price units in a particular bundle rather than quantities of goods A and B since the ranking of CE prices for the two goods varied across experiments. For each $\rho = 0.75$ versus $\rho = 0.25$ CE percentage comparison, a 2×2 χ-square statistic is presented. This test statistic assumes that each trading period is an independent binomial experiment where a buyer either obtains or fails to obtain the CE commodity bundle. The null hypothesis is that the population proportion of CE outcomes under $\rho = 0.75$ will be equal to the proportion of CE outcomes under $\rho = 0.25$.

It is clear from the information in Table 4 that, while the percentage of CE outcomes is larger with $\rho = 0.25$ in 8 of the 10 comparisons using all trading periods, the difference is statistically significant (in the predicted direction)

Table 4. Percentage of buyer quantity choices at CE

	Experience level 1					Experience level 2				
	6 buyer design			4 buyer design		6 buyer design			4 buyer design	
CE	3H,1L	2H,2L	1H,3L	4H,2L	2H,4L	3H,1L	2H,2L	1H,3L	4H,2L	2H,4L
All trading periods										
$\rho = 0.75$	47.8%	53.6%	43.5%	9.4%	25.0%	81.8%	45.5%	6.1%	50.0%	50.0%
	(33)	(37)	(30)	(3)	(8)	(27)	(15)	(2)	(18)	(18)
$\rho = 0.25$	58.0%	56.5%	52.2%	37.5%	40.6%	48.5%	48.5%	9.1%	47.2%	55.6%
	(40)	(39)	(36)	(12)	(13)	(16)	(16)	(3)	(17)	(20)
χ^2	1.047	0.0293	0.726	5.573*	1.134	6.673*	0	0	0	0.056
Trading periods 6 – end only										
$\rho = 0.75$	61.5%	66.7%	51.3%	11.8%	35.3%	88.9%	50.0%	5.6%	47.6%	42.9%
	(24)	(26)	(20)	(2)	(6)	(16)	(9)	(1)	(10)	(9)
$\rho = 0.25$	76.9%	61.5%	69.2%	52.9%	52.9%	61.1%	44.4%	16.7%	47.6%	52.4%
	(30)	(24)	(27)	(9)	(9)	(11)	(8)	(3)	(10)	(11)
χ^2	1.505	0.056	1.927	4.838*	0.4772	2.37	0	0.281	0	0.095

*Significant at $p \leq 0.05$ (direction not predicted)

only for the $(4H, 2L)$ pairing in the four-buyer design under experience level 1. However, a matched-pairs Wilcoxon test using the ten paired observations in the upper part of Table 4 yields $Z = 1.682$ (barely large enough to reject the null hypothesis of identical populations at the $p = .05$ level, direction predicted). It appears that the opportunity cost differential generated by the variation in ρ is simply too small to induce the predicted differences in behavioral outcomes with very much consistency. Pooling observations across buyer CE classes and experience levels, we find that 43.2% of the buyer quantity outcomes (191 of 442) are at the CE when $\rho = 0.75$ and 48% are at the CE when $\rho = 0.25$ (212 of 442). The difference is not statistically significant at the .05 level ($\chi^2 = 1.82$, $p = 0.177$).

From Table 4 we can also note that the frequency of CE outcomes is not consistently larger in the experience level 2 experiments relative to experience level 1. Pooling across designs and buyer classifications we find that under level 1 subject experience 46.3% of the buyer quantity outcomes are at the CE (252 of 542) compared with 44.4% under level 2 (151 of 341). The difference is not significant ($\chi^2 = 0.27$, $p = 0.6$).

Additional perspective on buyer performance relative to the CE prediction is gained from looking at the actual earnings of buyers relative to the CE prediction. Over all trading periods, buyers realized an average of 87.33% of the profit available at the CE. Deleting observations from the first five periods, the average earnings rises to 90.84% of CE earnings. All buyers obtained 100% of the CE profit in only 8.2% (14 of 170) trading periods. While deviations from the CE were obviously very common, buyers were generally "close" to the CE commodity bundle; frequently one high-price or low-price unit away. It is important to realize that buyers did not face a fixed price (at the CE or anywhere else) while making their purchase decisions. Given the uncertainty embodied in the problem they faced, deviations from the CE commodity bundle should not be interpreted as constituting deviations from "rational", utility-maximizing behavior.

4.3 Summary of experimental results

The two-commodity double auctions reported in this study tend to converge toward a behavioral price-quantity equilibrium that is near the competitive theoretic equilibrium. After about three to five trading periods, the mean price tends to stabilize slightly above the competitive prediction in the "low-price" market and slightly below the competitive prediction in the "high-price" market (where in our design $P_{hi}/P_{lo} = 2.077$ and $P_{hi} - P_{lo} = 2.80, 3.50$, or 4.20). Trading volume is slightly below the CE prediction on average, with the low-price market displaying somewhat higher volume than the high-price market. Around 45% of all end-of-period commodity bundles held by buyers coincide with the prediction from the competitive model. A smaller proportion of choices at the competitive equilibrium bundle is observed for buyers with an elasticity of commodity substitution of −4 (43.2%) versus −1.33 (48%). This result is consistent with the fact that the opportunity cost of deviating from the competitive equilibrium bundle is smaller for buyers with the higher elasticity of substitution (flatter indifference contours). The difference, however, is not statistically significant nor is it robust with respect to variations in buyer groupings. While deviations from the CE commodity bundle are quite common, buyers manage to earn an average of over 87% of the profits available at the CE.

References

Davis, D.D., Williams, A.W.: The Hayek hypothesis in experimental auctions: institutional effects and market power. Economic Inquiry **29**, 261-274 (1991)

Durham, Y., Rassenti, S., Smith, V.L., Wilcox, N.T., Van Boening, M.: Can core allocations be achieved in avoidable fixed cost environments using two-part price competition. Annals of Operations Research **68**, 61–88 (1996)

Forsythe, R., Palfrey, T.R., Plott, C.R.: Futures markets and informational efficiency: a laboratory examination. Journal of Finance **39**, 955-981 (1984)

Gjerstad, S.: Multiple equilibria in exchange economies, with homothetic, nearly identical preferences. Mimeo, University of Minnesota, Center for Economic Research, Discussion Paper No. 288 (1996)

Holt, C.A., Langen, L.W., Villamil, A.P.: Market power in oral double auction markets. Economic Inquiry **24**, 107-123 (1986)

Lian, P., Plott, C.R.: General equilibrium, markets, macroeconomics and money in a laboratory experimental environment. Economic Theory **12**, 21–75 (1998)

McCabe, K., Rassenti, S., Smith, V.: Designing a uniform-price double auction: an experimental evaluation. In: Friedman, D., Rust, J. (eds.) The double auction: institutions, theory, and evidence. Santa Fe Institute, Addison-Wesley (1991)

McCabe, K., Rassenti, S., Smith, V.: Designing call auction institutions: is double dutch the best? Economic Journal **102**, 9–23 (1992)

Plott, C.R.: Externalities and corrective policies. Economic Journal **93**, 106–127 (1983)

Smith, V.L.: Experimental auction markets and the Walrasian hypothesis. Journal of Political Economy **73**, 387–393 (1965)

Smith, V.L.: Experiments with a decentralized mechanism for public good decisions. American Economic Review **70**, 584–599 (1980)

Smith, V.L.: Microeconomic systems as an experimental science. American Economic Review **72**, 923–55 (1982)

Smith, V.L., Williams, A.W.: An experimental comparison of alternative rules for competitive market exchange. In: Engelbrecht-Wiggans, R., et al. (eds.) Auctions, bidding and contracting: uses and theory. New York: N.Y.U. Press 1983

Smith, V.L., Williams, A.W.: The boundaries of competitive price theory: convergence, expectations and transaction costs. In: Green, L., Kagel, J.H. (eds.) Advances in behavioral economics. Norwood, NJ: Ablex 1990

Smith, V.L., Williams, A.W.: Experimental market economics. Scientific American **276**, 116–121 (1992)

Van Boening, M., Wilcox, N.: Avoidable cost: ride a double auction roller coaster. American Economic Review **72**, 461–477 (1996)

Williams, A.W.: Computerized double-auction markets: some initial experimental results. Journal of Business **53**, 235–58 (1980)

Some factors affecting demand withholding in posted-offer markets★

Bradley J. Ruffle

Department of Economics, P.O. Box 653, Ben-Gurion University, Beer Sheva, 84105, ISRAEL
(e-mail: bradley@bgumail.bgu.ac.il; website: http://econ.bgu.ac.il/facultym/bradley)

Summary. Both oligopoly theory and experiments are concerned almost uniquely with sellers' behavior. Buyers' ability to exhibit non-trivial behavior in different market institutions remains unaddressed. This paper investigates the impact of three variables (number of buyers, surplus division at the market-clearing price and information revelation) on strategic and fairness-motivated demand withholding. Demand withholding and its ability to force lower prices increase as the number of buyers or the share of surplus earned by the buyers decreases. However, increasing the information revealed to subjects about the surplus inequality favoring sellers mildly facilitates collusion among sellers rather than provoking demand withholding as conjectured.

Keywords and Phrases: Strategic buyer behavior, Demand withholding, Fairness, Reference profit.

JEL Classification Numbers: C90, D43, L13.

1 Introduction

1.1 Motivation

The overwhelming majority of industries in all but retail sales are characterized by bilateral oligopoly: a handful of sellers confronting a handful of buyers.

★ This paper, based on chapter 3 of my Ph.D. dissertation at Princeton University, has benefitted from many helpful discussions with James Cox, Vernon Smith and Bart Wilson. I also thank my advisors, Avinash Dixit and Timothy Van Zandt, as well as Gordon Dahl, Mark Isaac, David Lucking-Reily, John Ramberg, Steve Rassenti, Stanley Reynolds, Reinhard Selten, Jason Shachat, an anonymous referee and one of the guest editors of this special issue for useful comments. Funding for this project was provided by a visiting John M. Olin Foundation Predoctoral Fellowship at the University of Arizona's Economic Science Laboratory.

Yet, industrial organization theory, laboratory oligopoly experiments and antitrust legislation all neglect the potential strategic, counteracting role of buyers in markets. Oligopoly theory treats buyers as price-takers where sellers are the only strategic players. Oligopoly experiments focus on seller behavior (e.g. signaling and collusion), often actively censoring the role of human buyers by employing a computer algorithm to simulate buyer behavior. And antitrust policy ignores the possibility that a few, powerful buyers may offset what appears to be an anticompetitive industry structure.

Using a posted-offer market, this paper initiates a line of research intent on exploring buyers' ability to exhibit non-trivial behavior in various market institutions. In the posted-offer institution, a period begins with each seller simultaneously choosing a price and a quantity to make available at that price. Buyers then proceed sequentially to make the purchases each desires. Thus, the only choice available to buyers is to accept or reject a posted price. Admittedly, this limited strategy space differs from that typically available to the large industrial buyers this research aims at understanding. The posted-offer institution has instead been noted to characterize the structure of retail markets.[1] Nonetheless, we begin by examining buyer behavior in this very structured bargaining setting as a first step towards more complicated, less structured settings.

1.2 Why would buyers withhold demand?

Buyers may be motivated to reject profitable purchases, that is to withhold demand, for fairness or strategic reasons. One interpretation of fairness is a concern for relative payoffs. Evidence from the experimental bargaining game literature suggests that people will reject unequal divisions of surplus. Ultimatum-game responders often reject offers less than one half and rejection rates increase as the offer decreases as a fraction of the available surplus.[2] The posted-offer institution is the natural, multi-player, market extension of the ultimatum game. Sellers post prices which buyers can then either accept or reject. Acceptance yields sellers a payoff determined by the difference between the price they post and their cost on each unit sold. Buyers earn the difference between their valuation and the purchase price on each unit bought. If a buyer rejects a particular seller's posted price, both buyer and seller receive zero surplus. The natural question to ask is, does the concern for relative payoffs in the ultimatum game extend to the posted-offer market? To answer this question, one of the variables studied is the surplus division between buyers and sellers. The supply and demand curves are constructed so that at the market-clearing price each seller earns considerably more than each buyer. To make this earnings inequality more salient to the buy-

[1] Ketcham, Smith and Williams (1984) provide a brief history of the posted-offer institution in retail trade.

[2] The ultimatum game involves two players, a proposer and a responder. The proposer's task is to divide an amount between the two players. The responder can accept or reject the proposed division. If the responder accepts the division, then players receive the amounts indicated by the division. If she rejects, both players receive nothing.

ers, the high level of the information variable (a second factor studied herein) reveals all subjects' profits at the end of each period in addition to providing full information with regard to the market configurations.

Buyers may also forego a profitable purchase with the intent to force sellers to lower their prices in future periods. This non-myopic, strategic purpose is likely to be more successful the more concentrated the buyers' side of the market. The number of buyers is the third factor explored.

The remainder of this paper proceeds as follows: Section 2 reviews previous posted-offer experiments which consider buyer behavior and fairness in markets. Section 3 details the experimental model and its theoretical predictions. Section 4 presents and analyzes the experimental results. Section 5 examines the strategic versus fairness motivation for the observed buyer withholding and its profitability. Section 6 concludes.

2 Related literature

2.1 Buyer behavior in previous posted-offer experiments

The role of buyers in markets remains unaddressed. A glance at previous posted-offer experiments, in particular, illuminates the overwhelming focus on seller pricing.[3] Human buyers are typically replaced by a computer algorithm that continues to purchase from the lowest-price seller as long as the price is less than (or equal to) the buyer's valuation.[4] Removing human buyers reduces subjects payments; furthermore, demand withholding is prevented, thus simplifying the interpretation of experimental results. Sellers may exploit all available gains from trade without fear of buyer repercussions.

There are nonetheless several studies which run at least one replication with human buyers and compare the results obtained with other replications involving simulated buyers. These include Cason and Williams (1990), Davis and Williams (DW) (1991) and Kruse (1991). DW and Kruse both observe that the mere presence of human buyers (i.e. the threat of demand withholding) has a disciplining effect on sellers. This paper takes these insights seriously by studying buyer behavior and seller responsiveness to it.

2.2 More on fairness in markets

Franciosi et al. (1995) use a posted-offer market to examine the willingness of buyers to accept price increases depending on whether they are justified by increases in costs or not. If subjects are made aware of the unequal division of profits favoring sellers, profit-seeking behavior is initially blunted. Prices begin lower than in other treatments where sellers' profits are not revealed; however,

[3] See Holt (1995) for a survey.

[4] In fact, this censorship of human buyers dates back to the earliest oligopoly experiments (Fouraker and Siegel, 1963; Friedman, 1963).

even in the profit-disclosure treatments, prices soon converge to the competitive equilibrium as sellers gradually increase their posted prices in the face of excess demand.

Researchers conducting ultimatum games have attempted to extrapolate subjects' concern for relative payoffs to posted-offer markets. After all, a posted offer by a seller is identical to a take-it-or-leave-it offer issued by the proposer in an ultimatum game. Context and (perhaps) the number of players are the only differences between the two environments. Hoffman et al. (1994) attempt to reduce contextual differences between the two environments by framing the ultimatum game as a posted-offer market. For instance, they conduct the ultimatum game with the players labeled "seller" and "buyer" rather than "person A" and "person B". This produces significantly lower offers without affecting rejection rates. Hoffman et al.'s finding indicates the sensitivity of offers in an *ultimatum game* to contextual variables. The experiments conducted here examine fairness directly in a posted-offer market.

3 Experimental procedure and design

3.1 Posted-offer institution

The experimental implementation of the posted-offer institution has been carried out using the NovaNet (PLATO) software.[5] Subjects are randomly assigned to the role of either buyer or seller. Each period consists of the following sequence of events. Sellers simultaneously decide on a price, followed by a quantity to make available at that price (up to their capacities). After all sellers have made a price and quantity choice for the period, each seller's price (but not quantity) is revealed to the buyers and all other sellers. Buyers then proceed in random order to make their desired purchases. Purchase decisions are made privately so that buyers are unable to observe the purchases (or sacrifices) of other buyers. When it is a buyer's turn to shop, she sees only the posted prices of each seller, displaced by the message "Out of Stock" when a seller has sold all available units. The period ends when the last buyer has finished shopping. Buyers earn the difference between their valuations and the price they pay on each unit purchased. Sellers earn the difference between the price they post and the cost of the unit for each unit sold. They do not incur costs for units offered that remain unsold.

3.2 Experimental design

The cube in Fig. 1 displays the $2 \times 2 \times 2$ experimental design. The three treatment variables are: number of buyers, surplus division, and information revelation. Each variable is explored at two levels. This initial investigation attempts to identify factors that affect buyer behavior. Since there is no *a priori* reason

[5] See Ketcham, Smith and Williams (1984) for a detailed exposition of the features of this software.

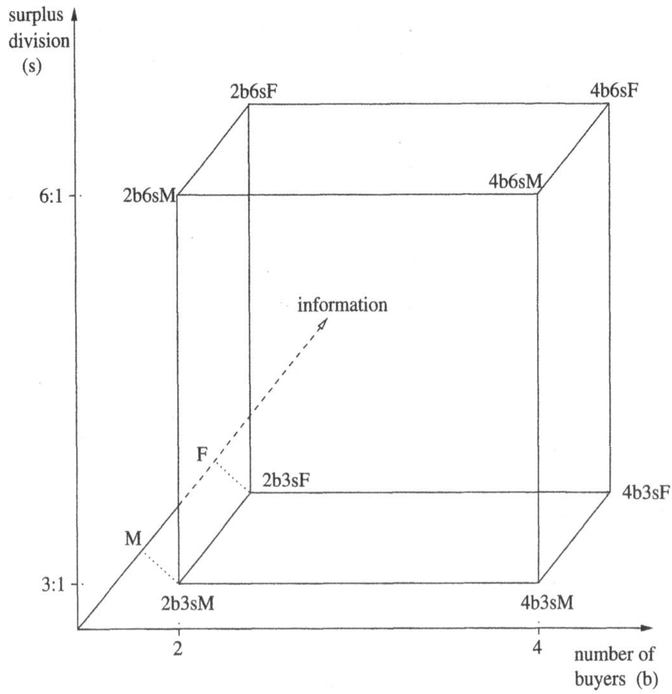

Figure 1. Geometric representation of the 2^3 factorial design

to suspect the relationship between any of the treatment variables and buyer withholding (or sellers' prices) to be non-monotonic, two levels per variable are sufficient to identify the major trend, if any. The specific levels for each treatment variable were chosen with the thought in mind that if the weaker level of each variable fails to stimulate withholding, then the stronger level should.

The first treatment variable is the number of buyers, b. Two (2b) and four (4b) are the numbers investigated here. It is important to keep in mind that the number of sellers is fixed at two for all treatments.

The surplus division variable, s, indicates the ratio of the profits of each seller to each buyer at the competitive price. The ratios of 3:1 (3s) and 6:1 (6s) were selected based on rejection rates in ultimatum, bargaining games. A vast number of ultimatum game experiments suggest that a 3:1 ratio of surplus inequality favoring the proposer is typically sufficient to elicit substantial rejection frequencies among responders. Does this finding extend to the market version of the ultimatum game? If not, perhaps a 6:1 earnings inequality will evoke demand withholding?

The third treatment variable of interest is the information revealed to subjects. The low-level information condition examined here provides subjects with full information about the market configurations (M); that is, all subjects are given each buyer's valuations and each seller's costs in tabular form. In the high-level information condition, in addition to the market configurations, the profits of each subject are made public at the end of each period. The intent is to make the earnings inequality more salient to buyers in an attempt to incite them to forego profitable purchases. I thus refer to the high-level information condition as the fairness condition (F).

Since we are interested in the ability of buyers to withhold demand, the model to be tested is constructed to hold constant across treatments the buyers' side of the market to the extent possible. In this manner, differences in observed levels of demand withholding (and seller pricing) across treatments can be attributed to changes in the levels of the treatment variables, as opposed to variations in the design in moving from one treatment to another. We thus hold fixed the aggregate demand curve across all treatments. To vary the number of buyers from four to two, the units of demand of the third and fourth buyers are redistributed to the first two buyers, as shown in Fig. 2. In so doing, this would double the earnings of each of the two buyers relative to their cohorts in the four-buyer treatments. To control for this, an exchange rate of two experimental dollars for one U.S. dollar is introduced in those treatments with two buyers. The end result is that, at the competitive price and full efficiency, all buyers earn the same amount ($0.40 per period) regardless of identity or treatment.

Changes in the surplus division are accomplished by shifts in the cost curve. Figure 2 displays three distinct cost functions. The middle one of the three, labeled "2b3s/4b6s", is relevant for four treatments, 2b3sM, 2b3sF, 4b6sM, and 4b6sF. To see this, consider the treatment 4b6sM, for instance. For the cost curve indicated, each of the two sellers earns six times as much as each of the four buyers at full efficiency at the competitive price. If we reduce the number of buyers to two, the ratio of the earnings inequality falls to 3:1, thus yielding 2b3sM. The three cost functions given therefore cover all eight treatments of interest.

3.3 Theoretical predictions

The ten-cent *vertical* overlap of the demand and cost configurations produces a unique competitive quantity prediction of eight units with a competitive range of prices. The midpoint of the $0.10 competitive tunnel is treated as the competitive price for ease of exposition. All subsequent prices are stated as deviations from this competitive price.

As discussed above, subjects were given full information regarding the market conditions. This serves two purposes. First, it speeds up subjects' learning process thereby quickening convergence, if any, to observed outcomes.

Secondly, full, symmetric information and public knowledge of the exchange rate (the payoff function) makes this a repeated game with complete information. We may therefore compute sellers' Nash equilibrium and cooperative strategies.[6]

[6] Note well that such a computation assumes buyers as price takers, that is, non-strategic players. Bilateral oligopoly remains unexplored in oligopoly theory.

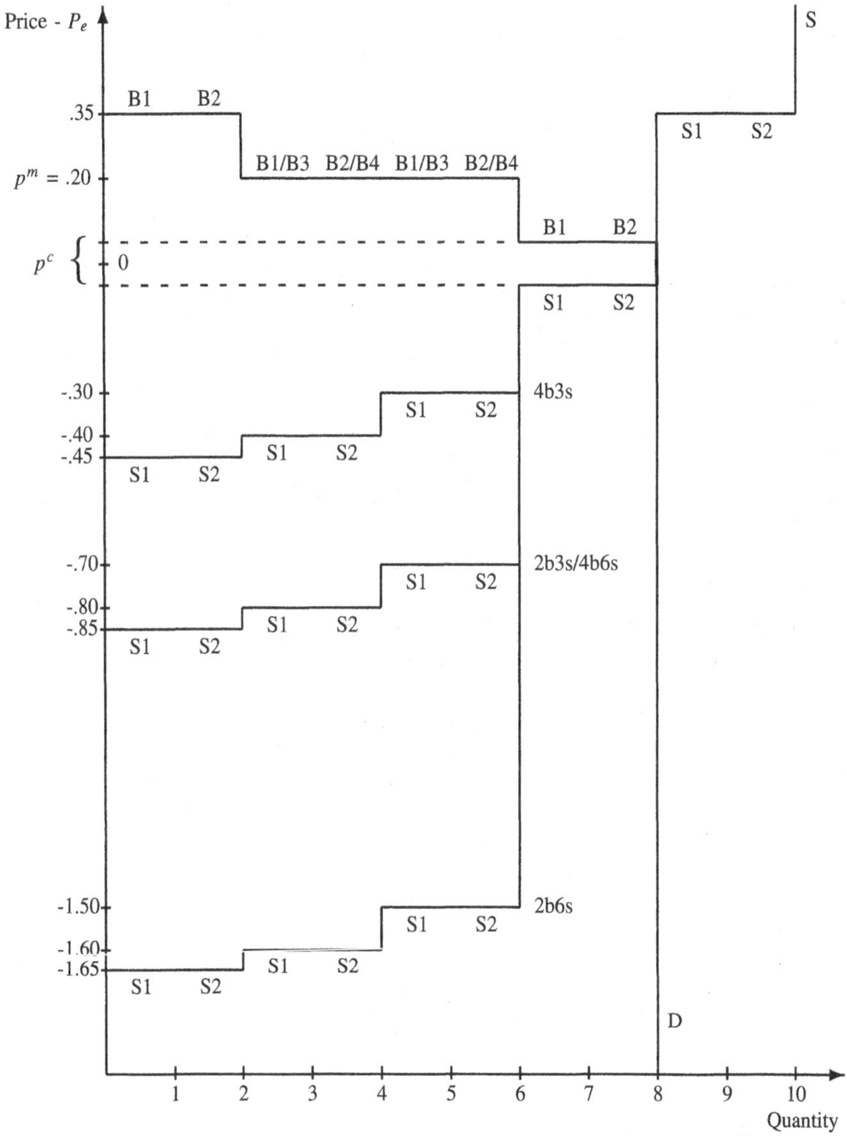

Figure 2. Market configurations for all eight treatments. The three distinct cost curves share common fourth and fifth steps at $p = -.05$ and $p = .35$, respectively. The labels above (below) the demand (cost) curve(s) refer to the buyer (seller) to whom the unit belongs. "B1/B3" indicates that buyer 1 (3) owns the unit in treatments with two (four) buyers

Seller i's best-response price function to seller j is provided by equation (1).[7] For simplicity of exposition, it assumes seller j chooses the optimal quantity

[7] This is easily derived for the two-buyer treatments. The extension to the four-buyer case involves a probabilistic assessment of the different possible orderings of the non-identical buyers. A proof that the best-response function remains unchanged is contained in an earlier version of this paper available for download from the author's website.

for a given price. The best-response function reveals that there does not exist a pure-strategy Nash equilibrium (PSNE) in prices for this model.

$$p_i^*(p_j) = \begin{cases} +.20, \; q_i^* = 4 \;\; \text{if} \; +.20 < p_j \\ p_j - .01, \; q_i^* = 4 \;\; \text{if} \; +0.05 < p_j \leq +.20 \\ +.20, \; q_i^* = 3 \;\; \text{if} \; p_j \leq +0.05 \end{cases} \tag{1}$$

where q_i^* is firm i's sales which are $\leq$ its choice of quantity.

To understand this result intuitively, note that seller i will price at +0.20 as long as he is able to sell 3 or more units of production. This, he achieves as long as his competitor, seller j, prices at, or below, the top of the competitive range ($p_j \leq +0.05$), or strictly greater than +0.20. Once seller j raises her price above the competitive tunnel but less than +0.20 she reduces seller i's sales to two units. Seller i's best response is then to undercut seller j by 0.01. With $p_j = +0.20$, seller i also responds optimally by charging +0.19 and selling four units. Given seller j's symmetric best-response function, there does not exist a pair of prices such that neither seller cannot increase its profits by changing its price. The non-existence of a PSNE is a product of the sequential ordering of buyers in the posted-offer institution. With the efficient-rationing rule where the highest valuation units are purchased first, a symmetric PSNE exists with prices set at +0.05 (and sales of four units for each firm).

There does exist a unique, symmetric, mixed-strategy Nash equilibrium for each treatment. It involves seller i choosing prices $p_i = (+.05, +.20)$ with probabilities $(4/5, 1/5)$, $(4/9, 5/9)$, and $(4/17, 13/17)$ for treatments 4b3s, 2b3s/4b6s, and 2b6s, respectively. Note that the expected price from playing the mixed-strategy equilibrium *increases* as the number of buyers decreases or the surplus inequality increases – a striking observation once we allow buyers to play a role.

The monopoly (or joint-profit-maximizing) outcome occurs at +0.20. As a point of reference, we may similarly compute the monopsonist price, as if this were a posted-bid auction. It turns out that buyers' profits are jointly maximized at a price equal to the third step on the sellers' cost curve with six units purchased.

4 Experimental findings

4.1 Subject pool

A total of 19 sessions were conducted at the University of Arizona's Economic Science Laboratory. Each of the 92 subjects participated in a single treatment as either a buyer or a seller. Buyers earned on average \$7.25, while sellers took away \$29.25, not including their \$5 participation fees. Reading the computerized instructions and one-page handout took up the first 30 minutes in all sessions. The shorter sessions (with two buyers and no revelation of profits) took about 1 hour in total while those with four buyers and revelation of profits required up to 1 hour and 40 minutes.

4.2 Observed outcomes

Unable to conduct multiple replications of all eight treatments due to a budget constraint, I faced the decision of which treatments to run and which ones to omit from the study. To resolve this problem, I adopt an experimentation strategy from the design of experiments literature in industrial engineering. This strategy (a fractional factorial) involves conducting a particular four of the eight available treatments at four replicates each. After a total of 16 runs, there still remained enough money to conduct three more experiments of 2b3sM – the cheapest of the four remaining treatments.[8]

Appendix A provides a graph of the mean contract price for the five treatments pooled across replications. The data from all five treatments is summarized in Table 1. The "Mean Efficiency" in the third column reports the amount of combined consumer surplus and producer surplus captured by buyers and sellers in the experiment as a percentage of the theoretical surplus available at the competitive price. "Sales Lost to Demand Withholding" indicates the additional units of production that would have traded if buyers made all purchases yielding a profit of at least $0.01.[9] It therefore provides a measure of the impact of demand withholding on sellers.

One feature common to all five treatments is that, when attempted, demand withholding persisted throughout the experiment. Buyers did not, for instance, withhold early on and then later abandon it. In many cases, buyers did not begin resisting sellers' prices until after five or six periods and withholding, where observed, continued at a more or less steady rate until the final couple of periods (19-20) at which point buyers resumed accepting all profitable prices.

4b3sM: Four buyers, 3:1 surplus division, full Market info

All four replications are similar in that prices converge to the midpoint or top of the competitive tunnel from above. Attempts at buyer withholding are limited to, at most, one or two units in a period. When attempted, withholding is ineffective in bringing down posted prices. In fact, prices in this treatment were higher than in any other. Of the total 80 periods played, at least one seller posted a price at the top of, or above, the competitive tunnel ($\geq$ +0.05) in 63 periods. This left buyers 1 and 2 no opportunity to make a positive profit on their fourth units of demand. Furthermore, the mean contract price was $\geq$ +0.05 27 times. As

[8] For those desirous of more detail of the advantages of fractional factorial designs, consult any Design of Experiments textbook, such as Box, Hunter and Hunter (1978), or an earlier version of this paper available from the author's website.

[9] Although the sales lost to demand withholding may be less than the actual number of units withheld, I use the two terms synonymously. Because the posted-offer software does not keep track of the random order in which the buyers proceed in each period, we sometimes cannot determine with exact precision the number of purchases withheld. Ambiguity arises in cases where a buyer makes no purchases in a period and sellers post different prices. In such instances, we sometimes are not able to determine if the buyer (having had the opportunity to purchase first) withheld demand or (having been last in the buyer queue) was unable to purchase profitably. Also, units passed up by one buyer in a given period may be subsequently purchased by another buyer later in the queue making the number of sales lost to withholding less than the number of units withheld.

Table 1. Summary statistics for the five treatments conducted where all replications within a treatment are pooled. All five treatments consist of four replications with the exception of 2b3sM which was conducted only 3 times. For all statistics each row reports the mean for the entire 20 periods, last 15 periods (pooled across replications) with the standard deviations for the respective means in parentheses beneath. For "Sales Lost to Demand Withholding", the number given is a total for the entire 20 periods, last 15 periods

Treatment	Mean contract price	Mean efficiency	Sales lost to demand withholding	% Buyer surplus captured	% Seller surplus captured
4b3sM	.047, .054	88.43, 88.29	27, 23	72.24, 70.14	99.22, 100.40
	(.045, .044)	(14.17, 14.23)		(17.75, 17.90)	(17.08, 16.27)
2b3sM	.023, .042	91.26, 93.91	4, 4	83.39, 78.81	93.89, 98.94
	(.082, .046)	(14.31, 11.64)		(30.95, 21.21)	(16.00, 10.50)
2b3sF	−.021, −.025	83.79, 85.01	73, 59	89.89, 91.77	81.76, 82.76
	(.065, .054)	(18.73, 18.89)		(27.19, 23.22)	(19.46, 19.76)
2b6sM	−.151, −.180	77.32, 73.98	113, 97	116.99, 120.12	70.71, 66.29
	(.144, .142)	(24.26, 24.31)		(57.78, 61.77)	(23.23, 21.94)
4b6sF	−.005, .017	93.92, 95.04	30, 24	93.67, 87.23	94.00, 97.64
	(.076, .034)	(10.21, 8.90)		(28.19, 15.94)	(14.05, 9.85)

a result of sellers' high prices, efficiency levels in this treatment were regularly the second or third lowest among the five treatments.

2b3sM: Two buyers, 3:1 surplus division, full Market info
Reducing the number of buyers from four to two causes sellers to price more cautiously initially, it would appear. In the three replications of this treatment, four of the six sellers' posted prices begin below the competitive tunnel. In the absence of buyer resistance, sellers unanimously increase their prices. Convergence to the competitive price from below is thus observed. Pricing and efficiency levels in the latter periods match those in 4b3sM.

2b3sF: Two buyers, 3:1 surplus division, Fairness
The results from this treatment reveal a tendency for prices to fall over time. In two of the four replications, prices start out at, or below, the bottom of the competitive tunnel and decay gradually. Slow convergence from above is observed in the other two replications. Overall, demand withholding is stronger and more persistent in this treatment: at least one unit of sales was lost to withholding in 35/80 periods run with three or more sales lost in 12 periods. Consequently, efficiency levels are notably less than in the aforementioned treatments.

2b6sM: Two buyers, 6:1 surplus division, full Market info
This treatment is exceptional in the intensity of buyer withholding and its ability to force lower posted prices. Prices did not converge to the Nash or competitive price. Instead, beginning in period 6, prices fell continuously throughout the session in all four replications. The mean contract price in the terminal period ranged from −.10 to −.42. Efficiency levels actually decrease throughout the experiment as withholding intensified and posted prices fell out of the competitive tunnel (thereby allowing a maximum of six units to trade).

What cannot be observed in the summary statistics or the summary graph of Appendix A is the differential impact on sellers' prices of early attempts at demand withholding versus later ones. It is in the early periods that sellers may establish a reference profit. That is, if sellers post prices at or near the competitive equilibrium early on and manage to sell all available units, they will be reluctant to offer price reductions in the face of later attempts at demand withholding. On the other hand, buyers who withhold demand early on reduce sellers' profits to considerably less than the available surplus. Sellers are therefore likely to respond by dropping prices in an attempt to encourage buyers to make purchases and *increase* their own profits.

The first and fourth replications of this treatment illustrate these divergent phenomena. In the first replication, buyer 1 boycotted the market entirely beginning in periods 1, 3, 5-8. This left the two sellers competing in price for the business of the remaining buyer. Prices fell steadily throughout the twenty periods with an average contract price of around −0.36 over the last five periods.

In the fourth replication, on the other hand, buyers did not even attempt to withhold demand until period 5. By this point, having previously enjoyed sales of at least three units each at similar prices, neither seller was willing to acquiesce. As a result, prices in the last five periods of this experiment were a less impressive −0.11. This contrast is particularly striking in view of the fact that total demand withholding in the fourth replication (39 units) significantly exceeded that in the first replication (26 units). What is more, in both of these sessions, the first-period contract prices and the overall efficiencies for the twenty periods (about 70%) were almost identical, yet the price paths differ markedly due to the *timing* of the withholding. The moral of the story is that if buyers are to have an impact on sellers' prices, they cannot afford to let the sellers get used to a comfortable profit level.

4b6sF: Four buyers, 6:1 surplus division, Fairness

Sellers again begin pricing very tentatively. In 3/4 replications, mean contract prices start out well below the competitive tunnel. However, attempts to withhold demand are either weak or non-existent signaling to sellers the potential to raise profitably their prices. By period 9, prices in all sessions are within the competitive range where they remain for the duration. Again (as in 2b3sM), despite an earnings imbalance in their favor, sellers push upward their prices when left unchecked by buyer withholding.

4.3 Data analysis

Let us now assess the quantitative impact of each of the three treatment variables and demand withholding on sellers' posted prices. Table 3 of Appendix B reports the results of four OLS regressions using the data from all 19 experimental sessions.

The buyer and surplus treatment variables are highly statistically significant no matter the collection of right-hand-side (RHS) variables. The impact of in-

Table 2. Support for various qualitative response models of seller pricing in treatments with profit revelation (F), 2b3sF and 4b6sF (N=300)

Qualitative response model		Observations			
Prediction	Condition	Support	Contrary	Neutral	Total
$p_{it} - p_{i\,t-1} > 0$	if $p_{i\,t-1} < p_{j\,t-1}$, $\Pi_{i\,t-1} < \Pi_{j\,t-1}$	42	3	15	60
$p_{it} - p_{i\,t-1} < 0$	if $p_{i\,t-1} > p_{j\,t-1}$, $\Pi_{i\,t-1} < \Pi_{j\,t-1}$	43	5	7	55
$p_{it} - p_{i\,t-1} > 0$	if $p_{i\,t-1} > p_{j\,t-1}$, $\Pi_{i\,t-1} > \Pi_{j\,t-1}$	13	15	34	62
$p_{it} - p_{i\,t-1} < 0$	if $p_{i\,t-1} < p_{j\,t-1}$, $\Pi_{i\,t-1} > \Pi_{j\,t-1}$	14	17	22	58
$p_{it} - p_{i\,t-1} >=< 0$	if $(p_{i\,t-1} - p_{j\,t-1})(\Pi_{i\,t-1} - \Pi_{j\,t-1}) = 0$	—	—	70	70

creasing the number of buyers from two to four is that each seller increases his per period posted price by an average of 2.2 cents.[10] Increasing the surplus inequality favoring sellers from 3:1 to 6:1 decreases each seller's posted price by 2.4 cents on average. The information variable is only found to be significant when the first five periods are dropped. It is typical for data from early periods to be quite noisy in market experiments as subjects figure out what to do and how to respond to the actions of other participants. Interestingly, the sign on the information variable is positive: revealing profits to subjects actually *increases* the prices sellers post in each period by approximately one cent. Treatment 4b6sF, for instance, reveals that sellers made use of the observability of each other's posted price and resultant profits to coordinate; they learned to post similar, if not identical, prices and extract most of the gains from trade.

Table 2 presents a series of qualitative response models for seller pricing in profit revelation treatments 2b3sF and 4b6sF. The level of support for the various models makes clear the logic behind seller price collusion: the seller whose profit was the lower of the two in period $t-1$ typically adjusts his period t price in the direction of the high-profit seller's period $t-1$ price (rows 1 and 2); whereas the aggregate data of the high-profit seller (rows 3 and 4) appear as if he plays a mixed strategy, keeping his period t price constant roughly half the time and the other half of the time increasing and decreasing his price with approximately equal frequency.

The use of four regressors for the quantity of sales lost by seller i due to demand withholding in period $t-1$, $W1_{i\,t-1}$, $W2_{i\,t-1}$, $W3_{i\,t-1}$, $W4_{i\,t-1}$, permits withholding to affect pricing non-linearly. The finding that $W1_{i\,t-1}$ is insignificant comes as no surprise in view of the market configurations: a sale lost at the competitive price costs a seller a mere \$0.05; whereas the steepness of the MC curve makes foregone sales due to withholding the second, third and fourth units of production very costly. The coefficient of -0.044 on $W2_{i\,t-1}$ indicates that if seller i loses two sales to demand withholding in period $t-1$, he lowers his period t posted price by 4.4 cents. The loss of the third and fourth units ruffles a seller even more, causing him to drop his subsequent period price

[10] The (normalized) market-clearing price in all treatments was \$1.90.

by 6.3 and 5.9 cents, respectively.[11] Interestingly, the decision to withhold three units dominates four units of withholding.

5 Fairness versus strategically motivated withholding

In this section we examine the motives underlying the observed withholding and its profitability.

For a given price, the punishment to sellers of rejecting a profitable purchase is greater the more extreme the earnings inequality. The observation that, for a given number of buyers, withholding is more frequent the larger the surplus inequality is therefore consistent with fairness. Consider, for instance, the four-buyer treatments 4b3sM and 4b6sF. Despite lower initial prices in 4b6sF than in 4b3sM, buyers in the former treatment withheld 30 units of demand, three more than in 4b3sM. Similarly, demand withholding was significantly higher in 2b6sM than the other two-buyer treatments explored. The frequent rejection of profitable prices in 4b6sF and especially 2b6sM reflects perhaps an intent to punish sellers' unwillingness to divide more equally the large, available surplus.

On the other hand, buyers' intensified withholding in treatments involving the 6:1 earnings inequality may be purely strategic.[12] Recognizing sellers' ability to post prices significantly below the competitive price, buyer withholding may be viewed as an attempt to force lower prices. From the regression analysis of Section 4.3 we learned that every sale lost to demand withholding lowered that seller's posted price in the subsequent period by just less than 2 cents. Still, was withholding profitable? Did the lower prices (at which buyers subsequently purchased) make up for the foregone gains from withholding?

The choice to withhold poses a free-rider problem for the buyers. Buyers benefit from withholding in the form of lower prices, but prefer others to do the withholding. The results of the regression on buyer earnings reported in equation (2) concretize the gains from having cohort buyers withhold.

$$\begin{array}{rcl} \Pi_k^{buyer} & = & \underset{(.000)}{5.460} + \underset{(.000)}{0.178}\, W_{-k} - \underset{(.000)}{7.716}\, \bar{P}_1 \\ & & \bar{R}^2 = .537 \qquad\qquad n = 54 \end{array} \qquad (2)$$

The W_{-k} variable represents the total number of units withheld in a session except those units withheld by buyer k. The coefficient of 0.178 (p-value=.000) indicates that every unit of demand withheld in a session by buyers other than k adds \$0.18 to buyer k's total earnings.

[11] The significance of these three withholding variables are robust to whatever data set is used, as are the significance of the own lagged price, $p_{i\,t-1}$, and other lagged price, $p_{j\,t-1}$, variables. Interaction variables composed of the withholding variables and each of the three variables turn out to be insignificant, as do withholding dummies for period $t-2$.

[12] It is only in the terminal period that we can attribute with certainty buyer withholding to fairness considerations. Of the 19 experimental sessions conducted, the sole instance of withholding in period 20 was a buyer in 2b3sF who withheld a single unit.

Further, there is evidence that on average buyers are made better off by withholding, their own included. In the treatment where withholding was greatest (113 units withheld in 2b6sM), buyer earnings were highest ($9.36 on average). Buyers in treatment 2b3sF withheld less and earned less, a total of 73 units withheld and $7.19 on average. In 4b3sM buyers earned still less, $5.42 on average, in conjunction with their withholding a mere 27 units of demand in total. While buyers in 4b6sF withheld only 3 additional units of demand (30 in total), initial prices well below the competitive equilibrium account for their appreciably higher average earnings of $7.50. The importance of initial prices on buyers' earnings is captured by the regression coefficient of −7.716 (p-value=.000) on $\bar{P}_1$: a $1.00 decrease in the first-period mean contract price increases a buyer's earnings by $7.72.[13] Initial prices below the competitive price in 2b3sM and 4b6sF provided a rare opportunity for significant single-period earnings for the buyers since prices in these treatments quickly rose to the competitive price and beyond.

This observed tendency of convergence to the competitive equilibrium from below in 2b3sM and 4b6sF speaks to the issue of fairness on the sellers' side of the market. Already enjoying almost three or even six times as much earnings as buyers, sellers' willingness to increase their initial prices when left unchecked by buyer withholding displays a disregard for sharing the available surplus more equally.

6 Conclusions

This paper establishes that buyers may not always play the passive, price-taking role allotted to them by oligopoly theory. Even in the restrictive posted-offer institution which limits buyers to accepting or rejecting proposed prices (i.e. buyers are unable to make a counter-offer or negotiate a better price), buyers appear able to form long-term strategies and forego myopically profitable purchases. As a result, sellers are forced to lower their prices in order to increase sales. Yet an active, or at least influential, role for buyers depends on their numbers and the surplus division between buyers and sellers at the competitive price.

Four-buyer treatments reveal buyers' ineffectiveness against sellers: sellers resist buyer attempts at demand withholding and maintain prices at, or above, the competitive price. Two buyers, on the other hand, often intensely withhold demand and successfully drive prices down well below the Nash and competitive equilibrium predictions. This is particularly true in the face of large earnings inequalities between buyers and sellers. Even in the four-buyer treatment involving extreme (6:1) earnings inequalities, buyers resist equilibrium prices early on, but eventually concede. Buyers lack of persistence in this treatment is an indication that, individually, they feel powerless when there are three others. Sellers, on the other hand, do not mind the earnings inequality. Their pricing behavior reveals

[13] None of the treatment variables nor a dummy variable for buyer identity, B1, B2, B3, B4, were found to be significant.

that as long as buyers are willing to buy, they will increase the price if it is profitable to do so.

7 Appendix A: Mean contract prices by treatment

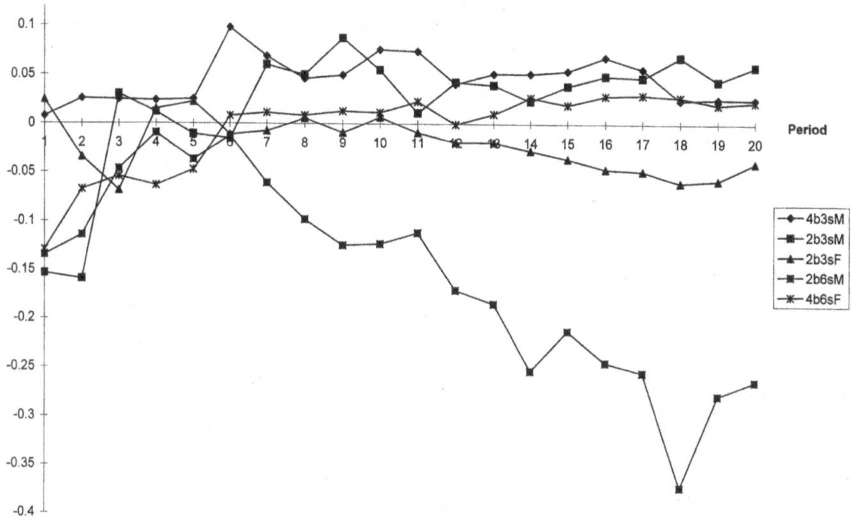

8 Appendix B: regression results

Table 3. Estimated OLS regression coefficients with p-values in parentheses. The treatment variables buyr, surp, and info are coded as "0" for 2, 3:1, and M, respectively, and "1" for 4, 6:1, and F. The indicator variables for withholding take on a value of "1" whenever the relevant number of units was withheld in period $t-1$, and "0" otherwise

$p_{i\,t}$	constant	buyr	surp	info	$W1_{i\,t-1}$	$W2_{i\,t-1}$	$W3_{i\,t-1}$	$W4_{i\,t-1}$	$p_{i\,t-1}$	$p_{j\,t-1}$	$\bar{R}^2$	Obs.
20	1.897	0.102	−0.122	0.009	—	—	—	—	—	—	.248	756
periods	(.000)	(.000)	(.000)	(.342)	—	—	—	—	—	—		
last 19	1.894	0.099	−0.116	0.011	—	—	—	—	—	—	.271	718
periods	(.000)	(.000)	(.000)	(.199)	—	—	—	—	—	—		
last 19	0.536	0.022	−0.024	0.007	−0.012	−0.044	−0.063	−0.059	0.417	0.301	.594	718
periods	(.000)	(.003)	(.002)	(.315)	(.293)	(.002)	(.002)	(.027)	(.000)	(.000)		
last 15	0.310	0.018	−0.020	0.013	−0.011	−0.026	−0.056	−0.059	0.465	0.369	.768	570
periods	(.000)	(.002)	(.001)	(.010)	(.206)	(.010)	(.000)	(.003)	(.000)	(.000)		

References

Box, G., Hunter, W., Hunter, S.: Statistics for experimenters: An introduction to design, data analysis and model building. New York, NY: Wiley 1978

Brown-Kruse, J.: Contestability in the presence of an alternate market: an experimental examination. RAND Journal of Economics **22**(1), 136–147 (1991)

Cason, T., Williams, A.: Competitive equilibrium convergence in a posted-offer market with extreme earnings inequities. Journal of Economic Behavior and Organization **14**, 331–352 (1990)

Davis, D., Williams, A.: The Hayek hypothesis in experimental auctions: Institutional effects and market power. Economic Inquiry **29**, 261–274 (1991)

Fouraker, L., Siegel, S.: Bargaining behavior. New York, NY: MacGraw-Hill 1963

Franciosi, R., Kujal, P., Michelitsch, R., Smith, V., Deng, G.: Fairness: Effect on temporary and equilibrium prices in posted-offer markets. Economic Journal **105**(43), 938–950 (1995)

Friedman, J.: Individual behavior in oligopolistic markets: An experimental study. Yale Economic Essays **3**, 359–417 (1963)

Hoffman, E., McCabe, K., Shachat, K., Smith, V.: Preferences, property rights, and anonymity in bargaining games. Games and Economic Behavior **7**, 346–380 (1994)

Holt, C.: Industrial organization: A survey of laboratory research. In: Roth, A., Kagel, J. (eds.) Handbook of experimental economics, chapter 5. Princeton University, Princeton, NJ: Princeton University Press 1995

Ketcham, J., Smith, V., Williams, A.: A comparison of posted-offer and double-auction institutions. Review of Economic Studies **51**(4), 595–614 (1984)

Firm-specific cost savings and market power*

Douglas D. Davis[1] **and Bart J. Wilson**[2]

[1] Department of Economics, Virginia Commonwealth University, Richmond, VA 23284-4000, USA (e-mail: dddavis@vcu.edu)
[2] Economic Science Laboratory, University of Arizona, McClelland Hall Room 116, P.O. Box 210108, Tucson, AZ 85721-0108, USA (e-mail: bwilson@econlab.arizona.edu)

Summary. We report a policy experiment that illustrates a potential problem of using historical pass-through rates as a means of predicting the competitive consequences of projected firm-specific cost savings in antitrust contexts, particularly in merger analysis. The effects of cost savings on welfare can vary vastly, depending on how the savings affect the industry supply schedule. In a capacity-constrained price-setting oligopoly, we observe that cost savings can overwhelm behaviorally salient market power incentives when the savings affect marginal (high cost) units. However, cost savings of the same magnitude on an infra-marginal unit leave market power unchanged.

Keywords and Phrases: Market power, Cost savings, Efficiencies and antitrust analysis, Pass-through rate.

JEL Classification Numbers: L1, L4, C9.

1 Introduction

Efficiencies recently have come to take on a more prominent role in the antitrust analysis of horizontal mergers. In April 1997, the U.S. Department of Justice and the Federal Trade Commission issued a revision of the Horizontal Merger Guidelines solely to articulate means for assessing the efficiency claims of merging parties. In contrast to the burden of showing an anti-competitive price effect

* Financial assistance from the National Science Foundation is gratefully acknowledged. We thank without implicating an anonymous referee, Tim Cason, Signe-Mary McKernan, Stan Reynolds, Bradley Ruffle, Chuck Thomas, and seminar participants at North Carolina State University, Virginia Commonwealth University, and the 1998 Regional Economic Science Association meetings. The data and a sample copy of the instructions are available upon request from the authors.

of a merger, which rests with the government, the Revised Guidelines implicitly impose on the merging parties the burden of demonstrating the magnitude and *merger specificity* of alleged efficiencies. Federal antitrust enforcement agencies will consider only those efficiencies likely to be accomplished with the proposed merger and unlikely to be accomplished otherwise to be a relevant factor in mitigating concerns about increased market power (see Revised Guidelines, p. 30).

The process of weighing potential increases in market power against alleged merger-specific efficiencies involves estimating the rate at which firm-specific cost reductions arising from the efficiencies are passed through to consumers in the form of lower prices.[1] In the oligopolistic market structures that are typically of antitrust policy interest, predicting pass-through rates is somewhat problematic. Such estimates require assumptions about the nature of oligopolistic interactions, in addition to assumptions about cost conditions and the nature of demand. Progress along these lines has been made for the case of constant marginal cost producers by Farrell and Shapiro (1990), who consider pass-though rates for homogenous-product Cournot competitors, and by Werden (1996), who discusses the case of differentiated product price-competitors. In the arena of antitrust policy, federal antitrust authorities recently have turned to estimating historical pass-through rates as part of the process of weighing merger efficiencies against the increased market power concerns. Historical pass-through rate estimates, for example, were a matter of considerable debate in the Staples/Office Depot merger litigation.[2] We wish to raise the issue that focusing on historical pass-through rates has the potential for errant policy conclusions, because this practice does not consider the way that the expected cost savings affect the cost schedule.

To appreciate the potential problems, consider a market where firms are capacity-constrained and have increasing marginal costs. A representative context might be the electric power industry, where the marginal cost of producing elec-

[1] As Yde and Vita (1996) point out, firm-specific and industry-wide cost reductions must be carefully distinguished when attempting to assess the extent to which cost reductions will be reflected in price changes. Consider a market with common, constant marginal costs and a linear demand curve. If the market is perfectly competitive, 100% of any industry-wide cost savings will be passed on to consumers. However, none of the cost savings realized by any particular firm will be passed on to consumers (provided each individually small firm is capacity-constrained), since in a competitive market the differentially efficient firm does not produce the marginal units that determine the market price. In fact, for consumers to realize benefits from firm-specific cost reductions, some market power is necessary. In the limiting case of a monopolist, for example, a firm-specific cost reduction will alter the intersection of marginal cost and marginal revenue. Under the presumed conditions of linear demand, a profit-maximizing monopolist will pass 50% of any firm-specific marginal cost reductions through to consumers in the form of lower prices.

[2] *Federal Trade Commission v. Staples, Inc.*, 970 F. Supp. 1066, 1090 (D.D.C. 1997). The merging firms claimed that two-thirds of cost reductions were historically passed-through to consumers in terms of lower prices. The Federal Trade Commission (FTC) argued that the firms failed to distinguish between industry-wide and firm-specific cost reductions, and that only the latter were relevant for estimating the pass-through rate for merger-specific efficiencies. The FTC estimated that the actual historical firm-specific pass-through rate was on the order of 15%. See Ashenfelter, Ashmore, Baker and McKernan (1998) for an explanation of the methodology used by the FTC.

tricity in different types of plants varies markedly. Nuclear-generated electricity, for example, has a much lower marginal cost than coal-generated electricity. When demand is high, the less efficient power plants are brought on-line. In such a context, cost savings that reduce the cost of marginal units can mitigate or even eliminate pre-existing market power. At the same time, a cost reduction of the same magnitude on infra-marginal units can leave unchanged pre-existing market power. When combined with a reallocation of units among sellers, cost reductions can, in a parallel way, either create market power or leave unaffected initially competitive conditions.

The intuition underlying these conclusions is that cost reductions affect the incentives to sell marginal units differently when firms have capacity constraints: cost reductions to marginal units make the sale of marginal units more profitable, and thus increase incentives to reduce prices. However, cost reductions on infra-marginal units do not affect pre-existing incentives to sell marginal units. Cost reductions on infra-marginal units, when combined with a reallocation of units to include some marginal units of capacity, may create market power, since the seller with the newly endowed marginal units may have little incentive to price low enough to sell the marginal units.

This paper uses controlled laboratory methods to examine firm-specific cost savings and market power in a capacity-constrained price-setting oligopoly. Our experimental design and the model on which it is based are admittedly stylized. The model involves transfers of plant capacity among firms in a way that is suggestive rather than a literal characterization of a complete merger. Nevertheless, the experiment is useful for policy purposes in that it illustrates in a straightforward way how cost savings can interact with market power in a both theoretically and behaviorally salient manner.[3] Section 2 below develops more formally the relationship between costs and market power. Section 3 presents the experiment design, and Section 4 reports the results. The paper concludes with a short discussion of the policy implications in Section 5.

2 Market power and cost savings

Market power is defined conventionally in the industrial organization literature as arising when a firm can set a price greater than marginal cost and still find buyers; i.e., when the firm has a downward sloping residual demand schedule. For the case of capacity-constrained price competitors, Holt (1989) articulates a game-theoretic variant of this definition: market power exists if one or more sellers can profit by unilaterally deviating from the competitive price. Thus, for

[3] The difficulty of generating closed-form Nash mixing predictions in a design that incorporates a horizontal consolidation and also varies power conditions and the type of cost savings symmetrically across treatments, deterred us from using a more complex environment with an explicit merger. Our inability to find closed form solutions makes it difficult to verify that the amount of power is held constant in different designs. Nevertheless, our design has obvious implications for merger policy and analysis, even without incorporating an explicit merger, since it allows insight into the effects of controlled and identifiable power and synergy interactions.

capacity-constrained price competitors, market power exists when the competitive price is not a pure strategy Nash equilibrium for the market viewed as a stage game.

Consider in particular a market for a homogeneous product served by an asymmetric triopoly. The firms each have a fixed production capacity, and they compete under posted-offer (Bertrand) rules; i.e., sellers post prices simultaneously, and production is to order. After all sellers post prices, a single buyer begins a shopping sequence. The buyer is assumed to reveal demand fully and to rotate purchases among the tied sellers in the case that one or more sellers post the same price.[4]

To see how market power arises in this context, compare the two supply and demand arrays shown on the left side of Figure 1. In each market the buyer will purchase five units at any price P_c or less and will purchase three units at any price above P_c, but not greater than the reservation price r. The seller identifiers $S1$, $S2$, and $S3$, printed below the cost steps on the market supply schedule, indicate unit cost allocations for the three sellers. Notice in market N that no seller can offer more than two units, and that an excess supply of two units exists at any price above P_c. Thus, no power exists in market N, since any unilateral deviation above P_c will reduce earnings for the deviating seller to zero. In this case the (unique) Nash equilibrium for this game has all sellers offering all available units at price P_c.[5]

Market P, shown at the bottom of Figure 1, is identical to market N, except that seller $S2$'s high cost unit has been reallocated to seller $S1$, as highlighted by the underlined identifier. This rather subtle change affects importantly the equilibrium for the market stage game, since seller $S1$ is now certain to sell at least one unit at reservation price r independent of the actions of the sellers $S2$ and $S3$. If seller $S1$'s earnings from selling a single unit at price r exceed $S1$'s earnings from selling three units at P_c, then P_c is no longer a pure strategy Nash equilibrium. Furthermore, no equilibrium in pure strategies exists for this game under these conditions. At a common price of r seller $S1$ has an incentive to cut price slightly and increase sales to three units. Seller $S1$ has similar incentives to shade on any common price down to a lower price $P_{\min}$, where $S1$'s earnings for selling three units just equals $S1$'s security earning available by posting r. Finally, $P_{\min}$ cannot be part of a pure strategy equilibrium strategy, since $S1$ will sell only a single unit at a common price of $P_{\min}$ and could increase earnings by posting the reservation price r. The equilibrium for the stage game involves mixing over the range $P_{\min}$ to r.

Due to Holt and Solis-Soberon (1992), the derivation of the equilibrium mixing distributions for this market is relatively standard, and is relegated to an

[4] Assuming a single buyer is equivalent to assuming that demand is rationed efficiently among the sellers, with the high-value buyers shopping first, followed by lower value buyers.

[5] Verifying the existence of the equilibrium at P_c is straightforward. The discussion in the text rules out unilateral price deviations above P_c. We complete the verification by ruling out unilateral price reductions below P_c, which follows since such deviations would only reduce earnings on units that are certain to sell. Uniqueness can be established readily by ruling out an equilibrium at any other possible price triplet.

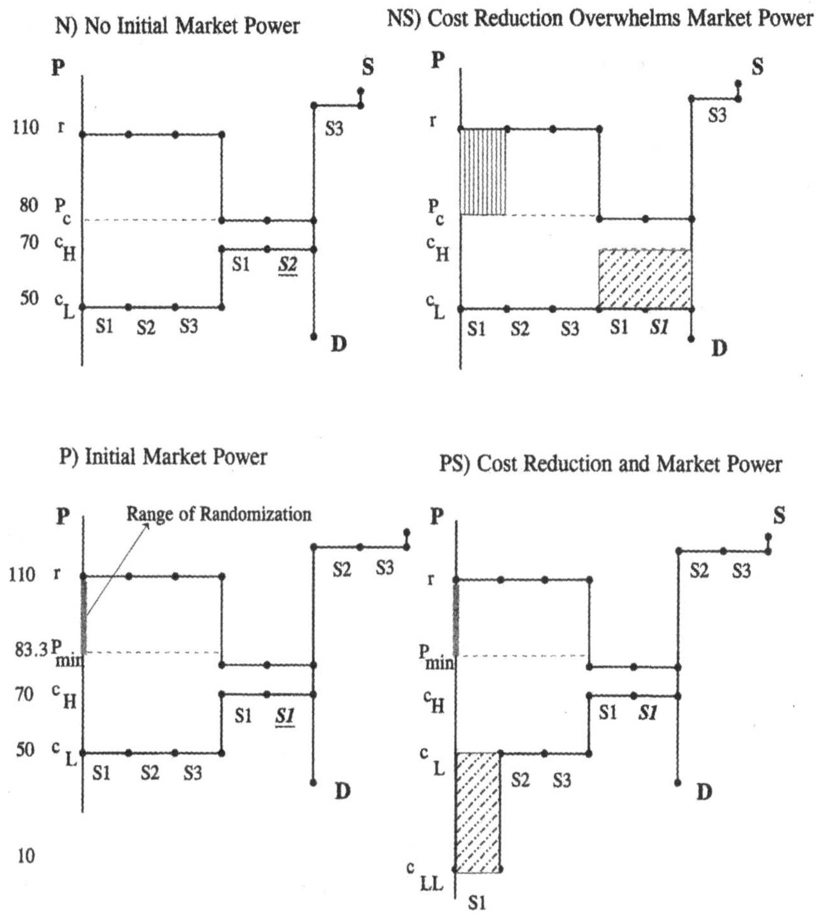

Figure 1. Experimental design. Starting with no initial power in market N, cost savings are combined with a reallocation of units that ensures seller $S1$ the sale of a unit at limit price r. When the cost savings affect the marginal units, no power is created, as seen in market NS. But the reallocation creates power when the synergy reduces costs on inframarginal units, as shown in market PS. Starting with initial power in market N, cost savings on marginal units eliminate initial power (market NS), while cost savings on invramarginal units leave initial power unaffected (market PS)

appendix which is available from the authors upon request. In an equilibrium, all three firms randomize over the support $[P_{\min}, r]$ where $P_{\min} = \frac{r+2c_H}{3} > P_c$. Seller $S1$'s cumulative mixing distribution is

$$F(p) = 1 - \frac{r + 2c_H - 3c_L^{\text{Small}}}{3p - 3c_L^{\text{Small}}}, \tag{1}$$

where c_L^{Small} is a small firm's cost of production. Notice that $F(r) = (2r - 2c_H)/(3r - 3c_L)$, which implies that the cdf for large seller $S1$ has a probability mass at r of $f(r) = (r + 2c_H - 3c_L)$. Setting $F(p) = .5$ and solving yields the median price of $S1$'s mixing distribution,

$$\tilde{p}_{\text{Large}} = \begin{cases} 2P_{min} - c_L & f(r) \leq .5 \\ r & f(5) > .5 \end{cases}.$$

In a symmetric equilibrium, the small sellers' cumulative mixing distribution is

$$G(p) = \frac{3p - 3c_H - r}{2p - 2c_H}. \tag{2}$$

$S2$ and $S3$ mix over the entire range $[P_{\min}, r]$ with no probability mass at either limit of the support. Setting $G(p) = .5$ yields the median for the small sellers of $\tilde{p}_{\text{Small}} = \frac{c_H + r}{2}$.

2.1 Market power and cost reductions

Consider now the interrelationships between cost savings and market power that are the primary focus of this paper. Suppose first that seller $S1$ initially has market power, as shown in market P in the bottom half of Figure 1. Reducing the costs for seller $S1$'s two marginal units from c_H to c_L can eliminate $S1$'s initial power, as seen by examining market P shown in the lower left panel and market *NS* in the upper right panel. Despite seller $S1$'s certainty of selling a single unit at price r in market *NS*, the cost reduction eliminates seller $S1$'s power, if the extra earnings from selling the two marginal units at price P_c (shown as the box with the diagonal lines) exceed the earnings forgone from selling a single unit at r (shown as the box with the vertical lines).

Cost reductions may also leave initial power unaffected, if the reduction affects only infra-marginal units. This situation is illustrated by comparing the supply and demand arrays shown as markets P and *PS* in the bottom panels of Figure 1. Reducing $S1$'s low cost unit from c_L to c_{LL} leaves initial power unaffected, since the extra earnings realized on the certain-to-sell unit does not affect seller $S1$'s decision to sell or to not sell his or her higher cost, marginal units at the competitive price. Notice that the power in market *PS* is identical to that in market P, in the critical sense that reducing the cost of $S1$'s infra-marginal unit from c_L to c_{LL} does not affect the equilibrium mixing distributions: The cost reduction does not affect profits for $S2$ or $S3$ and thus it does not alter the expression seller $S1$'s mixing distribution, $F(p)$ in equation (1). Further since seller $S1$ is certain to sell his or her infra-marginal unit c_L or c_{LL} do not appear in equation (2), the small sellers' mixing distribution.

Cost savings also can exert an impact on capacity reallocations that might create market power. To see this, consider again the supply and demand arrays shown in Figure 1 as market N. Comparing this initial no-power regime to markets *NS* and *PS* on the right side, the cost schedule for seller $S1$ changes in two ways. First, seller $S1$ acquires a high cost unit from seller $S2$. Second, seller $S1$'s unit costs are reduced, as was the case changing market P to markets *NS* and *PS*. The first change potentially creates power, since with the reallocation seller $S1$ is certain to sell at least one unit at a price of r. However, power interacts with the cost savings. In market *NS*, the costs of marginal capacity fall enough so that power is not created. In market *PS* the cost synergy occurs on seller $S1$'s infra-marginal units, and so the reallocation of units creates power, as specified in the mixing distributions in equations (1) and (2).

3 Experiment design and procedures

To evaluate the behavioral relevance of the predictions regarding cost savings and market power developed above, we conducted a market experiment. Importantly, the conformance of observed outcome to Nash prediction should not be regarded as a forgone conclusion in this design. In the experimental research most closely related to the present design (Davis and Holt, 1994; Wilson, 1998), the correspondence between Nash predictions and behavior was very poor, despite the general observation of predicted comparative-statics results. We had no reason *a priori* to believe that conformance to Nash mixing predictions would be any better here than in the previous studies, and thus it was for us an open question as to whether deviations from Nash predictions would dominate the comparative statics predictions here.

The experiment consisted of 16 laboratory markets using parameterized combinations of the markets shown in Figure 1. Each market lasted a total of 60 trading periods, with a regime change after period 30. Market design N with no initial power was used for the first 30 periods of eight sessions, and market design P with initial market power was used for the first 30 periods of the remaining eight sessions. For each initial condition, the period-31 regime change reduces the cost of marginal units for one-half of the sessions, either alone, or in conjunction with a unit reallocation from seller $S2$ to seller $S1$, to create the terminal condition without market power, shown as market design NS. For the remaining sessions, we reduce the cost of an infra-marginal unit for $S1$, either alone, or in conjunction with a reallocation of capacity from $S2$ to $S1$ to generate the terminal condition with power, shown as market design PS.

Value and cost parameters are set at $r = 110¢$, $P_c = 80¢$, $c_H = 70¢$, $c_L = 50¢$, and $c_{LL} = 10¢$. As summarized in Table 1, these parameters generate a clear separation of predictions in the power and no-power regimes, while still ensuring a reasonable saliency of payoffs for the small sellers. Inserting r and c_H into equation (1) generates the lower bound of the mixing distribution $P_{\min} = 83.33¢$, only slightly above $P_c = 80¢$. However, most of the pricing density is far above $P_{\min}$. Notice in the top row of P and PS columns in Table 1 that the mean of posted prices for seller $S1$ is 102.96¢, and the median price is 110¢. The mass point in $S1$'s mixing distribution at $r, f(r) = .55$, explains the very high prediction for $S1$'s mean and median posted price. The second row in Table 1 reports that the mean and median prices for small sellers are 92.06¢ and 90¢, respectively. As listed in the third row, the overall mean *transaction price* prediction for the power sessions is 94.61¢, almost exactly half the distance between reservation price $r = 110¢$ and $P_c = 80¢$.[6] Notice that in the lower half of Table 1 that the expected per period earnings range from 40¢ (lab currency) to 100¢ (lab currency) for seller $S1$, depending on power conditions, and the presence or absence of

[6] We numerically determine the mean posted price predictions for sellers $S1$, and $S2$ and $S3$. Calculating the overall mean transaction price is much more complicated and is estimated via simulation. Intuitively, this overall mean is far below the seller $S1$'s mean posted price, because the low-pricing sellers account for the bulk of the transactions in each period.

Table 1. Equilibrium predictions

Seller	Regime			
	No initial power (*N*)	Initial power (*P*)	No power with cost savings (*NS*)	Power with cost savings (*PS*)
	Price			
S1	80	102.96[a] (110[b])	80	102.96[a] (110[b])
S2 and S3	80	92.06[a] (90[b])	80	92.06[a] (90[b])
Overall (transaction prices)	80	94.61[a]	80	94.61[a]
	Earnings (per period)			
S1	40	60[c]	90	100[c]
S2 and S3	30	33.3[c]	30	33.3[c]

[a] Mean price. Note: Posted price means are determined numerically. Transaction price means are determined via the simulation of 5,000 periods where sellers priced according to equilibrium mixing distributions.
[b] Median price.
[c] Expected earnings.

cost savings. Expected earnings are either 30¢ or 33.3¢ (lab currency) for small sellers $S2$ and $S3$ in all treatments.

Individual sessions are identified by a combination of letters distinguishing the two treatment conditions and a session number in sequence, as listed in column (1) in Table 2. For example, session $N/NS\,1$, listed at the top row of column (1), indicates the first of four sessions conducted using the no-initial-power market design N for periods 1–30, and then the no-power-with-cost-savings market design NS for periods 31–60. Inspection of the entries in column (1) reveals the experiment's 2×2 incomplete-block design. This design allows for insight into the way that different types of cost savings interact with market power, both in markets with and without incipient market power incentives. Varying the initial power conditions is a crucial design consideration, since previous research indicates that the pricing history interacts prominently with induced incentives. Initial prices affect both the degree to which firms exert subsequent market power, and the competitiveness of markets without power (e.g., Davis and Holt, 1994; Wilson, 1998). However, we do not reverse the order of the cost savings treatment. This design choice is reasonable, since the issue of policy interest concerns only the realization and not the elimination of cost savings. Moreover, as will be seen below, restricting the cost savings treatment to the second session-halves permits some examination of interaction effects between cost savings and the introduction or elimination of market power.

A final interesting feature of this design is that it provides an alternative market environment for examining the power of mixed strategy equilibrium predictions to organize outcomes. In the relatively few market experiments for which

Table 2. Experiment design, observed mean transaction prices (standard deviations), and pass-through rates for the last half of each treatment condition

(1) Treatment	(2) N	(3) P	(4) NS	(5) PS	(6) PTR[a]
equilibrium price prediction	80 (0.00)	94.61 (5.39)	80 (0.00)	94.61 (5.39)	
$N/NS1$	80 (0.00)[b]		80 (0.00)		0%
$N/NS2$	82 (2.80)		102 (5.53)		−92%
$N/NS3$	80 (0.00)		81 (2.58)		0%
$N/NS4$	80 (0.00)		80 (0.00)		0%
$N/PS1$	80 (0.00)			107 (4.36)	−47%
$N/PS2$	80 (0.25)			97 (7.11)	−39%
$N/PS3$	82 (3.37)			95 (8.10)	−28%
$N/PS4$	80 (0.00)			101 (5.12)	−38%
$P/NS1$		90 (6.27)	80 (0.00)		51%
$P/NS2$		90 (4.89)	80 (0.00)		69%
$P/NS3$		88 (6.08)	80 (0.00)		71%
$P/NS4$		89.8 (4.63)	78 (3.33)		53%
$P/PS1$		95 (6.57)		86 (6.39)	12%
$P/PS2$		98 (6.21)		95 (4.37)	1%
$P/PS3$		91 (5.47)		94 (8.22)	−2.5%
$P/PS4$		93 (4.39)		100 (3.66)	−6%

[a] Pass-through Rate (PTR) is the percentage reduction in mean prices for units that experienced a cost reduction, divided by the percentage reduction in unit costs. To mitigate learning and transition effects, only data from the second half of the pre-synergy and post-synergy sequences of each session are used for the calculations. Formally,

$$PTR = \frac{(\overline{\text{Price}}_{46-60} - \overline{\text{Price}}_{16-30})/\overline{\text{Price}}_{16-30}}{(\text{Cost}_{46-60} - \text{Cost}_{16-30})/\text{Cost}_{16-30}} \times 100\%.$$

$PTR > 0$ implies that prices fell when the cost savings were introduced, and $PTR = 1$ implies that cost savings were passed fully through to consumers. Negative PTR values indicate that prices increased after the synergy.

[b] Each entry reports the mean and (standard deviation) observed in the last 15 periods of each treatment condition. The theoretical mean and standard deviation of transaction prices are estimated via the simulation of 5,000 periods, where sellers priced according to the Nash mixing prediction.

the equilibrium price distributions have been calculated, results vary widely. As mentioned at the outset of this section, mixing predictions organize outcomes very poorly in related environments with asymmetric sellers, despite the general observation of predicted comparative-statics results. Nevertheless, in markets with symmetric sellers, both Kruse (1993), and Kruse, Rassenti, Reynolds, and Smith (1994) report that the central moments of equilibrium mixing distributions organize observed behavior quite well. We discuss this further in the Section 4.

The experiment involved 48 undergraduate students who were recruited from economics courses at Middlebury College. Each subject participated in only one session, and no one had ever participated previously in a posted-offer market experiment.[7] Sessions were conducted manually with participants seated in visually isolated booths. At the outset of each session, a monitor read the instructions aloud as participants followed along on a printed copy. The instructions include a practice period from a market with a structure different from that used in any session. The role of the "advantaged" seller $S1$ was determined at the end of the instructions by eliciting optimal responses for a three-round alternating offer game. The most nearly optimal response was identified as the winner and was given the $S1$ folder.[8]

Within the session participants privately recorded price and output decisions each period. Once all decisions were complete, a monitor publicly announced the price and quantity decisions as well as the subsequent sales. Participants then recorded earnings prior to making decisions for the following period. (The monitor independently maintained the decisions and earnings on a spreadsheet.) The subjects made decisions in light of full information about the market demand schedule, but they were only given private information on costs and capacities.[9] Participants were told neither the total number of decision periods in the session, nor the number of periods in either regime. Rather, a monitor explained only that after a "large number of decisions in one regime, new cost schedules would be passed out, and a second regime would begin." Including the instructions, each session lasted between 100 and 120 minutes. Laboratory earnings were converted to US currency at a $3 laboratory equal $2 U.S. rate, and participants were paid

[7] Roughly half the participants had experience in an unrelated quantity-setting experiment.

[8] In the case of a tie, the $S1$ role was selected randomly from among the tied participants. Two considerations motivated our use of an alternating-offer game to assign positions. First, we wanted to mitigate the importance of fairness considerations as a behavioral motivation for the advantaged sellers. As shown by Hoffman and Spitzer (1982) and Hoffman et al. (1994), subjects tend to behave more in their own self-interest after they have won the right to be in an advantaged position. Second, we were interested in having a strategically astute player in the $S1$ position. The alternating-offer game was used instead of the more conventional selection devices, such a trivia-quiz, in order to increase the likelihood that the advantaged seller might appreciate the potential significance of his or her position. Finally, we are grateful to a referee for observing that using a game of skill rather than a game of chance to select an advantaged participant may, in some contexts, have undesirable consequences. The frustration of losing a game of skill may incite the disadvantaged players, both individually and collectively, to try to "beat" the advantaged player, even at some personal cost. Although we did see some scattered attempts on the part of small sellers to undermine the large seller here, these effects were not persistent and never resulted in any disciplining of the large seller. The very weak position of the small sellers in our design may explain the paucity of retaliatory efforts induced by selecting positions with a game of skill.

[9] Costs were not provided as public information in order to make it more difficult for sellers to attempt to maximize relative rather than absolute earnings. "Rivalistic" relative earnings maximization behavior has been observed in some full information oligopolies. See Davis (1999) and Rassenti, Reynolds, and Smith (1996). Even though the structure of pricing and contracting reveals very little information about costs in the posted-offer environment, our large sellers had little difficulty in identifying and exerting their market power, as is discussed in the results section. Perhaps this can be attributed to the limited number of market participants, and to the simple supply and demand parameters in our design. Potential sales quantities became obvious for each participant after a very few periods, despite a lack of information about earnings.

a $6 show-up fee plus their earnings from the session.[10] Per-participant earnings averaged $25.40 US.

4 Results

The four panels of Figure 2 illustrate the average transaction price paths for the four sessions in each treatment. Within each panel the contract price paths for individual sessions are shown as thin dotted, dashed, solid and dash-dotted lines, while the bolded line illustrates the mean price path for the four sessions in each treatment. (Individual session price paths are dithered slightly to help distinguish sessions about the competitive prediction.) We evaluate the results in light of the competitive prediction and the expected mean price for the market power sessions, shown as a solid and as dotted lines, respectively.[11] From the figure, it is apparent that power conditions affect performance markedly. For sequences conducted under a no-power condition (*N* or *NS* treatments), the mean price path tends to hover close to the competitive prediction. Higher and much more variable prices are observed in sequences conducted under a power condition (*P* or *PS* treatments). Further evident from the figure is that synergies do not exert any obvious effect on pricing performance. This can be seen by comparing price paths for sequences in a synergy condition (*NS* or *PS* treatments) with results for a pre-synergy condition, conducted under comparable power conditions (*N* or *P* treatments, respectively).

In what follows, our experimental results are summarized as a series of four findings. In addition to the qualitative results displayed in Figure 2, the transaction price means and standard deviations for the last half of each treatment sequence, shown in Table 2, are used as the basis for quantitative support. We begin with a consideration of pricing performance in the no-power treatments, which is largely a calibration result.

Finding 1: *Markets in the no-power treatments tend to stabilize on the competitive outcome. This is true both in the initial no-power treatment and in the terminal no-power-with-cost-savings treatment.*

Support: Consider first the initial no-power sequences of the *N/NS* and *N/PS* sessions, shown at the top of Figure 2. In each session prices collapse on the competitive prediction. As summarized in column (2) of Table 2, mean transaction prices for the last 15 periods of each sequence exactly equal the 80¢ competitive prediction in six of the eight instances, while mean prices for the remaining two instances are both just 82¢. Similarly, the standard deviation of transaction prices collapses on the competitive prediction of 0¢, in five of eight sequences, and a

[10] To reduce the possibility of having to cancel sessions due to participants failing to keep their appointments, an alternate was recruited in most sessions. The alternate was paid $10, and remained eligible to participate in future sessions.

[11] For purposes of brevity, the results are presented exclusively in terms of price outcomes. Results for efficiency, sales quantity and profit extraction rates (available from the authors on request) parallel our price observations.

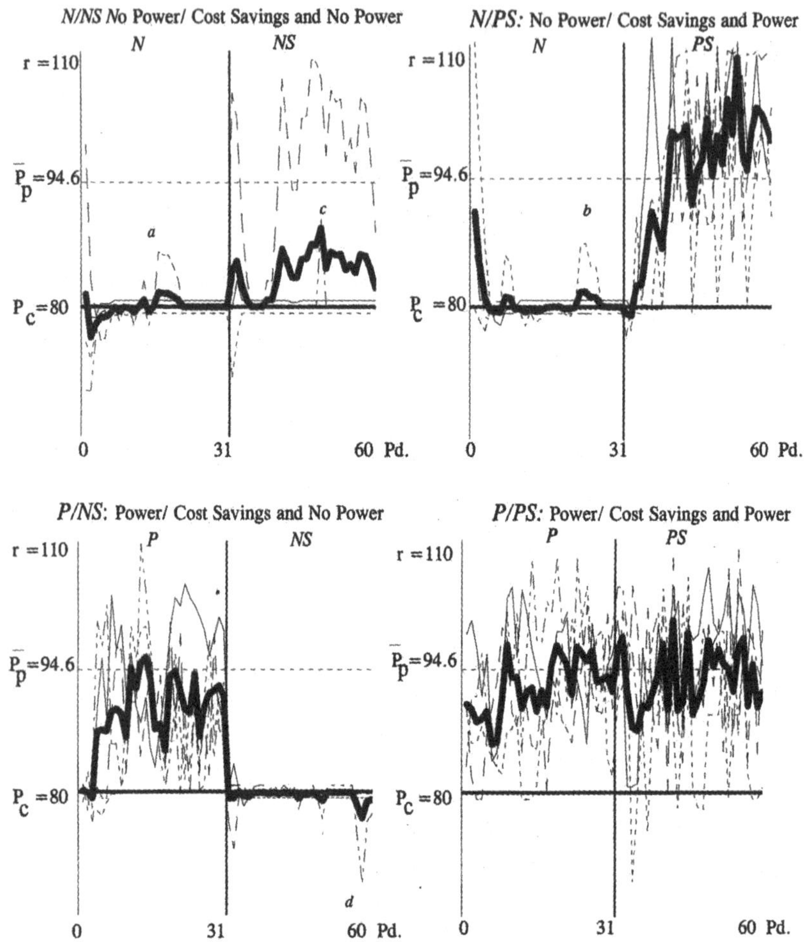

Figure 2. Mean contract price paths. Key: For each panel the mean contract price paths for the 4 sessions in each treatment are shown as thin lines (to facilitate distinction of the sessions prices are dithered slightly). The bolded line is the overall mean price. The solid line illustrates the competitive reference, P, while the dashed line illustrates the mean predicted price for the market power treatment, P. The italicized letters are reference points discussed in the text

standard deviation of 0.25¢ is observed in a sixth session. Further, single brief episodes of price increases explain the somewhat larger standard deviations in the N sequences of sessions $N/NS2$ and $N/PS3$. Notice in Figure 2 that session prices spike upward, as highlighted by the "a" and "b", when the seller-type $S1$'s fail to induce cooperation by repeatedly posting the reservation price.

The results are almost as marked for the sequences conducted in the no-power/cost savings treatment, as summarized the eight no-power with cost-savings session shown in the right half of the N/NS and P/NS panels of Figure 2, and in column (4) of Table 1. For five of the sequences, the mean transaction price equals the 80¢ competitive prediction, and in two of the remaining three sequences, deviations from the competitive prediction are two cents or less. These sequences are also fairly stable. Standard deviations of 0¢ are generated in five instances. Furthermore, notice that in two of the three remaining instances single-period price peaks cause the higher standard deviations. The "*c*" and "*d*" labels in Figure 2 identify these two cases.[12] □

Session $N/NS2$, however, is a distinct outlier. At 102¢, the mean observed price for the last 15 periods of the *NS* sequence of this session far exceeds the equilibrium mean prediction for the mixing distribution in the power sessions. This is also the second highest mean price observed for any 15 period sequence in the entire experiment. The observed price standard deviation of 5.53¢ matches very closely the 5.39¢ standard deviation predicted under equilibrium mixing in the power treatment. Seller $S1$ tried persistently to raise prices via signaling throughout this session. Due to a difference in the structure of the competitive equilibrium across the treatments, his price-increasing efforts are notably more successful in the *NS* market than in the *N* market. In the initial *N* design, sellers $S2$ and $S3$ can prevent $S1$ from selling any units by undercutting any price posting above P_c by seller $S1$. Thus, any attempt by seller $S1$ to signal a high price will result in *zero* sales for seller $S1$. However, seller $S1$'s signaling costs fall in the *NS* treatment. Sellers $S2$ and $S3$ each have only a single profitable unit to offer in the *NS* treatment, so seller $S1$ is always certain to sell at least one unit.

The effects of this change in signaling costs across treatments are displayed in Figure 3. In the figure, the vertically-aligned '♦', '+', and '×' symbols represent the price postings per period for sellers $S1$, $S2$ and $S3$, respectively. The vertical number of dots (•) at the bottom of the chart indicates the quantity of units sold by seller $S1$ per period. In the *N* sequence, shown in the left side of the figure, high price postings leave seller $S1$ out of the market for half of the first 20 periods. The seller largely reverts to competitive price postings after registering no sales in periods 15–18, 20 and 21. Seller $S1$'s efforts were costly; his earnings for the sequence averaged 24.1¢ per period, which is substantially less than the 40¢ per period earnings attainable by posting the competitive price.

In contrast, seller $S1$ sold at least one unit in each period of the *NS* sequence, shown on the right side of Figure 3. Due to this lower cost of signaling, seller $S1$ posted supra-competitive prices all but three of the 30 periods of the *NS* sequence, and in the each of the last 23 periods. Although seller $S1$ ultimately succeeded in raising the average transaction price in the *NS* sequence, his efforts were still costly. Seller $S1$'s average per period earnings for the entire *NS* sequence were 81.33¢, the lowest of the eight $S1$-type sellers in the *NS* sequences. Even toward the end of the sequence, when seller $S1$'s signaling efforts were beginning to generate supra-competitive profits, his mean per period earnings rose to only slightly above the competitive level. For example over the last 15

[12] Finding 1 is not supported with evidence from statistical tests, because such tests are relatively uninformative in this case. For each of the measures discussed above, the null hypotheses that mean prices or standard deviations do not deviate from competitive predictions cannot be rejected, even at very low levels of confidence. However, failure to reject the null hypothesis does not establish the truth of the null hypothesis. Rather, the best that can be done is to reject the predictions of alternative theories, as is done in the supports of Findings 2 and 3 below.

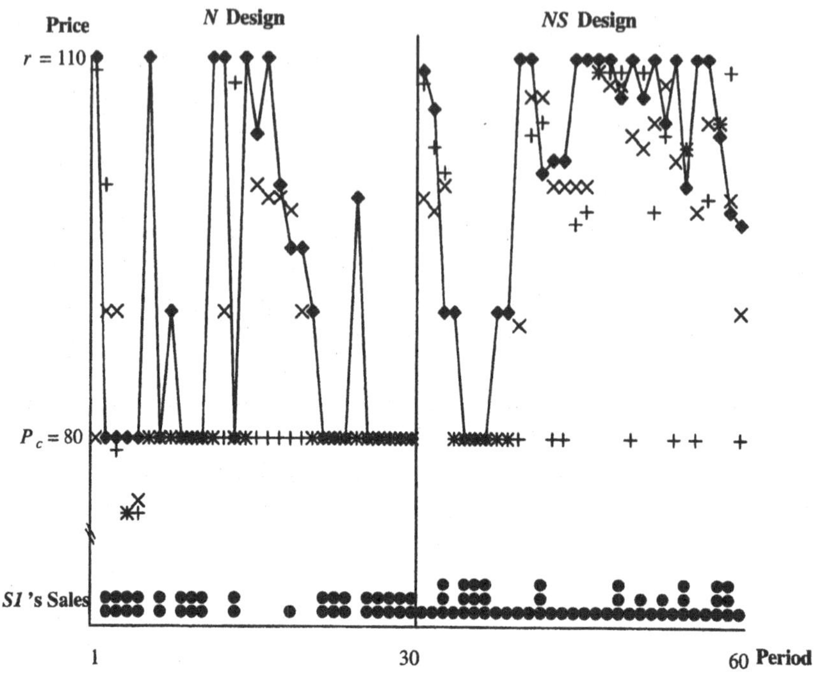

Figure 3. Sequence of Price Postings for Session $N/NS2$. Key: Vertically-aligned '♦', '+', and '×' markers are price postings per period for sellers $S1$, $S2$ and $S3$, respectively. The vertical number of dots (•) at the bottom of the figure indicates the quantity of seller $S1$'s sales per period

periods of the sequence, seller $S1$'s mean per period earnings are 91.87¢, a mere 2% above the 90¢ per period earnings at the competitive level.[13] Although $S1$'s behavior is atypical, and ultimately unprofitable, it highlights a potential dynamic consequence of cost savings even on marginal units, when markets are thin.[14]

Consider now the effects of introducing market power. We summarize our results in this respect as a second finding.

Finding 2: *Market power makes a difference. Holding other things constant, the introduction of power raises prices significantly.*

Support: Consider again the mean transaction price data shown in Figure 2. A comparison of the four sessions in the P/NS treatment with those of the N/PS treatment provides the cleanest test of the effect of market power. All prominent effects other than power are either held constant (e.g., the identity of the sellers), or controlled through blocking (e.g., order-in-sequence and the presence or absence of cost savings). These panels show that the power sequences generate higher and more variable prices than their counterpart no-power sequences. More formally, consider the mean prices for the last 15 periods of each session

[13] The seller knew that his signaling strategy was risky. After the session he expressed some regret for his strategy choice as he calculated his foregone earnings (about $7 U.S.).

[14] Although rare, signaling of this type has been occasionally observed in other laboratory markets. For example, Cason and Williams (1990) report a similar, albeit somewhat less intensive instance of signaling in a posted-offer market.

conducted in the power and no-power treatment and summarized in the second and third blocks of rows in Table 2. Using a Wilcoxon matched pairs test, mean prices are significantly higher in the power sequence than in the corresponding no-power sequence at a 99% confidence level.[15] Similarly, outcomes are more variable in the power sequences. Again, using a Wilcoxon matched pairs test, the standard deviation of prices is significantly higher in the power treatment than in the corresponding no-power treatment at a 99% level of confidence.[16] □

A third finding pertains to the effects of cost savings.

Finding 3: *Holding the degree of market power constant, cost savings do not affect mean prices in this environment. Price variability may be somewhat higher after the introduction of cost savings, but the evidence is mixed.*

Support: The eight sessions in the P/PS and N/NS treatments provide the evidence most pertinent to this finding. These sessions control for potential seller effects by maintaining constant seller identities within sessions as cost savings are introduced, and they control for potential market power effects by blocking power treatments across sessions. We find that the introduction of cost savings does not systematically reduce mean transaction prices. As shown in the upper block of rows in Table 2, prices in the N/NS sequence remain unchanged or nearly unchanged in three of the four sessions, and only rise sizably in session $N/NS2$ (discussed above). The more variable mean prices in the P/PS treatment sequence, summarized in the bottom block of rows in Table 2, fall somewhat with the cost savings in two sessions ($P/PS1$ and $P/PS2$) and increase slightly in the two remaining sessions ($P/PS3$ and $P/PS4$). Combined, we can not reject the null hypothesis that mean prices are unaffected by the introduction of cost savings, using the Wilcoxon matched pair test.[17]

However, these data suggest that the introduction of cost savings may affect price variability. The standard deviation of price increases with cost savings in two of the four sessions in the N/NS treatment sequence ($N/NS2$ and $N/NS3$). Although the standard deviation of transaction prices decreases in three of the four P/PS sessions ($P/PS1$, $P/PS2$ and $P/PS4$), the largest change in price variability observed in this treatment cell occurred in the remaining session ($P/PS3$). Using the Wilcoxon signed ranks test, the null hypothesis that cost savings do not

[15] The null hypothesis that power does not affect prices can be rejected at a 99% confidence level (Test statistic $W = 0$ is less than the critical value of 1 [8 degrees of freedom, 2-tailed test, $\alpha = .01$]).

[16] The null hypothesis that power does not affect the variability of prices can be rejected at a 99% confidence level ($W = 0$ is less than the critical value of 1, [8 d.f., 2-tailed test, $\alpha = .01$]). Comparing independent observations across treatments may also support these conclusions. Using a Mann-Whitney test, the null hypothesis that over the last 15 periods mean transaction prices for the eight P sequences do not differ from mean transaction prices for the eight N sequences may be rejected at a 99% confidence level ($U = 36$ is less than the critical value of 44 [8,8 d.f., 2-tailed test, $\alpha = .01$]). The parallel null hypotheses for the NS and PS treatments may similarly be rejected at a 99% confidence level ($U = 43$ is less than the critical value of 44 [8,8 d.f., 2-tailed test, $\alpha = .01$]). The null hypotheses of equal standard deviations may also be rejected for each comparison, at a minimum 99% confidence level. (Comparing the N sequences to the P sequences, $U = 36$ is less than the critical value of 44 [8,8 d.f., 2-tailed test, $\alpha = .01$]. Comparing NS sequences to the PS sequences, $U = 40$ is less than the critical value of 44 [8,8, d.f., 2 tailed test, $\alpha = .01$]).

[17] $W = 7.5$, which is much greater than the critical value of 2 [6 d.f. (2 ties), 2-tailed test, $\alpha = .10$].

affect price variability for all eight sessions can be rejected at a 95% confidence level.[18] Consideration of additional evidence in the N/PS and P/NS treatment cells casts some doubt on the hypothesis that cost savings increase price variability. Notice that the four N sequences in the N/PS treatment cell are independent from the four NS sequences of the P/NS treatment cell. Similarly independent are the four P sequences in the P/NS cell and the four PS sequences in the N/PS cell. When comparing sequences across treatment cells, we observe that the standard deviation of transaction prices exceeds zero cents in only one of the four NS sequences in the P/NS treatment, while the standard deviation of transaction prices exceeds zero in two of the four N sequences in the N/PS treatment cell. Transaction price standard deviations tend to be considerably larger in all the power sequences than in the no-power sequences, but a rank-ordering of the price standard deviations in the P sequences of the P/NS treatment cell with the PS sequences of the N/PS cell suggests no obvious difference in observed variability. Using a Mann-Whitney test, the null hypothesis that cost savings do not increase the standard deviation of transaction prices may not be rejected at any conventional level of confidence for either comparison.[19] □

Although observations across treatments are satisfactorily independent, interaction effects between the power and no-power conditions potentially confound comparisons of sequences across the P/NS and N/PS treatment cells. A critic could object to our observation that these data suggest that cost savings may not affect the variability of transaction prices by arguing that the change in power conditions somehow interacts with the introduction of cost savings to reduce transaction price variability. The absence of any obvious explanation for an interaction effect of this type leaves us reasonably comfortable with our above comments about price variability. Nonetheless, a comparison of mean transaction prices across the P/NS and N/PS cells suggests one potential interaction effect. Returning to Table 2, observe that mean prices for the P sequences of the P/NS treatment cell are uniformly lower than mean prices in the PS sequences of the N/PS cell. Even with the limited number of observations per cell available, the null hypothesis of no difference in mean transaction prices can be rejected at a 98% confidence level, using a Mann-Whitney test.[20] The reason for this difference in prices is unclear. Perhaps after suffering relatively low earnings in the no-power sequence of a session, neither the price-leading $S1$ sellers, nor the small $S2$ and $S3$ sellers are anxious to return aggressively to competitive pricing.

Inspection of pass-through rates for the cost savings induced in the various treatments of this experiment highlight the potential imprecision of pass-through rate (PTR) calculations as a predictor of the effects of future cost changes. The

[18] $W = 0$, which is less than the critical value of 1, [6 d.f. (2 ties), 2-tailed test, $\alpha = .05$].

[19] Comparing N sequences of the N/PS cell with NS sequences of the P/NS cell, $U = 16$ exceeds the critical value of 11 [4,4 d.f., 2-tailed test, $\alpha = .10$]. Comparing P sequences of the P/NS cell with PS sequences of the N/NP cell, $U = 16$ exceeds the critical value of 11 [4,4 d.f., 2-tailed test, $\alpha = .10$].

[20] $U = 10$, which is equal to critical value [4,4 d.f., 2-tailed test, $\alpha = .02$].

PTR calculations, shown in column (6) of Table 2 present relevant information. For each session *PTR* is the percentage reduction in mean prices for units underwent a cost savings, divided by the percentage reduction in unit costs. To mitigate learning and transition effects, only data from the last half of each pre- and post-cost savings sequence are used in the calculations. The *PTR* is not bounded either from above or below, but $PTR > 0$ implies that some of the savings were passed on in the form of lower prices and $PTR = 1$ implies that the pass-through was complete. Negative *PTR* values imply that prices rose as costs fell. The *PTR* calculations in Table 2 reveal far more about the change in power conditions that occurred with a synergy than the competitive state of the market: The large positive values in the P/NS sessions indicate that power was eliminated as the synergy was introduced, while the large negative values in the N/PS sessions indicate that power was injected along with the synergy.[21] With the exception of session $N/NS\,2$ (discussed above) *PTR* values are close to zero for the N/NS and P/PS sessions. By implication, these pass-through rates indicate nothing about the prospective effects of any further costs savings in these markets. In this context such effects may be predicted only with some information about the change in market power.

Finally, we turn attention to pricing behavior relative to Nash mixing predictions in the power treatment. As has been observed by previous commentators (e.g., Kruse, Rassenti, Reynolds and Smith, 1994), the obvious autocorrelation in prices due to sellers' inability to treat each stage game as independent and the limited information conditions characteristic of market environments invalidate any strict test of Nash mixing predictions in this environment. Nevertheless, a rather surprising result is that the central moments of the Nash mixing prediction characterize pricing behavior reasonably well in this *asymmetric* environment. This is our fourth finding.

Finding 4: *The central moments of Nash mixing predictions organize behavior quite well in this context.*

Support: Consider first the transaction price data for the power treatments summarized in columns (3) and (5) of Table 2. In 15 of the 16 sequences, mean transaction prices are closer to the 94.61¢ mean of the mixing distribution than to the 80¢ competitive prediction (the exception is the power-synergy sequence of $P/PS\,1$). The mean transaction prices are also closer to the mean of the mixing distribution than to the reservation price of 110¢ in all but one instance (the power sequence of session $N/PS\,1$, in column 5). Further, in 11 of the 16 sequences mean prices are within five cents of the mean price for the mixing distribution in the power treatment. Similarly, the standard deviation of observed transaction prices is uniformly closer to the 5.39¢ standard deviation predicted in the Nash mixing equilibrium than to the competitive prediction, and the observed standard deviations are within one cent of the prediction in 11 of the 16 half-sessions.

[21] Notice that the negative values in the N/PS sessions tend to be roughly half the magnitude of the P/NS sessions. This is an artifact of the calculations, since the cost reduction on the single unit in the *PS* sequence is twice the size of the per unit cost savings in the *NS* sequences.

Inspection of the price decisions allows more direct examination of behavior relative to Nash mixing predictions. The left and right panels of Figure 4 present the relevant empirical price distributions for seller $S1$ and for sellers $S2$ and $S3$, respectively. In each panel, the solid line illustrates the predicted pricing density, while bars indicate observed densities. The white bars indicate observed densities for the first 15 periods of each power sequence, and the shaded bars illustrate observed densities for the final 15 periods. Power sellers $S1$ exhibit a tendency to price at "nickel", and particularly "dime" price nodes of 80¢, 90¢, and 100¢. However, comparing shaded and clear bars, the pricing density at nickel and dime nodes diminishes in the second half of a session, and the pricing density at the reservation price increases as predicted under the equilibrium mixing distribution. Across session halves, the mean of posted prices for sellers $S1$ increases from 97.44¢ to 101.74¢, only slightly more than a penny away from the Nash prediction of 102.96¢.

Convergence toward the predicted posting densities for sellers $S2$ and $S3$ is somewhat less complete, as is seen in the right panel of Figure 4. In addition to a tendency to price at "nickel" nodes, considerable density lies below lower bound of the mixing distribution $P_{\min}$, particularly at 80¢. Nevertheless, when comparing the pricing densities across sequence halves, we observe that the price density at 80¢ falls considerably in later periods, and that the observed mean of the pricing distribution moves slightly closer to the predicted mean of 92.06¢, increasing from 89.38¢ to 93.74¢.[22] □

The equilibrium mixing distributions in Section 2 rely on the assumption of common and complete cost information. In this experiment, however, subjects were only given private cost information. Our finding that the sellers are behaving "as if" they have common cost information parallels the theoretical justification for such outcomes by Friedman and Ostroy (1995). Nevertheless, the organizing power of the Nash mixing distribution is surprising, given the very poor performance of Nash equilibrium mixing predictions in other asymmetric environments. Perhaps the nature of the asymmetry explains the improved performance. Here, a single seller with very considerable power is grouped with a pair of small sellers with no market power. A reasonable behavioral conjecture in this context is that sellers try to anticipate the actions of the others in making pricing decisions each period. This process becomes more difficult over time when only one large seller has market power. [See Davis and Wilson (1998) for a discussion on the decay of the price cycles in this and other posted-offer market power designs.] Eventually such a process could drive sellers to the pre-

[22] The capacity of Nash mixing predictions to organize pricing outcomes in this context should not be overstated. Even for the sample distribution that most nearly matches the theoretical mixing prediction (the second sequence-half for the type $S1$ sellers), the null hypothesis that observed prices are drawn from the theoretical mixing distribution can be rejected at a high level of confidence, using a Kolmogorov-Smirnov test. (For the type $S1$ sellers' pricing decisions in the second session halves, test statistic $K = .86$, just less than the critical value of .87 [240 d.f., 2-tailed test, $\alpha = .05$]. Empirical pricing distributions for type $S1$ sellers over the first session-halves, as well as for type $S2$ and $S3$ sellers in either session-half deviate from the equilibrium mixing distribution by a substantially larger margin.

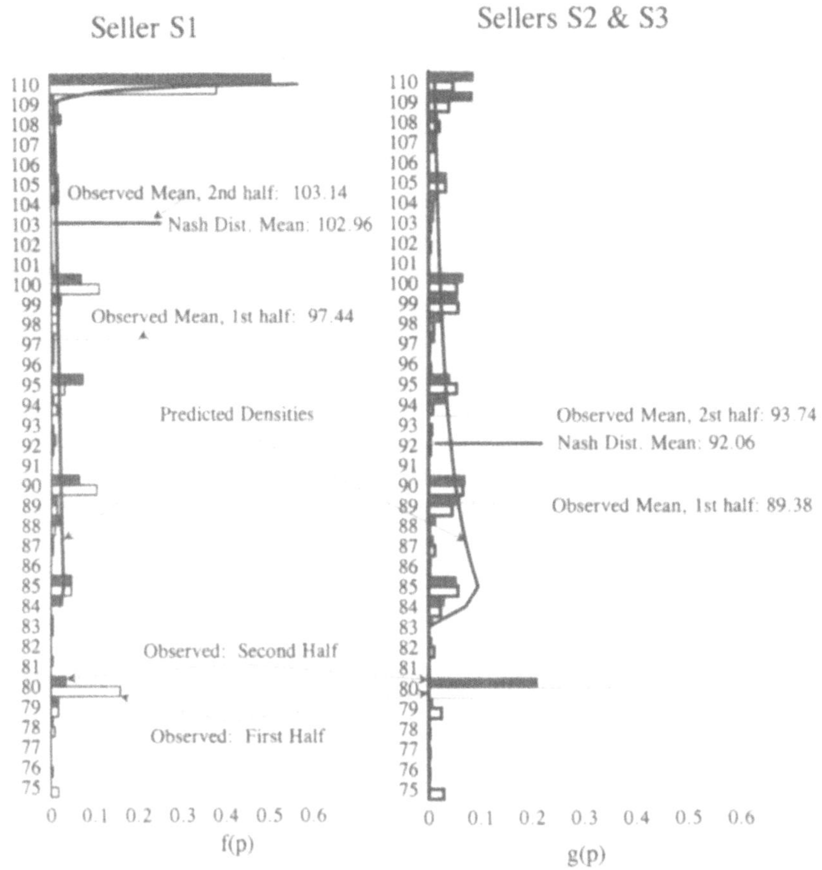

Figure 4. Predicted and observed pricing densities. In each panel the solid line illustrates the Nash equilibrium predicted density. The white and dark bars illustrate, respectively, the observed densities for the first half and last half all power treatment sessions

dicted equilibrium mixing distributions. In contrast, Nash predictions fared much worse in Davis and Holt (1994) and Wilson (1998) when multiple sellers had power and were symmetric to each other, but asymmetric with respect to the other sellers. Possibly, this particular type of design interacts with the sellers' decision-making process in a way that generates persistent deviations from static Nash-predicted outcomes. Sellers, for example, may generally make decisions with errors, but the error-process may generate persistent deviations from Nash-predicted outcomes only when the asymmetric sellers are sufficiently balanced. While our experiment provides evidence in support of a Nash equilibrium theory of market power, this question merits further investigation.

5 Discussion and policy implications

While interesting, the relation between observed prices and static Nash mixing predictions summarized in Finding 4 should not detract from our central results. The experiment reported here illustrates clearly the potentially important interrelationships that can exist between market power and firm-specific cost savings. With incipient market power, firm-specific savings may either leave power conditions unaffected or eliminate it altogether. Without initial market power, cost savings combined with a reallocation of units from a small producer to a large producer (as might be the case in a merger) may create market power, or it may maintain competitive outcomes. All of these outcomes depend on the way that the firm-specific savings affect the cost schedule. Another observation from this experiment may be pertinent to merger investigations. We detect that the no-power environment with cost savings is more susceptible to multi-period signaling and the concomitant higher transaction prices than the baseline no-power environment, because the opportunity cost of signaling is lower.

The policy implications of the experiment merit emphasis. Previous work on the study of post-merger pricing predictions with merger-specific cost savings has focused on the specific case of constant marginal costs. The application of this research to antitrust contexts has led to a focus on estimating historic pass-through rates for firm-specific cost savings. We have demonstrated that cost savings can have vastly different implications for post-merger welfare in a plausible case where firms have capacity-constraints and upward sloping marginal costs. Both competitive and anti-competitive outcomes can arise from cost savings, independent of a market's initial competitiveness. We do not conclude that the identification of historic pass-through rates is uninformative in assessing the competitive consequences of a merger. Rather, we contend that in conducting such an analysis some attention must also be devoted to assessing where on the market's supply schedule the alleged cost savings will be realized.

Further experimental work pertaining to this issue remains to be explored. In closing we suggest two directions for further research. First, in the environment used here the number of sellers was held constant, and capacity was reallocated among these sellers. It would be useful to conduct a series of sessions where mergers actually occurred, and the number of sellers was reduced as the cost synergy was realized. (The difficulty of generating Nash mixing predictions in this considerably more complex environment deterred us from using some variant of this design here.) However, it would, we think, be interesting to evaluate performance in this context even if closed-form solutions for the equilibrium Nash mixing predictions do not exist. Second, antitrust authorities appear to place much more emphasis on alternative institutional environments in their analysis of mergers. Cournot competition and differentiated-product Bertrand competition are of particular interest. Again, it may be useful to evaluate parallel results in these more complicated institutional contexts.

References

Ashenfelter, O., Ashmore, D., Baker, J., McKernan, S.: Identifying the firm-specific cost pass-through rate. FTC Working Paper #217 (1998)

Cason, T., Williams, A.: Competitive equilibrium convergence in a posted-offer market with extreme earnings inequities. Journal of Economic Behavior and Organization **14**, 331–352 (1990)

Davis, D.: Advance production and cournot outcomes: An experimental investigation. Journal of Economic Behavior and Organization (forthcoming) (1999)

Davis, D., Holt, C.: Market power in laboratory markets with posted prices. Rand Journal of Economics **25**, 467–487 (1994)

Davis, D., Wilson, B.: Mixed strategy nash equilibrium predictions as a means of organizing behavior in posted-offer market experiments. Manuscript, Virginia Commonwealth University (1998)

Farrell, J., Shapiro, C.: Horizontal mergers: An equilibrium analysis. American Economic Review **80**, 107–26 (1990)

Federal Trade Commission v. Staples, Inc.: 970 F. Supp. 1066, 1090 (D.D.C. 1997) (Hogan, J.)

Friedman, D., Ostroy, J.: Competitivity in auction markets: An experimental and theoretical investigation. Economic Journal **105**, 22–53 (1995)

Hoffman, E., Spitzer, M.: The Coase theorem: Some experimental issues. Journal of Law and Economics **25**, 73–98 (1982)

Hoffman, E., McCabe, K., Shachat, K., Smith, V.L.: Preferences, property rights, and anonymity in bargaining games. Games and Economic Behavior **7**, 346–80 (1994)

Holt, C.: The exercise of market power in laboratory experiments. Journal of Law and Economics **32**, S107–S130 (1989)

Holt, C., Solis-Soberon, F.: The calculation of mixed-strategy equilibria in posted-offer markets. In: Isaac, R.M. (ed.) Research in experimental economics, Vol. 5. Greenwich: JAI Press 1992

Kruse, J.: Nash equilibrium and buyer rationing rules: Experimental evidence. Economic Inquiry **31**, 631–446 (1993)

Kruse, J., Rassenti, S., Reynolds, S., Smith, V.L.: Bertrand-Edgeworth competition in experimental markets. Econometrica **62**, 343–371 (1994)

Rassenti, S., Reynolds, S., Smith, V.L.: Adaptation and convergence of behavior in repeated experimental Cournot games. Manuscript, University of Arizona (1996)

United States Department of Justice and Federal Trade Commission: Horizontal Merger Guidelines, Section 4 (revised) (1997)

Werden, G.: A robust test for consumer welfare enhancing mergers among sellers of differentiated products. Journal of Industrial Economics **44**, 409–413 (1996)

Wilson, B.: What collusion? Unilateral market power as a catalyst for countercyclical markups. Experimental Economics **1**, 133–145 (1998)

Yde, P., Vita, M.: Merger efficiencies: Reconsidering the "passing-on" requirement. Antitrust Law Journal **64**, 735–747 (1996)

An experimental comparison of two search models*

Eric Abrams[1], **Martin Sefton**[2], **and Abdullah Yavas**[3]

[1] Department of Economics, Hawaii Pacific University, 1060 Bishop Street, Suite 402, Honolulu, HI 96813, USA (e-mail: eabrams@hpu.edu)
[2] Department of Economics, University of Newcastle, Newcastle-upon-Tyne, NE1 7RU, UK (e-mail: martin.sefton@ncl.ac.uk)
[3] Smeal College of Business Administration, Pennsylvania State University, University Park, PA 16801, USA (e-mail: ayavas@psu.edu)

Summary. We report an experiment designed to investigate markets with consumer search costs. In markets where buyers are matched with one seller at a time, sellers are predicted to sell at prices equal to buyers' valuations. However, we find sellers post prices that offer a more equal division of the surplus, and these prices tend to be accepted, while prices closer to the equilibrium prediction are rejected. At the other extreme, sellers are predicted to sell at a price equal to marginal cost when buyers are matched with two sellers at a time. Here, we find prices are closer to, but still significantly different from, the equilibrium prediction. Thus, our results support theoretical comparative static, but not point, predictions.

Keywords and Phrases: Experimental search markets, Price dispersion, Diamond paradox.

JEL Classification Number: C9, D4, L1.

1 Introduction

We consider behavior in markets where search is costly. Such markets are commonplace, perhaps the norm, and yet analyses rooted in the theory of perfectly

* We acknowledge the Division of Research and Ph.D./M.S. Programs at Pennsylvania State University for their funding of the experiments. We appreciate comments from Timothy Cason (co-editor), an anonymous referee, Robert Forsythe, Eric Maskin, John Morgan, Tom Rietz, Bradley Ruffle, participants at the Amsterdam Workshop on Experimental Economics, 1997, and the Conference on Economic Theory, Antalya, Turkey, 1997.

competitive markets fail to address many of their salient features. Partly for this reason, a large and important theoretical literature has emerged, devoted to markets where search is costly. The modern approach begins with Diamond (1971), and has been developed by, for example, Salop and Stiglitz (1977), Burdett and Judd (1983), Stahl (1989), and Bagwell and Ramey (1992).

Two aspects of this literature are striking. First, some of the results obtained are sharp, and not always intuitive. For example, in markets where it is costly for buyers to search Diamond concludes that sellers will charge monopoly prices even when there are many buyers and sellers, and even when the search costs are small (the so-called "Diamond Paradox"). Burdett and Judd show that this result relies crucially on the assumption that buyers get one price quote per search. If buyers get more than one price quote per search, sellers will charge the competitive price. This is a strong comparative static prediction and seems, in principle, easy to test. However, this provokes our second observation: it appears to be very difficult to confront these theories with naturally-occurring empirical evidence.

These observations underlie our motivation to use experimental methods to investigate some of the pricing predictions of models with costly search. We designed an experiment to provide a direct test of the theoretical results outlined above. In our sessions a buyer can buy only one unit per trading period, and the buyer's valuation for this unit is common knowledge. A seller can sell any number of units in a trading period at zero cost. In each trading period, sellers post prices and buyers are initially matched with sellers. Buyers can then buy from a seller with whom they are initially matched, or can choose not to buy and search for another seller.

This setup resembles a posted offer market, except that buyers in our experiment do not observe all posted prices, but instead only see the prices of the sellers with whom they are matched. In sessions involving our "Diamond" treatment, buyers are matched with one seller at a time. Under this treatment, sellers are predicted to post a price equal to the buyers' valuation and buyers are predicted to accept the initial price. This implies that sellers in the Diamond treatment appropriate the entire surplus, and there is no search. In our "Bertrand" treatment buyers are matched with two sellers at a time. Here, the theoretical prediction is that sellers post zero prices, which buyers accept, and again there is no search. Thus, in theory, simply allowing two quotes per search, rather than one, enables buyers to get the entire surplus.

We also note that these predictions are independent of the size of the search cost. This is somewhat counter-intuitive because, other things equal, a higher search cost might be expected to deter search, enabling sellers to post higher prices in the initial period. Theoretically, this argument falls down because, in equilibrium, buyers would not benefit from searching and so are deterred by any positive search cost. Nevertheless, behaviorally, search costs may matter. Davis and Holt (1996), in a related experiment, find that doubling search costs leads to higher prices. We set search costs equal to 5% and 15% of buyers' valuations in our low and high cost treatments respectively to investigate search cost effects.

We conduct 12 sessions, using different subjects in each session. These sessions involve balanced combinations of the two treatment conditions: three sessions featuring each of Diamond-low cost, Diamond-high cost, Bertrand-low cost, and Bertrand-high cost. Comparisons of the outcomes of these sessions allow a test of the theoretical predictions outlined above.

We find that transaction prices in our Diamond treatment differ substantially from the equilibrium prediction. Transaction prices offer close to an equal division of the surplus. We note that in our Diamond treatment the seller's ability to post a price generates ultimatum game-type incentives and argue that this finding is quite consistent with results from ultimatum game experiments. However, while deviations from an equal split of the surplus favor the proposers (sellers) in ultimatum games, such deviations favor the buyers in our experiment.

For the Bertrand treatment, there is also a divergence between observed and predicted prices. Nevertheless, a comparison with the Diamond treatment generally supports the comparative static theoretical prediction. Observed transaction prices in the Bertrand treatment are significantly lower than in the Diamond treatment. These findings hold for either search cost, and in fact we find no significant search cost effects in our experiment. The remainder of the paper is organized as follows. In Section 2 we review some related experimental literature. In Section 3 we describe the experimental environment in more detail. In Section 4 we present results and in Section 5 we conclude.

2 Related literature

In experiments on individual search behavior (for example, Schotter and Braunstein, 1981; Hey, 1982; Harrison and Morgan, 1990) "choices appear to correspond closely to the theory of optimal sequential search" (Davis and Holt, 1993, p. 78). These experiments, however, study search behavior within an individual choice context, whereas ours uses a market context.

Our market context resembles a posted offer market, where sellers post prices and buyers, observing these, decide whether to buy or not. Previous experiments with posted offer markets, beginning with Williams (1973) and reviewed in Davis and Holt (1993), suggest that transaction prices converge to the relevant competitive price prediction. This is true even when the underlying costs and valuations induce extreme earnings inequalities (Cason and Williams, 1990). Our experiment modifies the standard posted offer environment by incorporating buyer search.

The experimental environment that most closely resembles ours is the "search treatment" of Davis and Holt (1996). They compare a posted-offer market, in which all posted prices are public, to a "search" market, in which buyers must pay to view a seller's price. Their first conclusion is that search matters: when posted prices are not public, transaction prices (and posted prices) are higher than when posted prices are public. Their second conclusion is that the Diamond-paradox monopoly price is not consistently observed in this environment. In fact, in more than half of their sessions, the average prices were closer to the competitive price than the monopoly price.

This latter finding contrasted with that of Grether, Schwartz and Wilde (1988), who found that transaction prices converged to the equilibrium prediction in a posted-offer market with buyer search.[1] Davis and Holt investigated one possibility for the divergent results. In their original treatments prices were associated with a seller number whereas in the Grether et al. experiment sellers' identities were concealed from buyers. Thus, sellers in Davis and Holt experiment could build a reputation for selling at lower prices, thereby attracting more buyers through search. However, Davis and Holt conducted further sessions in which seller identities were concealed and found this not to be the case and so they concluded that seller reputations did not lower prices. Of course, numerous other differences in design could account for the differing results of these two experiments.[2]

A key aspect of our experiment is the information available to buyers when they meet sellers, and the matching technology by which they meet. In our environment buyers observe only the prices of sellers with whom they are matched, and, according to the treatment, see either a single price or two prices at a time.[3]

Thus in our Diamond treatment transactions take place within the context of a bilateral monopoly game. In this game the ability to search gives buyers an (uncertain) outside option, and the ability to post prices gives sellers a take-it-or-leave-it power somewhat similar to that of proposers in ultimatum game experiments [see Güth, Schmittberger and Schwarz (1982) or Forsythe, Horowitz, Savin and Sefton (1994)]. In theory, sellers should offer buyers just enough to induce them to buy now rather than take the outside option. However, it should be noted that in ultimatum game experiments such predictions fail: proposals almost always imply a division of the surplus that is closer to the equal-split than the predicted split.

In our Bertrand treatment transactions take place within the context of a three-player game, where a sole buyer chooses between buying from one of two sellers, or taking two (again uncertain) outside options. Here, competing sellers make initial offers, and so they may be substantially better for the buyer than the outside options. In fact, the only equilibrium has sellers offer the entire surplus to buyers, despite the sellers' ability to post prices.

This contrast is examined in Roth, Prasnikar, Okuno-Fujiwara and Zamir (1991), where outcomes from ultimatum game experiments are compared with

[1] In fact, four sessions employed this treatment and in three of them prices converged to the predicted price. The non-convergence in the first session was unexplained.

[2] For example, in Grether et al., but not in Davis and Holt, buyers were shown the entire distribution of prices prior to deciding whether to buy from the person with whom they were originally matched or whether to search.

[3] Note that buyers only observe the prices offered by sellers with whom they are matched, and do not observe the entire distribution of prices. While the Diamond-Burdett-Judd models assume buyers draw prices from a known distribution of prices, we preferred the design in which price information is only acquired via search, as in Davis and Holt. In either case the theoretical predictions are the same. It is not clear, a priori, whether one choice or the other facilitates convergence to equilibrium.

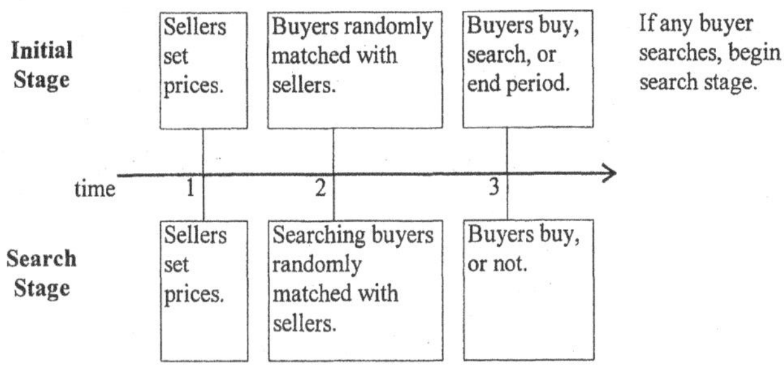

Figure 1. Market trading institution

those from a market experiment. In the ultimatum treatment, a "buyer" proposes a price and a randomly matched "seller" can accept or reject this price. If the seller rejects, both earn zero; if the seller accepts, the seller receives the price offered and the buyer receives a known valuation minus this price. In the market treatment, multiple (ten, to be exact) buyers submit offers to a single seller. The seller can accept or reject the highest offer. If accepted, the seller gains an amount equal to the price, the winning buyer gains the difference between the buyer valuation and the price, and all other buyers gain zero. If the seller rejects, all participants receive zero.

Roth et al. find that in the market environment transaction prices converge quickly to the buyers' common valuation (which is the equilibrium price prediction). Thus, almost all the surplus is competed away to the monopolist seller. This is in stark contrast to the ultimatum treatment, where proposed prices are close to half the buyers' valuation.

Our Diamond treatment is strategically similar to the ultimatum treatment of Roth et al., except that the search stage gives buyers a disagreement payoff that may differ from zero. Our Bertrand treatment differs from the Roth et al. market treatment in the same respect, and also in that we allow two, rather than ten, traders to compete in prices.

3 A market with costly search

3.1 Description of trading periods

Our experimental market is designed to capture important aspects of the Diamond-Burdett-Judd models. These in turn should be viewed as highly stylized versions of imperfectly competitive markets. All 12 sessions of our experiment were divided into 25 identical trading periods; next we describe the structure of a trading period in a session employing the Diamond treatment. The structure for each session is represented in Figure 1.

In each trading period there are eight buyers and eight sellers. Each seller can costlessly produce an unlimited number of goods so that for each unit sold the seller receives a payoff equal to the transaction price. Each buyer can buy at most one unit in a period, and receives 120 points less the transaction price when they purchase a unit. In addition, a buyer may incur search costs, as will be explained shortly. All traders were informed of these payoff features.[4]

In the initial stage of a trading period, sellers simultaneously post prices. (Sellers never observe the prices posted by other sellers.) Each buyer is then randomly matched with a seller, each seller being equally likely to be drawn independently of previous matchings or decisions; this means that a seller may be matched with none, some, or (very improbably) all of the buyers. The buyer observes the price posted by the seller with whom he or she is matched, and chooses whether or not to buy a unit from him or her. If the buyer chooses to buy, the round ends for this buyer and he or she earns a number of points equal to his or her valuation (120 points) less the price. If the buyer chooses not to buy, he or she then chooses whether or not to search. If the buyer chooses not to search, the trading period ends for this buyer and he or she earns zero points. If the buyer chooses to search, he or she incurs a search cost (6 points in the low search cost condition, 18 in the high search cost condition) and enters the search stage of the trading period. (If none of the buyers choose to search the search stage is omitted.)

At the beginning of the search stage the sellers again simultaneously post prices. (Sellers are not required to post the same price that they posted in the initial stage.) Each buyer that enters the search stage is then randomly matched with a seller (possibly with the same seller he or she was matched with in the initial stage). The buyer observes the price posted by the seller with whom he or she is matched, and chooses whether or not to buy a unit from that seller. If the buyer chooses to buy, his or her payoff for the period is the buyer's valuation, less the price, less the search cost. If the buyer chooses not to buy, his or her payoff is minus the search cost. This completes the trading period for all buyers and sellers.

For the Bertrand sessions, the structure is identical except buyers are matched with two sellers at a time. Thus, buyers observe two prices in the initial stage and can choose which, if any, to accept. If they choose not to accept either, and choose to search, they incur the search cost and are rematched with two more sellers in the search stage. Again, they can choose which, if any, of the two prices to accept.

3.2 Experimental procedures[5]

The experiment was conducted at the University of Iowa in Spring Semester 1997. Participants in the experimental sessions were recruited from courses in the College of Business, and were informed that they would earn between \$5 and \$15 for a session lasting less than 75 minutes. Each session involved 16 subjects,

[4] This contrasts with much of the experimental literature on posted offer markets, where sellers' costs and buyers' valuations are typically private information.

[5] Experimental instructions are available from the authors upon request.

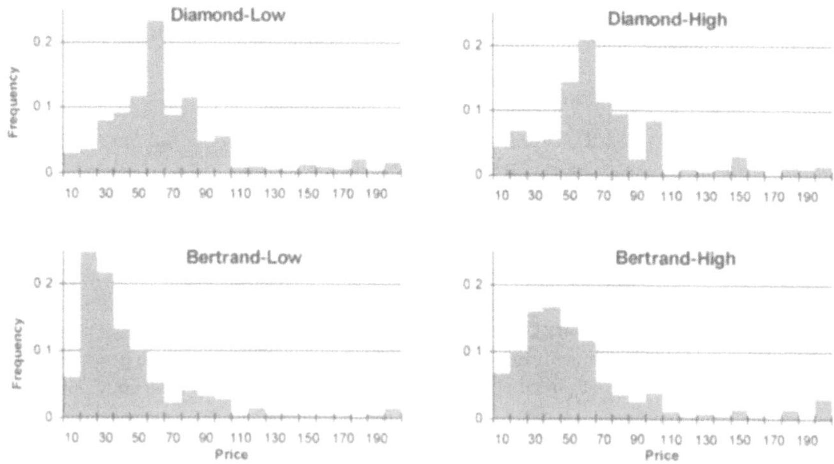

Figure 2. Histogram of initial stage posted price by treatment

who sat at visually isolated terminals. No subject was allowed to participate in more than one session. At the beginning of the session subjects were given a set of instructions which were then read aloud by the experimenter. Subjects were then given an opportunity to ask questions, and then the decision-making part of the experiment began.

Throughout the session no communication between subjects was permitted. All decisions were transmitted via computer terminals. Sellers chose prices by entering a whole number between 0 and 200 inclusive. If a seller did not enter a price within 30 seconds of receiving the prompt, the computer chose a price of 200 on their behalf. Buyers observed the seller identification number of the seller they were matched with, and the price posted by this seller. To choose not to buy, the buyer entered 0; to choose to buy, the buyer entered the identification number of the seller they wanted to buy from. If the buyer did not make a decision within 30 seconds of receiving the prompt, the computer chose not to buy on their behalf.[6]

Note that these trading restrictions permit sellers to choose prices in excess of buyers' valuations, and permit buyers to buy at these prices. Also, in the Bertrand sessions, buyers are free to buy at the higher of two prices if they wish. All of these behaviors are difficult to interpret, but we did not want to prohibit them by constraining the traders' strategy spaces.

The extreme earnings predictions associated with each treatment prompted us to give subjects an initial balance of 1200 points. This ensured that a subject who earned nothing from the trading periods would still earn $5 from the session. The initial balance also served to reduce potential losses from fruitless search or from purchasing at prices above valuations. Earnings from the trading periods were added to the initial balance and at the end of the session subjects received

[6] The time constraint was binding a total of 35 times (out of 12 sessions each with 25 periods, 16 subjects and up to 2 stages). This occurred no more than 7 times within a single session.

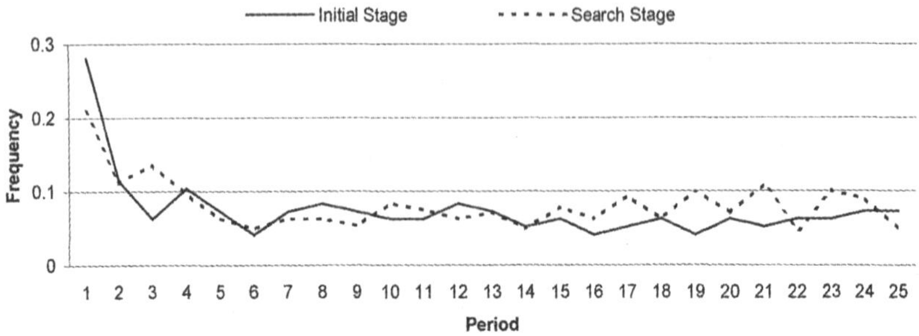

Figure 3. Proportion of posted prices exceeding buyer valuation. Note: In the initial stage the number of posted prices is 96 each period (12 treatments × 8 sellers per period). However, the number of posted prices is not constant in the search stage because in some treatments in some periods there is no search stage. The number of posted prices used in the search stage ranges from 40 to 96.

$1 for every 240 points accumulated. All sessions were completed in less than 75 minutes, minimum earnings were $5.75 and maximum earnings were $15.75. Average earnings were $10.97.

4 Results

4.1 Overview

Figure 2 presents histograms of prices posted in the initial stage of each trading period for each treatment.[7] Relatively few prices correspond to the equilibrium prediction and there is considerable variability in prices. In fact, observed prices fill the entire range of legitimate choices. In all treatments, some posted prices even exceed the buyer's valuation (120 points). Figure 3 shows how these excessive prices are distributed across trading periods. In the first period, almost 30% of initial stage posted prices and over 20% of search stage posted prices exceed the buyers' valuation. However, the proportion quickly falls to less than 10% for both stages and in a typical session almost all of the excessive prices are due to a single seller.

As is apparent from Figure 4 posted prices in the Diamond treatment tend to be higher than in the Bertrand treatment. However, the difference between posted prices in the low and high search cost treatments is not as noticeable. In Figure 4 we present the average initial stage posted price over periods. Each panel represents a treatment combination and in each panel there are three price series, one for each session.[8] In the first trading period the average posted price is similar in all four treatments. This average price of approximately 100 is the midpoint of the range of legitimate choices. In later trading periods average prices decline and differences between treatments emerge. Table 1 shows the

[7] Histograms for the search stage posted prices are nearly identical to histograms for the initial stage posted prices. For example, a 95% confidence interval for the difference between mean prices in initial and search stages is [−2.70, 2.18].

[8] It should be noted that the average price includes excessive prices.

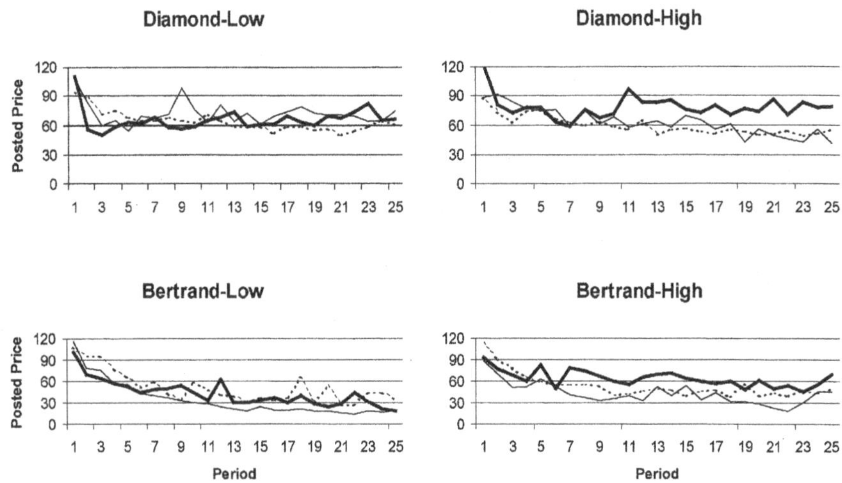

Figure 4. Average initial stage posted price by session

Table 1. Posted price averaged over 5 periods

	Initial stage				Search stage			
	Diamond		Bertrand		Diamond		Bertrand	
Periods	Low	High	Low	High	Low	High	Low	High
1–5	73.42	81.15	78.28	74.38	75.38	77.39	75.95	61.81
	(38.89)	(40.00)	(39.67)	(45.57)	(40.93)	(40.30)	(41.41)	(41.21)
6–10	66.92	65.41	44.94	52.73	65.40	58.95	41.65	58.29
	(42.87)	(43.13)	(34.92)	(36.35)	(41.93)	(39.14)	(30.35)	(38.33)
11–15	65.29	67.35	33.49	51.86	64.97	63.65	34.24	48.63
	(35.12)	(40.40)	(24.61)	(44.29)	(39.86)	(45.37)	(27.28)	(29.86)
16–20	64.35	60.83	31.98	45.94	65.18	63.68	36.20	44.83
	(38.21)	(35.63)	(29.96)	(35.31)	(42.70)	(39.99)	(33.38)	(40.59)
21–25	65.48	59.17	27.09	44.43	69.32	64.32	31.04	48.97
	(39.97)	(38.60)	(30.35)	(45.14)	(42.22)	(43.48)	(40.45)	(46.91)
All	67.09	66.78	43.16	53.87	68.18	65.79	45.20	54.74
	(39.10)	(40.26)	(37.15)	(42.81)	(41.52)	(41.79)	(38.41)	(39.11)

Note: Standard deviations are in parentheses

initial stage posted prices averaged across the first five periods, the next five, and so on.

For the Diamond low cost treatment initial stage posted prices appear to stabilize quite quickly as the averages for the last 4 groupings are quite similar to one another. The other treatments are slower to converge. Note, however, that none of the treatments converge to the equilibrium prediction.

The separation between Diamond and Bertrand treatments persists in Figure 5, which presents average initial stage transaction prices across periods for each session. Table 2 breaks down transaction prices by averaging across groups of five

Table 2. Transaction price averaged over 5 periods

	Initial stage				Search stage			
	Diamond		Bertrand		Diamond		Bertrand	
Periods	Low	High	Low	High	Low	High	Low	High
1–5	60.72	66.47	50.58	48.78	73.60	62.05	46.75	37.14
	(34.29)	(27.04)	(23.43)	(28.89)	(23.25)	(26.48)	(17.70)	(15.25)
6–10	51.42	49.14	25.13	31.87	54.73	49.92	27.80	31.45
	(27.31)	(25.20)	(9.29)	(20.56)	(22.32)	(31.35)	(8.82)	(17.61)
11–15	49.93	54.71	19.70	30.49	61.58	59.68	22.67	38.00
	(19.78)	(21.83)	(9.47)	(16.71)	(26.78)	(41.04)	(10.66)	(14.90)
16–20	50.23	48.18	18.08	29.08	52.56	57.59	21.00	28.17
	(22.09)	(24.55)	(7.60)	(12.23)	(18.85)	(43.25)	(7.42)	(8.61)
21–25	49.91	44.56	15.16	25.05	50.96	50.94	13.50	48.25
	(19.84)	(22.70)	(7.98)	(14.13)	(21.23)	(37.19)	(4.38)	(51.37)
All	52.26	52.24	25.23	32.73	58.73	55.93	29.04	35.73
	(25.18)	(25.27)	(17.86)	(20.71)	(23.89)	(35.72)	(16.44)	(20.03)

Note: Standard deviations are in parentheses

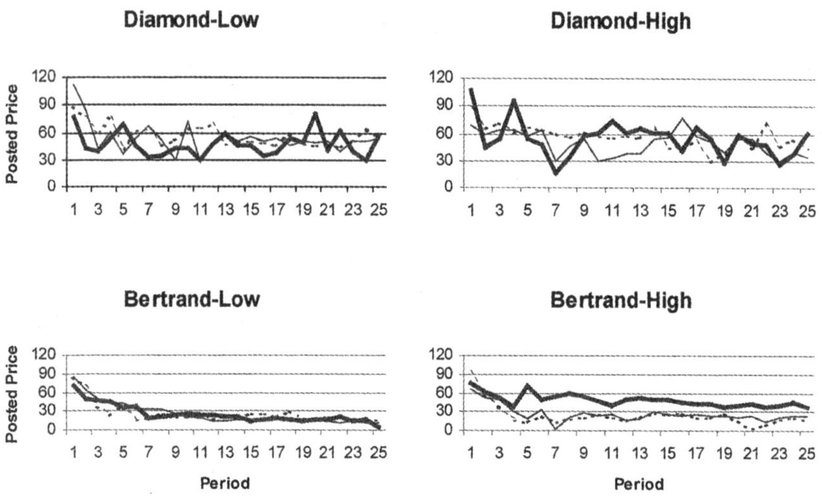

Figure 5. Average initial stage transaction price by session

periods each. The Diamond price is always above the Bertrand price, however neither price reaches the equilibrium prediction.

Average transaction prices are lower than average posted prices for two reasons. Most obviously, in the Bertrand treatment, the transaction price is generally the minimum of the two posted prices that the buyer observes. Second, in both treatments, rejected prices are included in the average posted price but not in the average transaction price, and as might be expected, rejected prices tend to be relatively high.

Finally, note that with constant costs and constant valuation, trading efficiencies can be calculated from knowledge of when each transaction occurs, if at all. Total profit extracted is calculated by multiplying the total number of trans-

Table 3. Aggregate buyer decisions

	All periods				Last 5 periods			
	Diamond		Bertrand		Diamond		Bertrand	
Description	Low	High	Low	High	Low	High	Low	High
Initial transactions	411	451	521	546	87	101	112	115
"No search"	15	19	1	7	1	1	0	1
Search transactions	135	104	59	34	24	16	8	4
Search rejections	39	26	19	13	8	2	0	0
Trading Efficiency	89.6%	89.3%	96.0%	95.5%	91.2%	95.3%	99.7%	98.7%

Note: There are a maximum of 600 transactions (8 buyers × 25 periods × 3 sessions per treatment) for all periods and a maximum of 120 transactions for the last 5 periods.

actions (from both stages) by the buyers' valuation (120) then subtracting the search cost by the number of searches that took place. Trading efficiency is then calculated by dividing the total profit extracted by the maximum available. The maximum available profit is 72,000 points (600 transactions × 120 points per unit) per treatment for all periods. Overall trading efficiency and its components are listed in Table 3.

As shown in the last row of Table 3, the Diamond treatment attains trading efficiency of 89.6% (low cost) and 89.3% (high cost). In the Bertrand treatment, trading efficiency is 96.0% (low cost) and 95.5% (high cost). Thus, search costs appear to have relatively little effect on efficiency, though efficiency appears to be enhanced by increasing the number of prices the buyers observe.

As the traders gain experience trading efficiency increases in all treatments. The greatest increase is in the Diamond treatment under a high search cost. Nevertheless, even in this case, efficiency remains higher in the Bertrand treatment throughout the sessions. For the Bertrand treatment, considering both high and low cost sessions, near perfect trading efficiency is observed in the last 5 periods, as all but one unit was traded, and there is very little search.

It is clear from Table 3 that a reasonable amount of price rejections occurred throughout the Diamond treatments. Table 4 displays the rejection rates for different prices at various intervals in the Diamond sessions. As expected, the rejection rate is higher for higher posted prices.

4.2 Treatment effects

Formal tests of treatment effects are based on permutation tests. Let S_i represent any summary statistic based on data from session i of treatment configuration S, and T_i be the same summary statistic based on data from session i of treatment configuration T. These statistics can be ordered, and under the null hypothesis that behavior is unaffected by the various treatments each ordering is equally likely. We use theoretical considerations to identify a predicted ordering: for example, if the summary statistic is average posted price, S is the Diamond-low treatment, and T is the Bertrand-low treatment, theory suggests $S > T$. Our rejection criteria

Table 4. Rejection rates for diamond treatment

	Low search cost									
	Initial stage Periods					*Search stage* Periods				
Prices	*1-5*	*6-10*	*11-15*	*16-20*	*21-25*	*1-5*	*6-10*	*11-15*	*16-20*	*21-25*
<41	0.10	0.03	0.04	0.00	0.00	0.00	0.00	0.00	0.00	0.00
	21	32	28	31	33	3	10	10	7	8
41-80	0.50	0.25	0.28	0.24	0.29	0.25	0.19	0.13	0.06	0.17
	62	48	71	75	76	16	21	15	18	18
>80	0.70	0.78	0.81	0.86	1.00	0.33	0.70	0.33	0.80	0.83
	37	40	21	14	11	15	10	12	5	6
	High search cost									
	Initial stage Periods					*Search stage* Periods				
Prices	*1-5*	*6-10*	*11-15*	*16-20*	*21-25*	*1-5*	*6-10*	*11-15*	*16-20*	*21-25*
<41	0.31	0.04	0.00	0.00	0.02	0.33	0.09	0.00	0.00	0.00
	13	28	23	33	42	3	12	5	6	6
41-80	0.08	0.15	0.14	0.16	0.05	0.15	0.00	0.00	0.00	0.00
	63	61	71	64	62	20	14	11	12	8
>80	0.64	0.84	0.65	0.70	0.94	0.78	0.67	0.57	0.50	0.50
	44	31	26	23	16	9	6	7	8	4

Note: First number in each cell is the rejection rate while the second number (the number with the smaller font) is the number of times the corresponding price was offered.

is to reject the null hypothesis of no treatment effect if $\min\{S_i\} > \max\{T_i\}$. With three sessions conducted under each treatment configuration, there are 6! ways these statistics can be ordered, and of these $3!^2$ ways satisfy our rejection criteria. Thus the probability of rejecting the null when it is true is 1/20, and so this procedure delivers an exact 5% significance test (against a directional alternative).

Based on this procedure, we reject the hypothesis that transaction prices, averaged over all trading periods, are the same in the Diamond and Bertrand treatments. This conclusion holds for both levels of search cost. Thus, the comparative static prediction holds up in our experiment, even though the sharp equilibrium prediction clearly does not. Focusing on transaction prices is clearly favorable to the theoretical prediction: if buyers buy at the lower price, then even if the distributions of posted prices are identical under the alternative treatments we should expect to observe lower transaction prices, on average, in the Bertrand treatment.

This observation motivates our second hypothesis, which is based on the choices of sellers in the initial stage of each trading period. We test the hypothesis that average posted prices are the same in the Bertrand and Diamond treatments. We reject this hypothesis for the low cost treatment where average posted prices are 55% higher in the Diamond treatment than in the Bertrand treatment (an average price of 67.09 versus 43.16). In fact, the difference between the two treatments increases over the course of the sessions, so that this differential understates the differential in later rounds. For the high cost treatments, we cannot reject the hypothesis that average posted prices are the same in the Bertrand and Diamond treatments.[9]

Finally, the effect of the search cost is not significant. We do not find significant differences between average prices in low cost versus high cost sessions for either treatment. This is true whether we consider posted or transaction prices and is consistent with the theoretical prediction about the effect of search costs.

5 Conclusion

Markets with costly search operate quite differently from perfectly competitive markets. This is true both at the theoretical level and the laboratory level. However, we find that the differences go far beyond those predicted by standard theory. As a very general rule, theory works well in predicting laboratory outcomes in other market experiments, but not so well in our markets with costly search. Although theory delivers comparative static predictions that are supported in our laboratory markets with costly search, point predictions perform poorly.

To understand these experimental results we think our experimental environment might be better understood as a series of bilateral monopolies, in the case of the Diamond treatment, or as a series of duopolies in the case of the Bertrand treatment, rather than as a market with many buyers and many sellers.

In the case of the bilateral monopoly, our trading rules allow the seller to obtain all the gains from trade, at least in theory. In practice, sellers set prices that allow buyers to share some of the gains from trade. Sellers that set prices close to the predicted price find they are unable to make sales: buyers appear to regard these prices as exorbitant. These results are clearly difficult to reconcile with the theory, but on the other hand are quite consistent with the results of ultimatum games that have a similar structure.

Given that our subjects play 25 rounds of the same game, and possibly get matched with the same opponent(s) more than once, it is interesting to ask whether their behavior can be explained by strategic rather than fairness motivations. For example, in our Diamond treatment could a seller repeatedly offer prices below the monopoly price in all but the last period in an attempt to induce the buyers to reject other offers in the initial stage and search for that seller? Similarly, could the rejection of high prices by buyers in all but the last period be an attempt by buyers to lower sellers' prices? We are doubtful that strategic factors offer a complete explanation for the experimental outcomes. First, subjects would have found these strategies difficult to implement given the information conditions of our experiment. We did not reveal the complete price distribution ex post to either buyers or sellers. This would make it difficult for single sellers to set prices to undercut rivals, and would make it difficult for multiple buyers to coordinate and withhold demand to strategically manipulate price. Second,

[9] Although the latter result should be interpreted cautiously: for high cost sessions we can reject the hypothesis that median prices are the same, or that modal prices are the same, in the Bertrand and Diamond treatments.

Ruffle (1997) studied posted-offer markets where, in equilibrium, the division of the surplus is heavily skewed toward the sellers, and found that buyers' attempts to force lower prices by withholding demand were ineffective when there were more than three buyers. He found that "buyers resist equilibrium prices early on, but eventually concede". Third, and finally, we do not observe a significant endgame effect: average prices in rounds 24 and 25 are roughly the same, and buyer responses are roughly the same.[10]

This is not to say that strategic considerations are irrelevant in our experiment. As can be seen from the last row of Table 2, transaction prices yield buyers more than half the surplus in each of the Diamond treatment sessions. Average transaction price was less than 50% of the buyer's redemption value in all six Diamond sessions. Note that buyers in our sessions receive a somewhat more favorable division than is observed in ultimatum games where buyers (receivers) almost always receive half or less of the surplus. We attribute this difference to the extra bargaining power afforded to buyers in the Diamond treatment by the presence of an option to search. Further, turning to the Bertrand treatment, we observe lower prices than in the Diamond treatment, as theory predicts. Even here, however, the division of surplus is not as extreme as predicted. In theory the buyers should get the entire surplus. In fact, sellers charge prices in excess of equilibrium prices, effectively demanding, and sometimes getting, a substantial part of the surplus.

In conclusion, restricted mobility of traders and the concealment of price information necessarily characterize markets with costly search. In turn, this places traders in a context where non-strategic, as well as strategic, factors play a large role in determining outcomes. The main role that search plays in this context is to offer an outside option, with an uncertain payoff, to buyers who would otherwise be facing a monopolist or duopolist.

References

Bagwell, K., Ramey, G.: The diamond paradox: a dynamic resolution. Discussion Paper No. 1013, Northwestern University (1992)

Burdett, K., Judd, K. L.: Equilibrium price dispersion. Econometrica **51**, 955–969 (1983)

Cason, T. N., Williams, A. W.: Competitive equilibrium convergence in a posted-offer market with extreme earnings inequities. Journal of Economic Behavior and Organization **14**, 331–352 (1990)

Davis, D. D., Holt, C. A.: Experimental economics. Princeton, NJ: Princeton University Press 1993

Davis, D. D., Holt, C. A.: Consumer search costs and market performance. Economic Inquiry **34**, 133–151 (1996)

Diamond, P. A.: A model of price adjustment. Journal of Economic Theory **3**, 156–168 (1971)

Forsythe, R., Horowitz, J. L., Savin, N. E., Sefton, M.: Fairness in simple bargaining experiments. Games and Economic Behavior **6**, 347–369 (1994)

Grether, D. M., Schwartz, A., Wilde, L.: Uncertainty and shopping behavior: an experimental analysis. Review of Economic Studies **55**, 323–342 (1988)

Güth, W., Schmittberger, R., Schwarz, B.: An experimental analysis of ultimatum bargaining. Journal of Economic Behavior and Organization **3**, 367–388 (1982)

[10] For example, a 95% confidence interval for the difference in mean initial stage posted prices between rounds 24 and 25 is [−5.71, 3.98].

Harrison, G. W., Morgan, P.: Search intensity in experiments. Economic Journal **100**, 478–486 (1990)

Hey, J. D.: Search for rules of search. Journal of Economic Behavior and Organization **3**, 65–81 (1982)

Roth, A. E., Prasnikar, V., Okuno-Fujiwara, M., Zamir, S.: Bargaining and market behavior in Jerusalem, Ljubljana, Pittsburgh, and Tokyo: an experimental study. American Economic Review **81**, 1068–1095 (1991)

Ruffle, B. J.: Do buyers always act as passive price-takers? Demand withholding in experimental posted-offer markets. Working Paper, Ben-Gurion University (1997)

Salop, S., Stiglitz, J.: Bargains and ripoffs: a model of monopolistically competitive price dispersion. Review of Economic Studies **44**, 493–510 (1977)

Schotter, A., Braunstein, Y. M.: Economic search: an experimental study. Economic Inquiry **19**, 1–25 (1981)

Stahl, D., O. II: Oligopolistic pricing with sequential consumer search. American Economic Review **79**, 700–712 (1989)

Williams, F. E.: The effect of market organization on competitive equilibrium: the multi-unit case. Review of Economic Studies **40**, 97–113 (1973)

Intertemporal pricing in laboratory posted offer markets with differential information*

Aldo Rustichini[1] **and Anne P. Villamil**[2]

[1] Department of Economics, Boston University, Boston, MA 02215, USA
[2] Department of Economics, University of Illinois, Champaign, IL 61820, USA

Summary. This paper analyzes intertemporal seller pricing and buyer purchasing behavior in a laboratory retail market with differential information. A seller posts one price each period that a buyer either accepts or rejects. Trade occurs over a sequence of "market periods" with a random termination date. The buyer and seller are differentially informed: The seller's cost of producing a unit of a fictitious good is known and constant in all periods, but the buyer's value for the good (demand) is a random variable governed by a Markov Process whose structure is common knowledge. At the beginning of each period the unit's value is determined by "nature" and is privately revealed only to the buyer. The market termination rule is a binary random variable. We conduct 32 laboratory experiments designed to study intertemporal pricing by human subjects in the Posted Offer Institution when demand follows a stochastic process. There are four series of experiments: 8 with simulated buyers, 8 with inexperienced subjects, 8 with once experienced subjects, and 8 with twice experienced subjects.

Keywords and Phrases: Intertemporal pricing, Differential information.

JEL Classification Numbers: C91, D82, D83, D40, C61, L16.

1 Introduction

In this paper we study seller intertemporal pricing strategies and buyer intertemporal purchasing strategies in a laboratory market where demand follows a Markov Process. The market is organized as a Posted Offer institution, where the seller posts a single "take-it" or "leave-it" price which a buyer either accepts or rejects. There is no bargaining or price revision.[1] This institution has

* We gratefully acknowledge financial support from the University of Illinois.

[1] See Davis and Holt (1998) for a recent analysis of price revision in the posted offer market.

been studied extensively in laboratory markets because it corresponds to trade in many naturally occurring retail markets. Indeed, the Posted Offer institution is the most common market structure in developed economies.[2] We extend experimental work on the Posted Offer institution by considering an explicitly intertemporal setting with differential information where demand follows a Markov Process. Most previous work has considered static, deterministic supply and demand conditions with private information about seller costs and buyer values for a fictitious good. In contrast, in our study the buyer's value for the good (i.e., the induced demand curve) follows a Markov Process whose structure is common knowledge, but the current realization is privately revealed only to the buyer at the beginning of each "market period." The precise date that the market ends is determined randomly and is independent of the stochastic demand process. This termination procedure corresponds to an intertemporal discount rate.

Intertemporal pricing and purchasing relationships are clearly important in many naturally occurring markets. In addition, many markets are subject to recurrent cyclical fluctuations (e.g., seasonal and/or business cycle movements). To our knowledge, the only experimental papers that consider fluctuating demand conditions in the Posted Offer institution are by Hoffman and Plott (1981) and Davis, Harrison, and Williams (1992). However, the fluctuations in these models are perfectly predictable (e.g., follow a high, low, high, low pattern, etc.). In contrast, in our experiments demand fluctuations are uncertain but follow a well defined stochastic process. Extending experimental work in this direction is important because a main contribution of experimental research has been to establish that certain market institutions drive sparsely informed agents to equilibrium in static, deterministic environments. Indeed, many economists have argued that this is why market economies have been so successful (relative to non-market economies).[3] A crucial unanswered question in this literature is — Are markets as effective in stochastic, intertemporal environments where agents problems' are more formidable? This paper analyzes this question.

The paper is organized as follows. In Section 2 we discuss the theoretical model we will test and specify the equilibria. In Section 3 we discuss the testable implications of the theory. In Section 4 we discuss our experimental design and the results of 32 laboratory experiments. Finally, Section 5 contains concluding remarks.

2 The model and equilibria

Consider a Posted Offer market with two risk neutral traders, a buyer and a seller, who may trade over a sequence of periods. In each period, the seller may produce one unit of an indivisible good. The seller's production cost is fixed and common knowledge. In contrast, the buyer has a reservation value for the

[2] See Plott (1989) for a survey of the large experimental literature on this institution. See also Davis and Holt (1996).

[3] See Smith (1982) for a discussion of markets as "economizers of information."

good each period v which follows a Markov Process that is common knowledge. The unit's value v takes on one of two possible values, h (high) or l (low). The Markov Process describes the serial correlation in the unit's value over the sequence of market periods. The structure of the process is common knowledge, but the current realization of v is privately observed by the buyer each period. In particular, all agents know that given the unit's value in the previous period, the probability that the unit's value is the same in the current period is $1 - \alpha$, and the probability that it changes (i.e., either from high to low or from low to high, given the previous state) is α. The unit's value to the buyer in the first period (i.e., the initial state) is a random variable, drawn from a known uniform distribution. We restrict attention to the case of positive serial correlation (i.e., $0 < \alpha < 1/2$).[4]

Each period trade occurs according to the following sequence of events. At the beginning of the period the unit's value is determined by "nature." The value of the first unit is drawn randomly from a known distribution. The value of each subsequent unit is determined by the stationary Markov process. A complete description of the process is common knowledge of both agents, but the unit's current value is privately observed by the buyer only. The seller posts a single price offer that the buyer can either accept or reject. If the buyer accepts, the unit is traded at the price posted by the seller. The buyer's profit is the difference between the unit's value and the posted price $v - p$, and the seller's profit is the difference between the posted price and the cost $p - c$. If the buyer rejects the offer the unit is not produced and is not traded, and both traders make zero profit. This concludes the market period. The probability that the market continues in each subsequent period is given by $\delta \in (0, 1)$. For risk neutral traders, this is equivalent to assuming a discount factor of δ. These procedures are repeated in each period until the market ends.[5]

We focus on equilibria in *stationary Markov* strategies. Given a distribution of values for the initial unit, a history of past seller price offers, and buyer answers, the seller forms a belief (for any equilibrium of the game) about the buyer's value for the current unit of the good. Each period the seller considers only this belief about the current unit's value to the buyer, and the buyer considers only his/her own (known) value and the seller's belief about the value. The sequence of previous price offers and answers is reflected in the seller's current belief. To derive equilibrium strategies, we assume that the seller's initial belief is common knowledge.[6] A stationary strategy for the seller is a map from the seller's belief to the space of price offers, where the seller's belief is a number $w \in [0, 1]$ which denotes the seller's subjective probability that the buyer's value for the current unit of the good is *high*. A stationary strategy for the buyer is a map from the

[4] Three other cases are possible: if $\alpha = 0$ the initial draw determines the unit's value for all periods; if $\alpha = 1/2$ the unit's values are independent; and if $1/2 < \alpha \leq 1$ the unit's values exhibit negative serial correlation.

[5] See Villamil (1999) for a discussion of alternative termination procedures.

[6] This assumption is induced in the experiments by drawing the first unit's value randomly from a fixed distribution that is known by subjects.

unit's value for that period, the seller's belief, and the seller's price offer to a binary decision variable which indicates the buyer's answer.

Rustichini and Villamil (1996) prove that equilibria in stationary strategies are characterized by a triple (w^*, p_l, p_h), which indicates the *seller's critical belief*, the *low price offer* and the *high price offer*, respectively. Equilibrium strategies for each agent are a pair $(p(w), A(p, v))$, where

$$p(w) = \begin{cases} p_h & \text{if } w \geq w^*; \text{ or} \\ p_l & \text{otherwise;} \end{cases} \qquad (P_s)$$

and

$$A(p, v) = \begin{cases} Y & \text{if either } p \leq p_l \text{ or } p \leq p_h \text{ and } v = h; \\ N & \text{otherwise.} \end{cases} \qquad (P_b)$$

(P_s) is a state dependent seller price offer, and (P_b) is a state dependent buyer answer where Y denotes "yes" and N denotes "no."

Rustichini and Villamil (1996) show that in equilibrium, strategies (P_s) and (P_b) solve the following dynamic programming problems:[7]

The Seller's Problem: A stationary strategy for the seller is a pricing function p. The seller knows his/her own beliefs, w, and value function, $V_s(\cdot)$. Let $\mathbf{1}_{\{A=Y\}}$ denote the indicator function of the set $\{A = Y\}$, which is a random event from the seller's perspective, and c denote the seller's (known) cost of producing each unit sold.[8] The seller's discounted dynamic programming problem is to choose a pricing function to maximize the following functional equation:

$$V_s(w) = \max_{p \geq 0} E_w \{p\mathbf{1}_{\{A=Y\}} + \delta V_s(w'(p, A, w))\},$$

where E_w denotes the expectation over w and δ is the discount factor.

The Buyer's Problem: A stationary strategy for the buyer is an answer function $A \in \{Y, N\}$. The buyer knows the realization of the unit's current value, $v = \{h$ or $l\}$, takes the seller's price as given, and knows his/her own value function, $V_b(\cdot)$. The buyer's discounted dynamic programming problem is to choose an answer function to maximize the following functional equation:

$$V_b(w, p, v) = \max_{A \in \{Y, N\}} E_v \begin{cases} v - p & \\ +\delta V_b(w'(p, Y, w), p'(w'(p, Y, w)), v') & \text{if } A = Y; \\ \delta V_b(w'(p, N, w), p'(w'(p, N, w)), v') & \text{if } A = N. \end{cases}$$

The expectation is taken over v, the stochastic process for the unit's value, where v is the current value.

Rustichini and Villamil (1996) show that the equilibria of the game are given by the triple (w^*, p_l, p_h). A crucial step in completing the analysis of the equilibria is to provide a rule that the seller follows to form beliefs, both on and off the equilibrium path. Let w denote the seller's belief that the buyer's value for the current unit is high, and $w'(\cdot)$ denote the seller's belief that the buyer's value

[7] "Primes" denote next period's value of a variable.

[8] The seller's cost plays no role in our analysis so we normalize $c = 0$.

for *next period's* unit is high. If the seller receives no additional information in a period, the seller's next belief is given by:

$$\hat{w}(\cdot) = w(1-\alpha) + (1-w)\alpha.$$

This equation indicates that the value for next period's unit may be high for two reasons: the current unit's value was high and did not change (the first term) or the unit's value was low but changed states (the second term).

The stationary strategy of each agent is binary. There are five possible belief situations on and off the equilibrium path:

(i) $w'(p, Y, w) = 1 - \alpha$ for any $p_h \geq p > p_l$: If the seller posts a price higher than the low price but less than the high price, and the buyer accepts, then the seller believes that the unit's value was high.

(ii) $w'(p, N, w) = \alpha$ for any $p > p_l$: If the seller posts a price higher than the low price and the buyer rejects, then the seller believes that the unit's value was low.

(iii) $w'(p, Y, w) = w(1-2\alpha) + \alpha \equiv \hat{w}$ for any $p \leq p_l$: If the seller posts a price lower than or equal to the low price and the buyer accepts, then the seller has no useful new information and revises his/her belief according to $\hat{w}$.

(iv) $w'(p, N, w) = \hat{w}$ for any $p < p_l$: We assume that if the seller posts a price less than the low price and the buyer rejects, then the seller updates as in the case where the buyer accepts.

(v) $w'(p, Y, w) = \hat{w}$ for any $p > h$: If the seller posts a price higher than the high reservation value and the buyer accepts, then we assume that the seller again updates via $\hat{w}$.

Beliefs (i), (ii), and (iii) are equilibrium path strategies that follow directly from $\hat{w} = w(1-\alpha) + (1-w)\alpha$. When the buyer accepts the seller's high price offer, this is a perfect signal that $v = h$; thus $w = 1$ and w' is given by (i). When the buyer rejects the seller's high price offer, this is a perfect signal (in equilibrium) that $v = l$; thus $w = 0$ and w' is given by (ii). When the buyer accepts the seller's low price offer, this action is *not* perfectly revealing; thus w' is given by (iii). Finally, (iv) and (v) are "off the equilibrium path" strategies, so we must attribute to the agent some belief to complete the belief specification rule. We assume that if the buyer rejects the seller's low price offer the seller believes $v = l$ so $w = 0$ and $\hat{w} = \alpha$. This is plausible because when $v = l$ the buyer loses nothing by rejecting p_l but if $v = h$ the buyer foregoes substantial profit. If the buyer accepts a price higher than his/her reservation value, the buyer loses profit on the trade but this gives the seller no new information. Thus the seller updates via $\hat{w}$.

3 Testable implications of the theory

Rustichini and Villamil (1996) prove that strategies (P_s) and (P_b) solve the respective seller and buyer problems.[9] In our intertemporal Posted Offer market the seller moves first by posting a price, but does not observe directly the actual realization of the unit's value in any period. The buyer moves second and responds to the seller's price offer—after the unit's current value has been privately revealed. When $0 < \alpha < 1/2$ and the seller is rational (i.e., uses all available information and Bayes rule), equilibrium strategy (P_s) indicates that the seller behaves as if he/she forms a belief, w, about the unit's value to the buyer, and then decides whether to offer a high or a low price by comparing the current belief with a critical belief, w^*. The seller posts a high price (p_h) if $w \geq w^*$, and a low price (p_l) otherwise. The seller uses this strategy both to maximize revenue *and* to acquire information. Equilibrium strategy (P_b) indicates that the buyer should accept the seller's price offer, regardless of whether it is high or low, if the unit's value is high; but should accept only the seller's low price offer if the value is low.

Recall that an equilibrium is identified with the triple (w^*, p_l, p_h). Rustichini and Villamil (1996) characterize each component of this triple. The following results from their analysis are relevant for experimental investigation of the theory. If (w^*, p_l, p_h) describes all equilibria, then $p_l = l$ and $l < p_h \leq h$. Intertemporal price paths are determined as follows. The probability that the unit's value is high (or low) is $1/2$, and this is the seller's limit belief when no new information about the unit's value to the buyer is acquired.[10] The relationship between critical belief w^* and $1/2$ determines the equilibrium price path.[11]

Rustichini and Villamil (1996) derive parametric restrictions on the model which lead to only two types of equilibrium price paths. First, when $\alpha < w^* < 1/2$, equilibrium prices follow a cyclical pattern for any realization of value v. Suppose $w < w^*$, and the seller's equilibrium price offer is p_l. The sequence of future beliefs is given by $\hat{w}^i$, until the first time i_0 that $\hat{w}^{i_0} > w^*$ (where i_0

[9] Appendix B in Rustichini and Villamil (1996) proves existence of incentive compatible equilibria in stationary strategies. The buyer's *equilibrium* strategy sometimes truthfully reveals information to the seller because (P_b) involves an essential trade off: If the buyer accepts price p_h this action reveals information to the seller, *but* gives the buyer an immediate reward for telling the truth (i.e., profit from the trade). If the buyer lies by rejecting p_h when $v = h$, this distorts the seller's belief but "costs" the profit foregone on the rejected trade. Thus, the buyer faces a trade off between current profit and manipulating the seller's beliefs (to obtain higher future profit)—in a sequential game with a random termination date and an oscillating value sequence. Seller equilibrium strategy (P_s) takes this trade off into account.

[10] The $\lim_{i \to \infty} \hat{w}^i = 1/2$ for every w, where $\hat{w}^i$ is the ith iterate of the equilibrium Bayesian belief formation rule $w' = w(1-\alpha) + (1-w)\alpha$. Iteration shows that w' converges to $1/2$ for every w when $0 < \alpha < 1/2$.

[11] Rustichini and Villamil (1996, Section 4) show that if $h > \frac{2-\delta}{1-\delta}$, then $w^* < 1/2$ and equilibrium prices have persistent cycles. This result is essential for experimental tests of the theory, and is consistent with the following intuition: When the seller believes the unit's value is high (and $h > \frac{2-\delta}{1-\delta}$), he/she believes there is a large amount of consumer surplus available for extraction. Thus, the seller posts a high price to increase revenue *and* to acquire information (i.e., learn) about the unit's value.

is finite) when the seller's price offer becomes p_h. If the unit's value is low, the buyer refuses the offer, the seller sets the new belief to α, and the process begins again. If the unit's value is high, the buyer accepts the offer, the seller sets his/her new belief to $1 - \alpha$ and maintains a price offer of p_h (and a belief of $1 - \alpha$) until the unit's value becomes low. When the unit's value becomes low, the buyer rejects the offer, and a new period of low price offers begins. The average length of periods in which the seller makes low price offers is constant and given by:[12]

$$L \equiv \min_i \{\hat{\alpha}^i \geq w^*\}.$$

Thus when the seller's beliefs oscillate about w^*, price cycles are optimal. That is, the model predicts that persistent oscillations between high and low prices will be observed. In equilibrium, the length of the low price phase of the cycle is given by L, and the high price phase of the cycle should persist until p_h is rejected.

Second, when $w^* \geq 1/2$, equilibrium price cycles converge to p_l. Intuitively, suppose that the seller's initial price offer is p_h. If the buyer accepts p_h the seller continues with this price until the first time the price is rejected. The seller then sets $\hat{w}^i$ to α. However, since w' converges to $1/2$ in the limit, it will always be the case that the seller's belief is less than the critical value w^*, and (P_s) indicates that when $w < w^*$ the seller should post p_l. Thus, when $w^* \geq 1/2$ the seller's belief $\hat{w}^i$ can never "build-up" enough to make it optimal to try the high price so the seller always posts p_l (except for perhaps an initial phase).

4 Experimental design and results

The trading rules of the Posted Offer Institution are reported in Ketcham, Smith, and Williams (1984). Our market has a seller, a buyer, a sequence of "trading periods," and one unit of a fictitious good each period. The trading rules specify a "two-step" decision procedure: First, the seller privately makes a price decision and posts a "take-it" or "leave-it" offer. Second, the buyer either accepts or rejects the offer. We assign costs and values for the good each period in accordance with the procedures described in Smith (1976). The seller's cost of producing each unit is known and equal to zero (i.e., $c = 0$). The buyer's value for each unit (v) is either 1 or h in any period, where h is a treatment variable in the experiments. The unit's value is determined randomly each period as follows:

(i) The initial v is drawn from a known equal distribution: the probability it is high (i.e., h) is $1/2$ and the probability it is low (i.e., 1) is $1/2$.

[12] Acceptance of p_h by the buyer perfectly reveals that $v = h$. Further, if the buyer follows *incentive compatible* equilibrium strategy (P_b) the seller also implicitly learns when p_h is rejected that the state has switched from h to l. However, learning is imperfect because there is no way to signal a state change when the seller offers p_l. That is why the seller uses rule L: it tells him/her how many periods to use the (certain to be accepted but uninformative) low price before switching to the (informative but costly–in terms of lost trades) price p_h. Our parameter choices imply $L = 6$ when h = \$2.50.

(ii) All subsequent v's are determined by the first order Markov Process: $P(v = 1|v = 1) = P(v = h|v = h) = 1 - \alpha = 0.9$, and $P(v = h|v = 1) = P(v = 1|v = h) = \alpha = 0.1$, where $P(\cdot|\cdot)$ is a conditional probability. Thus, the probability that any subsequent unit's value is the same as last period's is 90 percent (so the probability v has changed is 10 percent).[13]

Cyclical pricing is optimal whenever $h > \frac{2-\delta}{1-\delta}$. In all the experiments we set $l = \$1$, $\delta = 0.05$, and $\alpha = 0.10$. The level of subject experience and h are treatment variables. It follows immediately that cyclical pricing is optimal for $h > \$2.05$ and a "flat" pricing strategy is optimal otherwise.

The *value determination rule* was publicly announced, and was induced in the experiment by the following procedure: The experimenter rolled a 20-sided die at the beginning of each market period. In period 1, if the outcome was an 11 through 20 the *first unit's* value was high; otherwise it was low. In periods $2, \ldots, end$, if the outcome was a 1 through 18, the *current unit's* value was the same as last period's value; otherwise it changed. Examples of stochastic processes were shown to subjects in the Instructions. In addition, subjects were told that (i) there would be period-to-period dependence in the unit's value over the course of the experiment; and (ii) on average about half the values would be high and half would be low, if the experiment lasted for many market periods.

Both the seller and the buyer knew the seller's cost (i.e., $c = 0$), and both knew the value determination rule. The unit's current value was privately revealed only to the buyer at the beginning of the period, but was never revealed directly to the seller at any time during the experiment. The fact that agents' had differential information was public knowledge. The termination rule used in all experiments was stochastic: Subjects were told the experiment would last between twenty minutes and three hours because the final market period would be determined by the roll of a 20-sided die at the end of every period. If the outcome was a 1 the experiment would end; otherwise it would continue. This termination procedure corresponds to a discount factor of $\delta = 0.05$. Subjects were undergraduate and graduate students at the University of Illinois at Urbana-Champaign.

We conducted the following series of experiments: Eight experiments with inexperienced subjects and simulated buyers denoted as series (a) experiments. Eight experiments with inexperienced buyers and sellers denoted as series (i) experiments. Eight experiments with once experienced buyers and sellers denoted as series (ii) experiments. Eight experiments with twice experienced buyers and sellers denoted as series (iii) experiments. In the series (a) experiments h was a treatment variable. Six experiments were conducted in which $h > 2.05$ thus cyclical pricing was predicted by the theory and two experiments were conducted in which h was less than this amount and thus "flat" pricing was predicted. In the series (i), (ii) and (iii) experiments experience was a treatment variable. Thus, each subject participated in a sequence of three experiments with the same parameters but different counterparts (and of course, different realizations of the

[13] In general, when α is small there is persistence in the process so information about v is valuable to the seller.

Markov process and termination period). The experience profiles of all subjects are reported in Appendix A.

The series (a) experiments were designed to investigate the seller's pricing strategy when the seller knew the buyer would always follow equilibrium strategy (P_b). Thus, in these experiments the seller was told that the buyer would accept the seller's price offer (regardless of whether it was high or low) if the value was high, but would accept only the low price if the value was low. These experiments are designed to establish baseline empirical support for pricing policy (P_S) when there is no strategic uncertainty about the buyer's behavior. In the first 6 experiments in this series parameters were chosen so that the theory predicts equilibrium price cycles.[14] In the remaining 2 experiments in this series parameters are chosen so that the theory predicts equilibrium prices which converge to p_l.[15] These latter two experiments are an important check on the model's theoretical consistency, though this parametric case is unlikely to be relevant in most naturally occurring markets.

The series (i), (ii), and (iii) experiments focus on cyclical pricing when there is strategic uncertainty about the behavior of one's counterpart. That is, both the seller and the buyer were human subjects with no information about how their counterpart would behave other than the information contained in the Instructions and their experience in the current and previous (if any) markets. In all of these experiments, h = \$2.50.

Tables 1 and 2 report summary statistics for each series. Specifically, the Tables report *measures of predictive success* (cf., Selten (1991)) for testable implications of the theory. The theory predicts a "high or low" seller pricing strategy (P_s), and a "yes" or "no" buyer answer strategy (P_b). The measure of predictive success assesses the statistical usefulness of an area theory.[16] Let m denote the measure of success, r denote the *hit rate* of the theory (the relative frequency of correct predictions), and a denote the area (relative size) of the predicted subset compared with the set of all possible outcomes. Define $m = r - a$, where $m \in [-1, +1]$ in general.

Figures 1a–8a and Figures 1-24 report price and answer data for the 32 experiments. Figures 1a–8a report the results of the eight series (a) experiments with simulated buyers; Figures 1-8 report the results of the eight series (i) experiments with inexperienced buyers and sellers; Figures 9-16 report the results of the eight series (ii) experiments with once experienced buyers and sellers; and Figures 17-24 report the results of the eight series (iii) experiments with twice experienced buyers and sellers.

Table 1 reports measures of predictive success (i.e., m's) for the eight series (a) experiments with simulated buyers. In Experiment 1a measure m for (P_s) is computed as follows: $r = 32/37$, where 32 is the number of successful predictions

[14] Price cycles occur when $h > \frac{2-\delta}{1-\delta}$, or $h >$ \$2.05 given $\delta = 0.05$. We set h = \$2.50 in Experiment 1a and h = \$2.20 in Experiments 2a–6a (see Figures 1a–6a).

[15] p_l is optimal when $h < 2 - \delta$. We set h = \$1.50 in these experiments (see Figures 7a and 8a).

[16] Area theories for the prediction of experimental results delineate regions of predicted outcomes within the set of all possible outcomes.

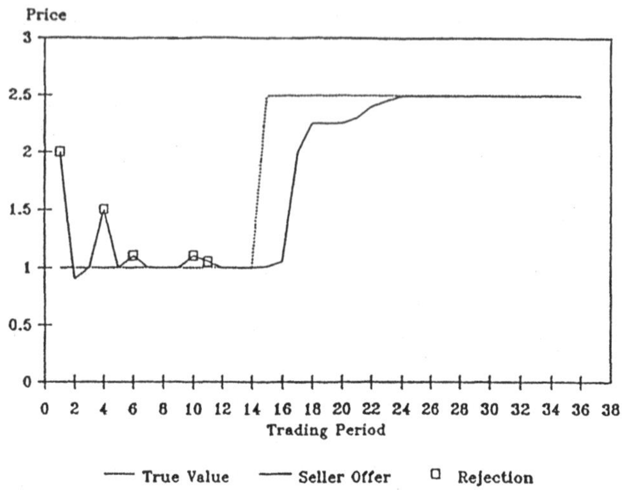

Figure 1a

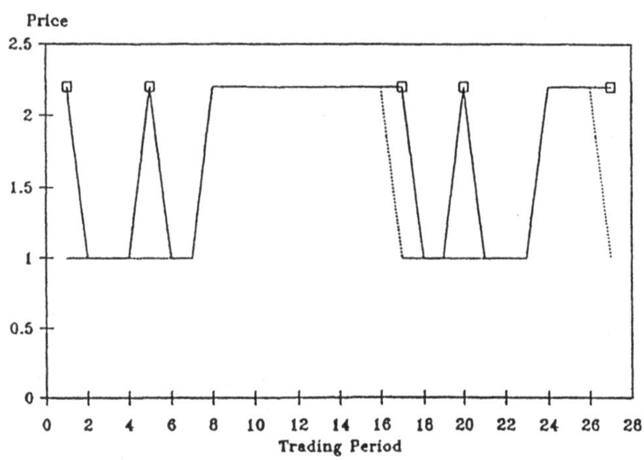

Figure 2a

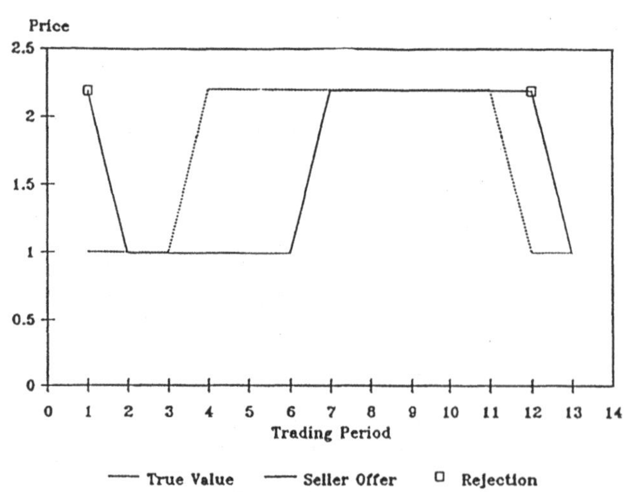

Figure 3a

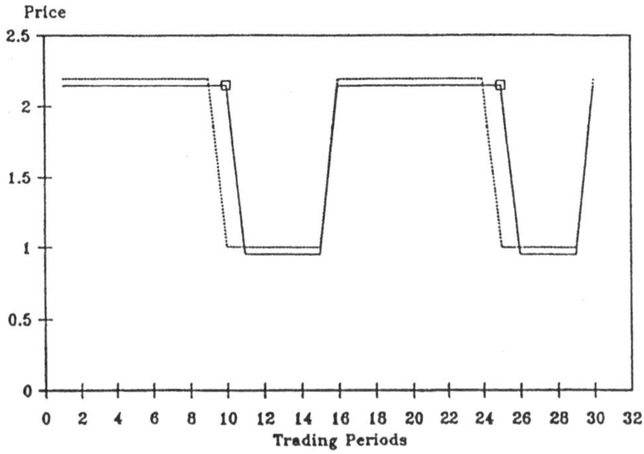

Figure 4a

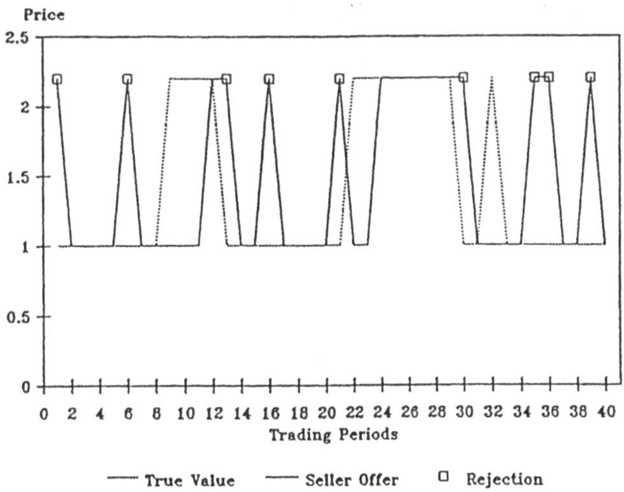

Figure 5a

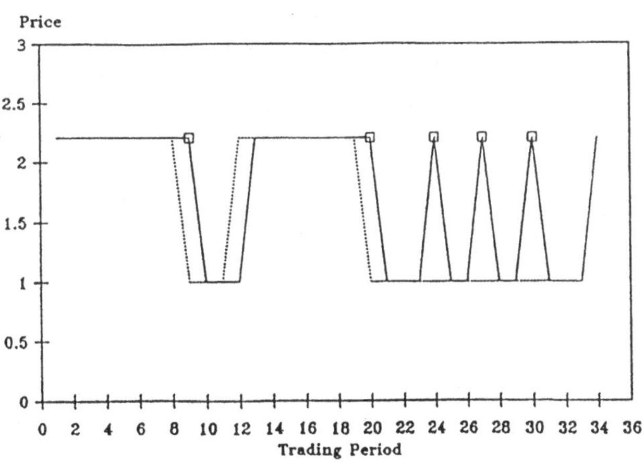

Figure 6a

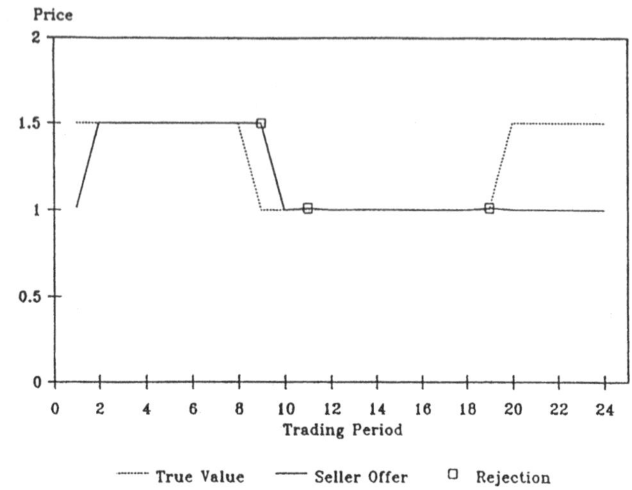

Figure 7a

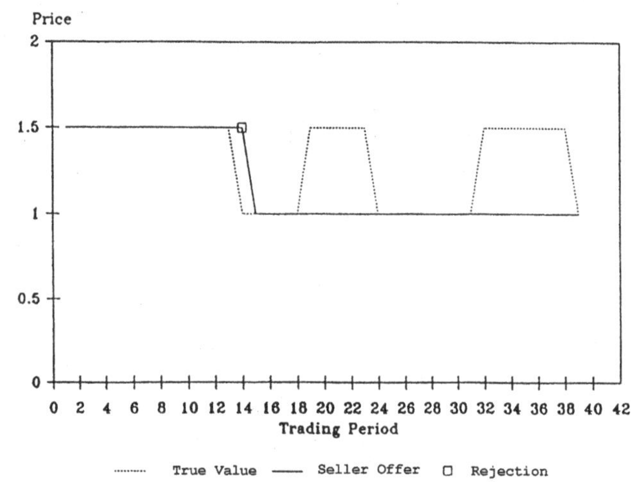

Figure 8a

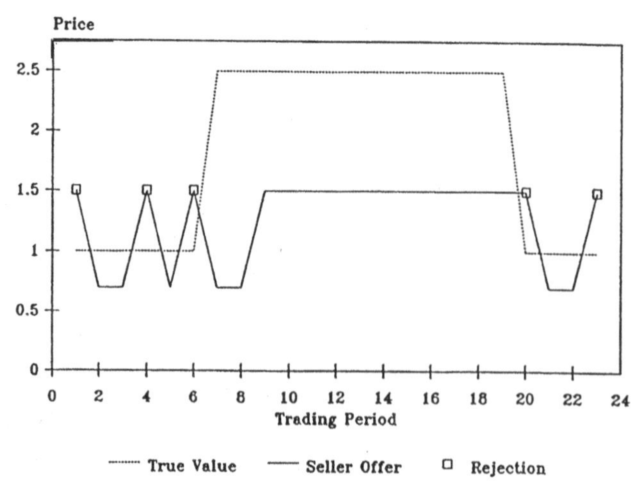

Figure 1

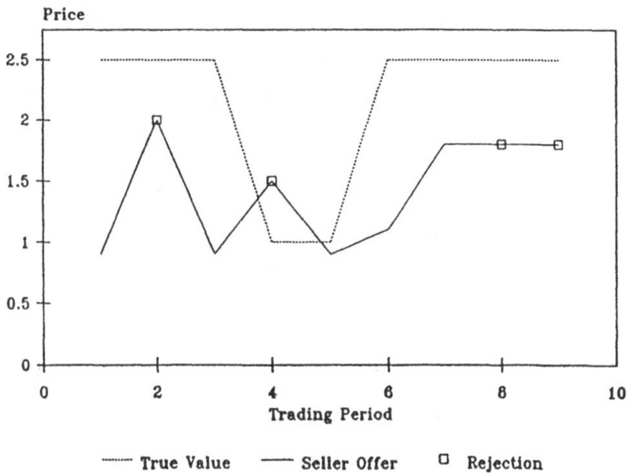

Figure 2

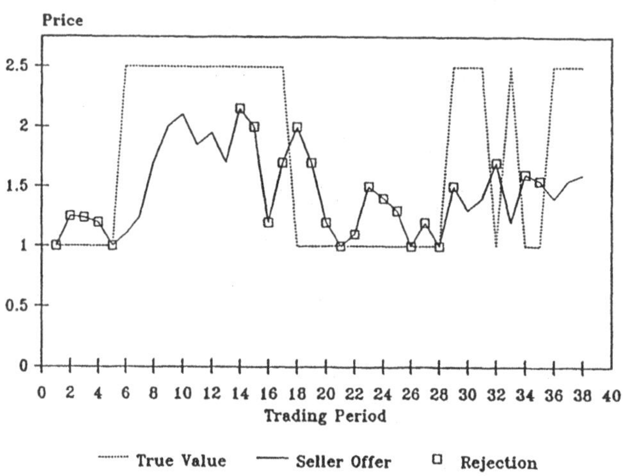

Figure 3

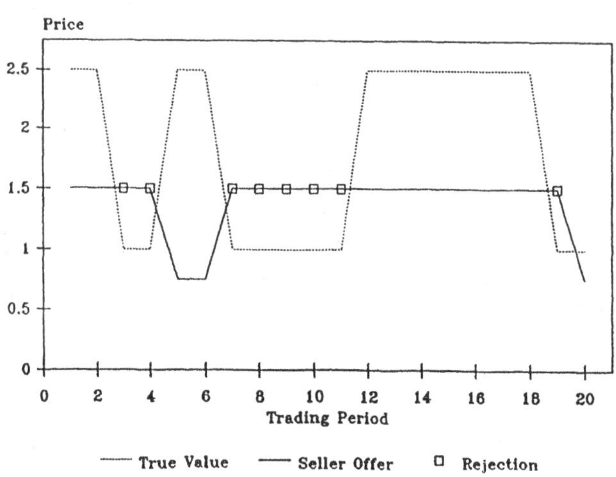

Figure 4

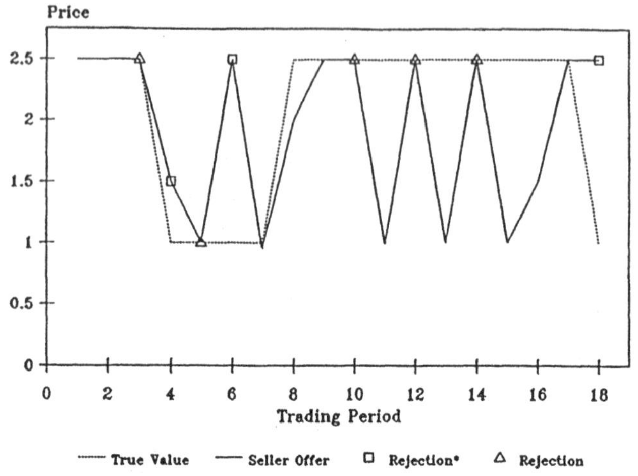

Figure 5

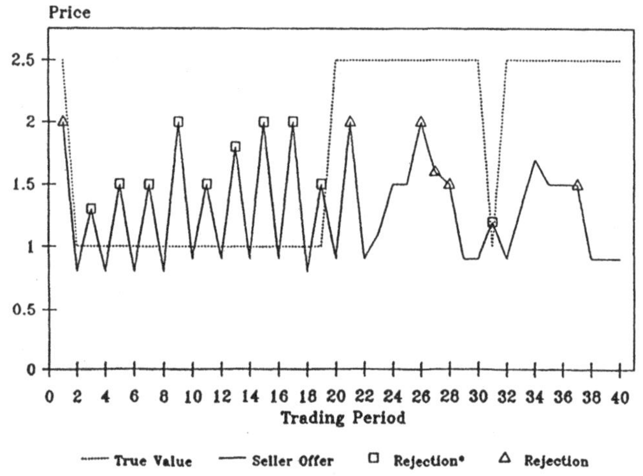

Figure 6

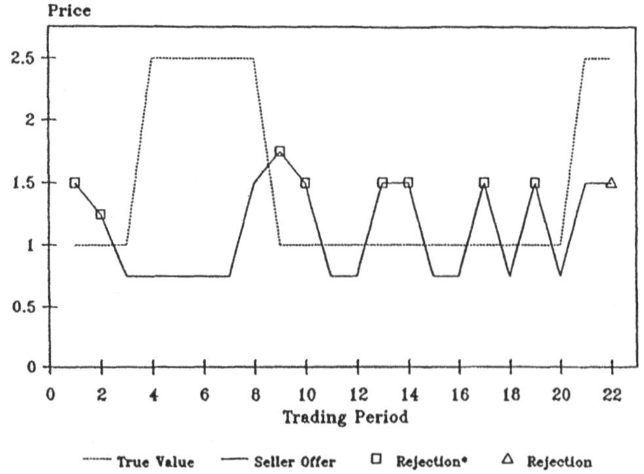

Figure 7

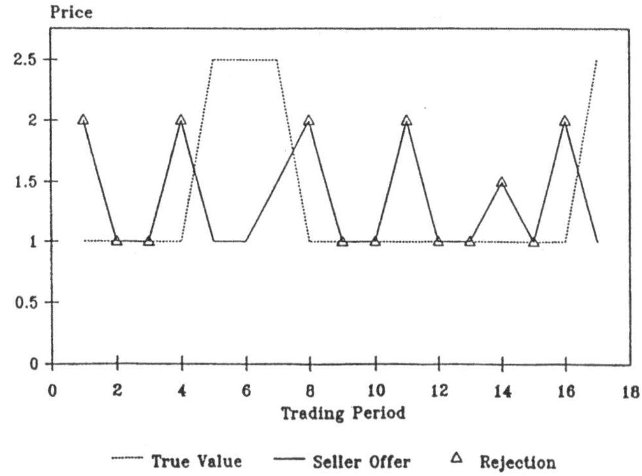

Figure 8

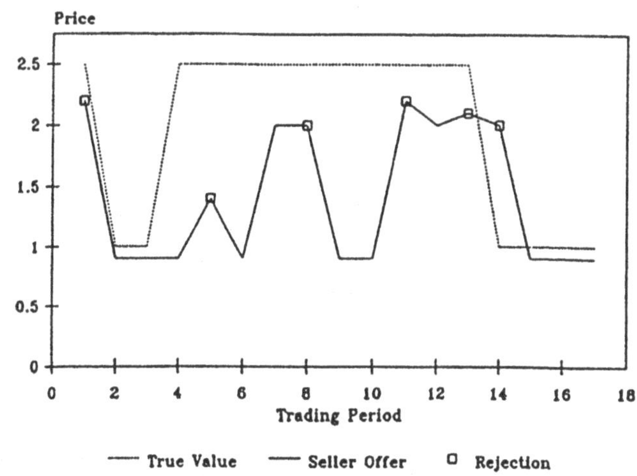

Figure 9

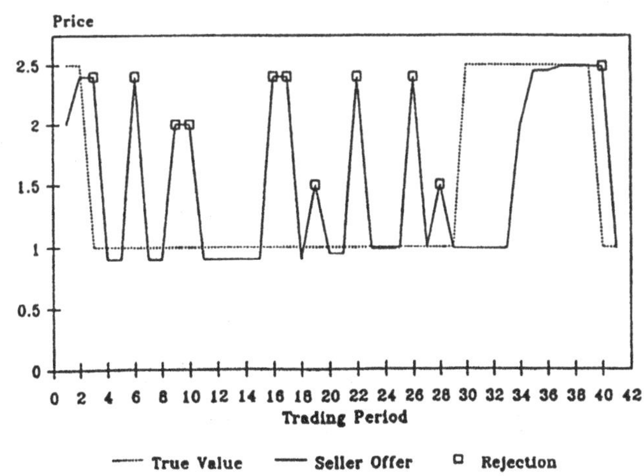

Figure 10

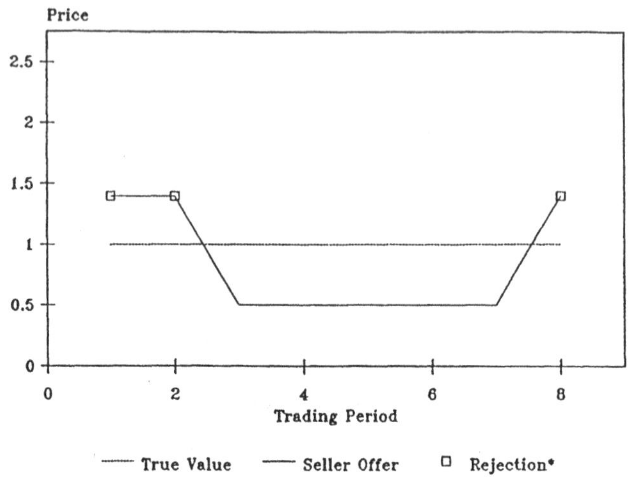

Figure 11

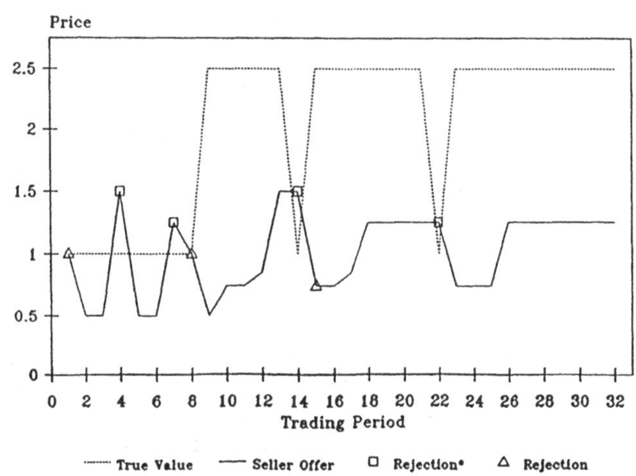

Figure 12

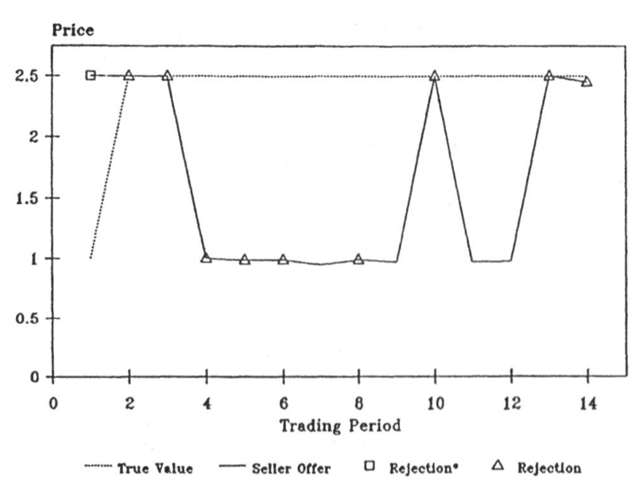

Figure 13

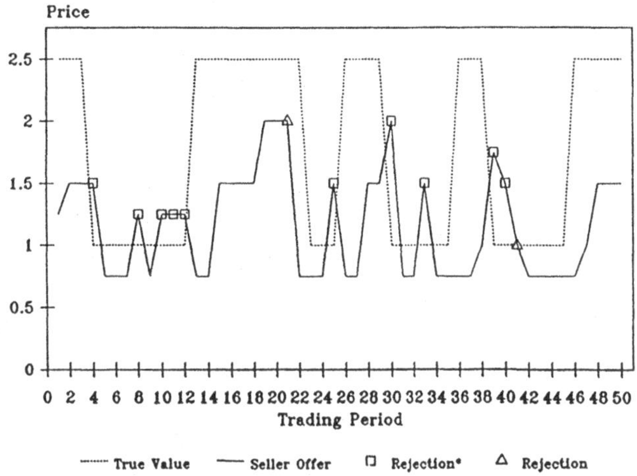

Figure 14

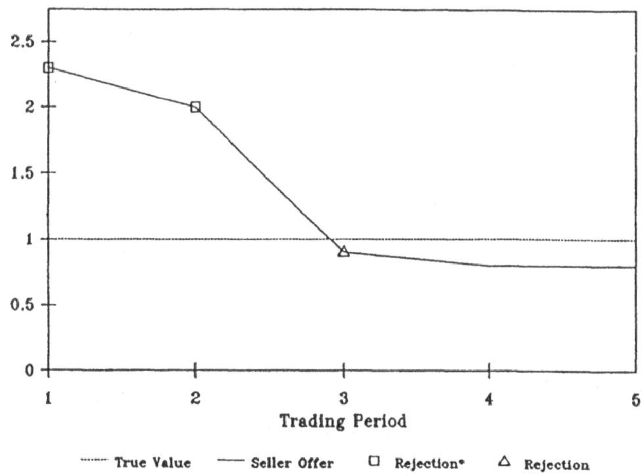

Figure 15

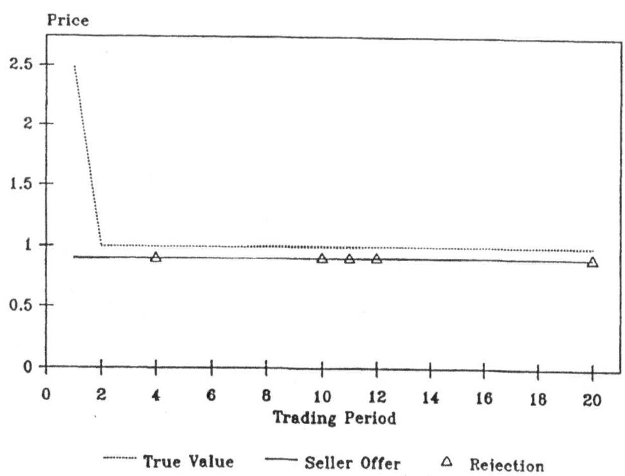

Figure 16

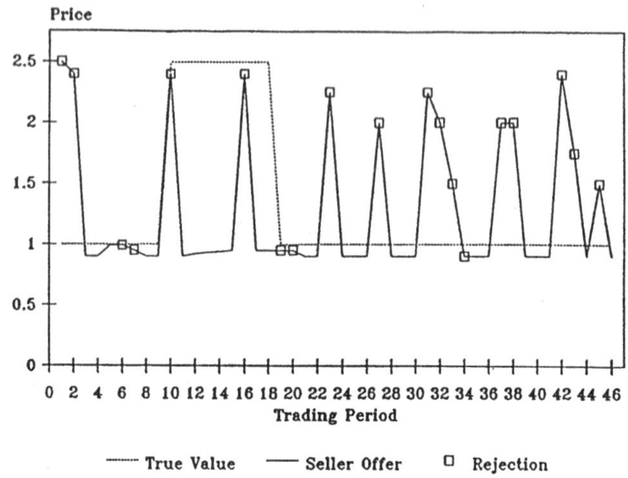

Figure 17

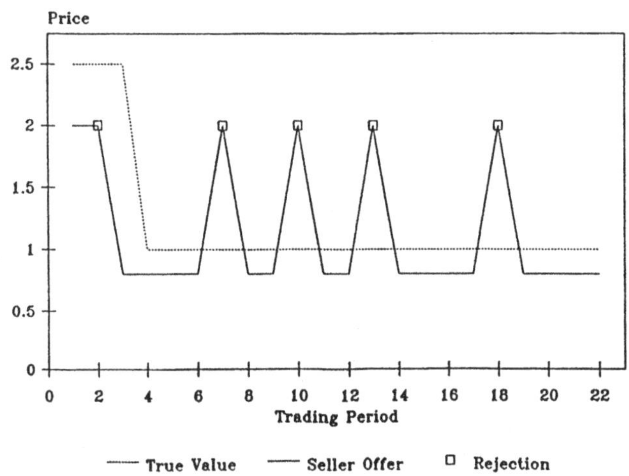

Figure 18

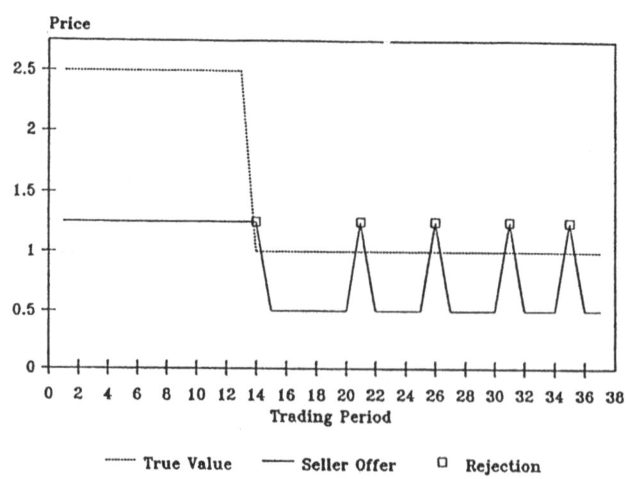

Figure 19

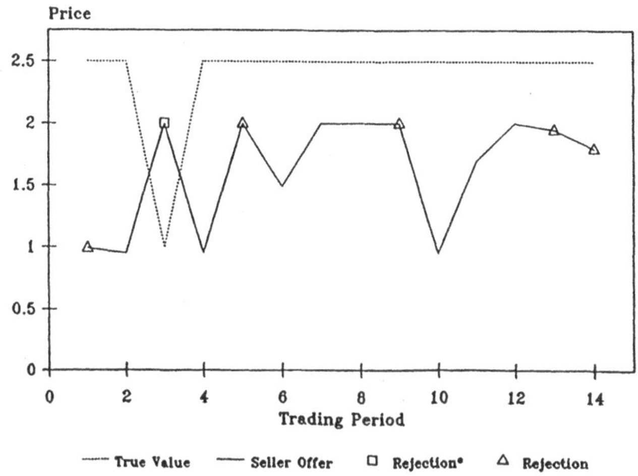

Figure 20

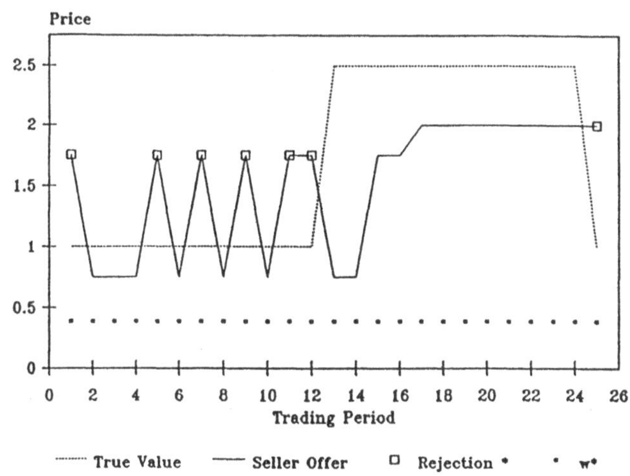

Figure 21

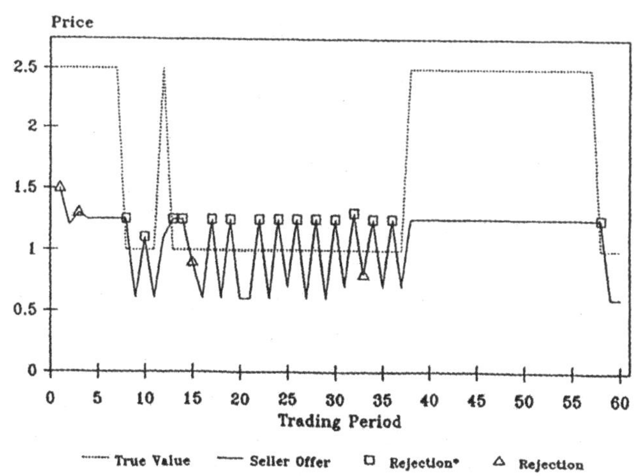

Figure 22

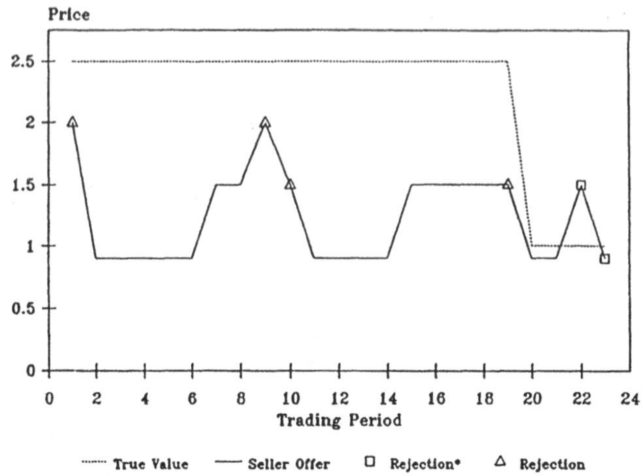

Figure 23

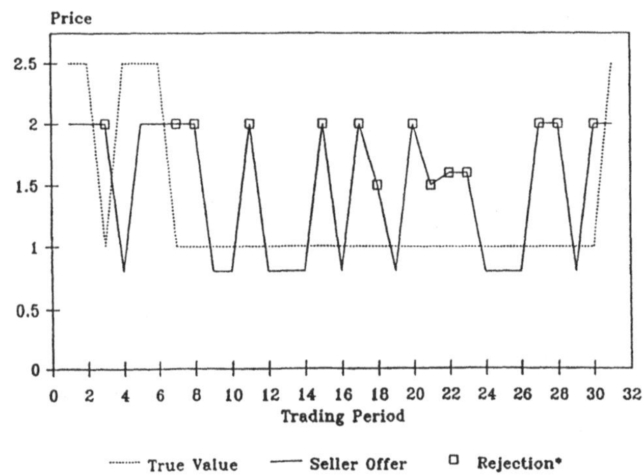

Figure 24

and 37 is the total number of outcomes; and $a = 150/251$, where 150 is the number of possible high prices $p_h \in [1.01, 2.50]$ and 251 is the total number of prices in $[0, 2.50]$. For (P_s) the theory's hit rate is high but the accuracy is low. When $h = \$2.50$ (Experiment 1a) the maximum possible m for (P_s) is 0.30. When $h = \$2.20$ (Experiments 2a, 3a, 4a, 5a, 6a) the maximum possible m for (P_s) is 0.46. When $h = \$1.50$ (Experiments 7a, 8a) the maximum possible m is 0.67. Most m in Table 1 are close to the maximum possible value. Since buyer behavior was simulated, m for (P_b) is not applicable. This is indicated by *NA* in column (P_b) of Table 1.

Table 2 reports summary statistics for the twenty four series (i), (ii), and (iii) experiments. For example in Experiment 1, $m = r - a = 0.27$ for (P_s), which is computed as follows: $r = 20/23 = 0.87$, where 20 is the number of

Table 1

Experiment	P_s	P_b	max m_s
1a	0.26	NA	0.30
2a	0.31	NA	0.46
3a	0.37	NA	0.46
4a	0.39	NA	0.46
5a	0.23	NA	0.46
6a	0.31	NA	0.46
7a	0.59	NA	0.67
8a	0.67	NA	0.67

Table 2

Series i			Series ii			Series iii		
Experiment	P_s	P_b	Experiment	P_s	P_b	Experiment	P_s	P_b
1	0.22	0.50	9	0.16	0.21	17	0.18	0.35
2	0.07	0.17	10	0.20	0.50	18	0.22	0.45
3	−0.13	0.24*	11	0.27	0.50	19	0.32	0.50
4	0.05	0.50	12	0.24	0.41	20	0.04	0.14
5	0.07	0.22*	13	0.11	−0.14*	21	0.16	0.50
6	0.02	0.35	14	0.22	0.46	22	0.13	0.43*
7	0.08	0.45	15	0.20	0.30	23	0.27	0.28
8	0.10	0.09*	16	0.40	0.25	24	−0.05	0.05
max m	0.40	0.50	max m	0.40	0.50	max m	0.40	0.50
$\mu_s^i = 0.06$	$\mu_b^i = 0.315$		$\mu_s^{ii} = 0.23$	$\mu_b^{ii} = 0.311$		$\mu_s^{iii} = 0.17$	$\mu_b^{iii} = 0.450$	

*Higher if zero-π trades are excluded.

successful predictions and 23 is the total number of outcomes (see Figure 1); and $a = 150/251 = 0.60$, where 150 is the number of $p_h \in [1.01, 2.50]$ and 251 is the total number of prices in $[0, 2.51]$. Since $h = \$2.50$ in all experiments, the maximum hit rate of (P_s) is high ($r = 1$ if the seller follows (P_s) perfectly) but the accuracy is low ($a = .6$). Thus in all 24 experiments the maximum possible m for (P_s) is 0.40 in Table 1. The maximum possible m for (P_b) is 0.50 since $r = 1$ if the buyer follows (P_b) perfectly and $a = .5$ since there are two possible answers (yes or no).

In the series (i), (ii), and (iii) experiments subjects have different levels of experience in each series. Thus, Table 2 reports the mean measure of predictive success for the eight experiments in the series i, ii, iii for the seller and the buyer, respectively. These means are denoted by μ_r^e, where $e = i, ii, iii$ denotes the experiment series and $r = b, s$ denotes buyer or seller. In the series (i), (ii) and (iii) experiments, $\mu_s^e = 0.06, 0.23$, and 0.17, for the seller in $e = i, ii, iii$. These means are all clearly below the maximum possible m. Most interestingly, however, is the fact that the means increase between the series (i) and (ii) experiments, which we interpret as evidence that sellers are learning (P_s), *but* decrease between the series (ii) and (iii) experiments.

The results in Table 2 suggest that experience affects sellers and buyers differently. Our a priori conjecture was that it would make subjects more likely to behave in accordance with the equilibrium strategies. The data indicate that buyers use equilibrium strategy (P_b) more frequently when they are more experienced: the maximum possible m was 0.50 and the respective mean measures of predictive success for (P_b) were $\mu_b^e = 0.315, 0.311$, and 0.45 for $e = i, ii, iii$. Observe that by the time buyers were twice experienced with the *institution* they were remarkably close to the maximum measure of predictive success for (P_b) (i.e., $m = 0.50$). However, seller's use of (P_s) declined in series (ii) to (iii) experiments: μ_s decreases from 0.23 to 0.17 and the variance of (P_s) increases from $\sigma_s^{ii} = 0.086$ in series ii to $\sigma_s^{iii} = 0.127$ in series iii. The variance of (P_s) in the series i experiments was $\sigma_s^i = 0.097$. The variance of (P_b) in the experiments was $\sigma_b^{ii} = 0.214$ and $\sigma_b^{iii} = 0.132$.

The increase in the variance of (P_s) suggests that there may have been something peculiar about some individual series (iii) experiments. Table 2 indicates that the measures of predictive success for the seller in each experiment in this series are $m_s^{17} = 0.18$ in Experiment 17, $m_s^{18} = 0.22$ in Experiment 18, $m_s^{19} = 0.32$ in Experiment 19, $m_s^{20} = 0.04$ in Experiment 20, $m_s^{21} = 0.16$ in Experiment 21, $m_s^{22} = 0.13$ in Experiment 22, $m_s^{23} = 0.27$ in Experiment 23, and $m_s^{24} = -0.05$ in Experiment 24. The maximum $m_s = 0.40$. Experiments 20 and 24 account for most of the increased variance, thus we now discuss the four series *iii* experiments with the worst seller performance in detail, with special emphasis on these two high variance cases.

In Experiment 20, buyer B20 (i.e., Subject 4 with experience profile S2, B12, B20: see Appendix A) frequently deviated from equilibrium strategy (P_b), resulting in $m_b^{20} = 0.14$. This buyer behavior is interesting in view of the fact that in Experiment 12 this individual was again a buyer (B12) with $m_b^{ii} = 0.41$, indicating that he consistently followed equilibrium strategy (P_b). Indeed in Experiment 12 this buyer had only 3 non-equilibrium rejections, and 2 of these 3 rejections were for zero-profit trades. Experiment 12 lasted for 32 periods. In contrast, in Experiment 20 this buyer had 5 non-equilibrium rejections in 14 periods, and only 1 of the 5 was a zero profit trade. The main difference between the buyer's behavior in Experiments 12 and 20 appears to be that in Experiment 12 the seller never offered a price higher than \$1.50 and by the end of the experiment p_h was consistently \$1.25. However, in Experiment 20 the seller consistently offered prices above \$1.50 (often \$2.00) and the buyer appeared unwilling to accept such a skewed profit split. The buyer may have been "punishing" the seller (and himself) in the hope of attaining a lower price for future trades. The seller was very slow in discerning this implicit lesson in the buyer's strategy, despite the fact that he had been a seller in Experiment 3. Although the measures of predictive success in Experiments 3 and 20 did show improvement (i.e., $m_s^3 = -0.13$ vs. $m_s^{20} = 0.04$), this subject had persistent trouble with the seller role but not with the buyer role (i.e., this subject was a buyer in Experiment 11 and $m_b^{11} = 0.50$).

In Experiment 21 the buyer followed equilibrium strategy (P_b) perfectly. This subject had two previous experiences as a seller (S5 and S13). The experiment lasted for 25 periods. The seller, however, had been a buyer twice previously (B8 and B14). This subject displayed no understanding of L, the optimal number of low price offers, and this is reflected in $m_s^{21} = 0.16$. This experiment suggests to us that S21's lack of prior experience with the seller role may have been more significant than B21's lack of prior experience with the buyer role, perhaps because buyers are much more passive in the Posted Offer institution than sellers.

In Experiment 22 the buyer followed equilibrium strategy (P_b) quite closely, and had previous experience as a buyer and seller (B7 and S14). The experiment lasted for 60 periods. The seller had been a seller once before (S6), and a buyer once before (B13). In Experiment 13 the buyer had an unusually long string of high values (13 out of 14 units). When this subject became a seller again in Experiment 22, he chose a relatively low p_h of \$1.25 (apparently to induce the buyer to tell the truth in the high state) but displayed no understanding of L. Perhaps his experience as a buyer in Experiment 13 (when he saw an unusually long string of high values) impeded his learning of L and resulted in $m_s^{22} = 0.13$.

In Experiment 24 the buyer followed equilibrium strategy (P_b) perfectly, and had two previous seller experiences (S7 and S16). The experiment lasted for 32 periods. The seller had been both a buyer and a seller previously (B5 and S15), however Experiment 15 lasted for only 5 periods so this subject's experience with the seller role was very limited. Interestingly in Experiment 15, $m_s^{15} = 0.20$ and in Experiment 24 $m_s^{24} = -0.05$ (for this same subject). Thus, this seller appears to have "unlearned" from his brief seller experience! In fact, closer inspection of Figure 24 reveals that a combination of an unusually large number of low realizations and imperfect comprehension of L (perhaps due to limited experience with the seller role) accounts for the extremely low $m_s^{24} = -0.05$.

The detailed search for anomalies in Experiments 20, 21, 22, and 24 indicates a common factor — experience with the buyer/seller role (not just with the institution) appears to be important when demand evolves according to an exogenous stochastic process with serial correlation. That is, whether a subject has been a buyer or a seller, and his/her particular experiences in the role are important when subjects are differentially informed and demand flutuates with some predictability. Subjects must not only attempt to learn the demand process, a seller must first form an opinion as to whether or not the buyer is revealing information about v truthfully.

Unusual prior experiences seem to affect sellers more than buyers in these experiments. Clearly, the buyer's role is easier to learn in the Posted Offer institution (i.e., the buyer simply accepts or rejects the price posted by the seller). In contrast, the seller must both choose a price *and* choose the optimal number of periods in which to make the low price offer (L). This dual task reflects the seller's more complicated problem: The seller is using price to both earn current profit on a trade and to learn the buyer's demand process (in the hope of making higher future profit). However, because sellers are informationally disadvantaged and realize that buyers may reject mutually profitable current trades in order to manipulate the seller's beliefs, subjects must learn about the institution, their own role, *and* their counterpart's role. This observation suggests that in future experiments it may be desirable to control for experience with only a particular role. That is, conduct a series of experiments where subjects are either buyers or sellers in all three experiments. These results can be compared with the results reported in Appendix A where role assignment was determined randomly over the course of the three experiments.

5 Concluding remarks

This paper studies intertemporal pricing and purchasing strategies in Posted Offer laboratory markets with demand values that follow a Markov Process. The following features of the model are important: First, differential information reduces market efficiency relative to full information when cyclical pricing is optimal because agents systematically forego trade on units for which the seller posts p_h to learn about the unit's current value when the value is low. Second, the buyer prefers the differential information (price cycle) solution to the full information solution because in general it allows the buyer to obtain some exchange surplus. Third, *in equilibrium information is revealed truthfully* but the equilibrium is *not* fully revealing (i.e., when the seller offers p_l the buyer's answer is uninformative).

The results from 32 laboratory experiments indicate that the Posted Offer institution performs reasonably well in revealing information in stochastic, intertemporal settings despite some pronounced inefficiencies (relative to full information) that are inherent in its structure. In particular, pricing patterns that appear to be "sticky," as well as periodic forgone trades, are part of an optimal intertemporal equilibrium pricing strategy. Since intertemporal pricing patterns are inextricably connected to agents' beliefs about the underlying nature of uncertainty that governs the system, laboratory experiments are useful in this setting because they allow researchers to observe and control the probability structures on which agents' actions are based. Thus, they may prove to be a useful tool for testing intertemporal, stochastic theories. However, the our results indicate that the nature of subject experience (i.e., with the institution versus with the role) matters much more in this setting than in previous static, deterministic Posted Offer experiments.

6 Appendix A: Experience profiles

All subjects in the series (a) experiments summarized in Table 1 were inexperienced. That is, they had never before participated in a Posted Offer experiment with stochastic demand. The experience profiles of the 16 subjects in the series (i), (ii) and (iii) experiments are given below. Subject 1 was the inexperienced buyer in experiment 1 (i.e., B1), the once experienced seller in experiment 11 (i.e. S11) and the twice experienced seller in experiment 19 (i.e., S19). After each role indicator, the associated measure of predictive success for that subject from Table 2 is reported. This provides a very rough profile of the subjects' learning over the course of the experiment. Following this notation, the experience and performance profiles for each subject are:

Subject experience and performance profile

Subject 01: B1 ($m_1^b = .50$), S11 ($m_{11}^s = .27$), S19 ($m_{19}^s = .32$).
Subject 02: S1 ($m_1^s = .22$), B10 ($m_{10}^b = .50$), S18 ($m_{18}^s = .22$).
Subject 03: B2 ($m_2^b = .17$), S09 ($m_9^s = .16$), B17 ($m_{17}^b = .35$).
Subject 04: S2 ($m_2^s = .07$), B12 ($m_{12}^b = .41$), B20 ($m_{20}^b = .14$).
Subject 05: B3 ($m_3^b = .24$), S12 ($m_{12}^s = .24$), B19 ($m_{19}^b = .50$).
Subject 06: S3 ($m_3^s = -.13$), B11 ($m_1 1^b = .50$), S20 ($m_{20}^s = .04$).
Subject 07: B4 ($m_4^b = .50$), S10 ($m_{10}^s = .20$), S17 ($m_{17}^s = .18$).
Subject 08: S4 ($m_4^s = .05$), B09 ($m_9^b = .21$), B18 ($m_{18}^b = .45$).
Subject 09: B5 ($m_5^b = .22$), S15 ($m_{15}^s = .22$), S24 ($m_{24}^s = -.05$).
Subject 10: S5 ($m_5^s = .07$), S13 ($m_{13}^s = .11$), B21 ($m_{21}^b = -.05$).
Subject 11: B6 ($m_6^b = .35$), B15 ($m_{15}^b = .30$), B23 ($m_{23}^b = .28$).
Subject 12: S6 ($m_6^s = .02$), B13 ($m_{13}^b = -.14$), S22 ($m_{22}^s = .13$).
Subject 13: B7 ($m_7^b = .45$), S14 ($m_{14}^s = .22$), B22 ($m_{22}^b = .43$).
Subject 14: S7 ($m_7^s = .08$), S16 ($m_{16}^s = .40$), B24 ($m_{24}^b = .50$).
Subject 15: B8 ($m_8^b = .09$), B14 ($m_{14}^b = .46$), S21 ($m_{21}^s = .16$).
Subject 16: S8 ($m_8^s = .10$), B16 ($m_{16}^b = .25$), S23 ($m_{23}^s = .27$).

References

Davis, D., Harrison, G., Williamson, A.: Convergence to nonstationary competitive equilibria: an experimental analysis. Discussion Paper (1992)

Davis, D., Holt, C.A.: Price rigidities and institutional variations in markets with posted prices. Economic Theory **9**(1): 63–80 (1996)

Davis, D., Holt, C.A.: Conspiracies and secret price discounts. Economic Journal **108**: 1–21 (1998)

Hoffman, E., Plott, C.: The effect of intertemporal speculation on the outcomes in seller posted-offer auction markets. Quarterly Journal of Economics **96**: 223–241 (1981)

Ketcham, J., Smith, V.L., Williams, A.: A comparison of posted-offer and double auction pricing institutions. Review of Economic Studies **51**: 595–614 (1984)

Plott, C.: An updated review of industrial organization: applications of experimental methods. In: Schmalensee, R., Willig, R.D. (eds.) Handbook of industrial organization, Vol. II. Amsterdam: Elsevier Publishers, North Holland 1989

Rustichini, A., Villamil, A.P.: Intertemporal pricing in markets with differential information. Economic Theory **8**: 211–227 (1996)

Selten, R.: Properties of a measure of predictive success. Mathematical Social Sciences **21**: 153–167 (1991)

Smith, V.L.: Experimental economics: induced value theory. American Economic Review **66**: 274–279 (1976)

Smith, V.L.: Markets as economizers of information: experimental examination of the 'Hayak hypothesis'. Economic Inquiry **20**: 165–179 (1982)

Villamil, A.P.: Termination procedures in laboratory markets. University of Illinois Discussion Paper (1999)

Dividend timing and behavior in laboratory asset markets*

Vernon L. Smith[1], **Mark van Boening**[2], **and Charissa P. Wellford**[3]

[1] Economic Science Laboratory, McClelland Hall, University of Arizona, Tucson, AZ 85721, USA
[2] School of Business Administration, 220 Holman Hall, University of Mississippi, University, MS 38677, USA (e-mail: mvanboening@bus.olemiss.edu)
[3] Federal Trade Commission, Bureau of Economics S-5013, 600 Pennsylvania Avenue NW, Washington, DC 20580, USA

Summary. This paper investigates the effect of dividend timing on price bubbles and endogenous expectations in twenty-six laboratory asset markets. In ten "A1" markets, a single dividend is paid at the end of the trading horizon. In nine "A2" markets, dividends are paid at the end of each trading period. In seven "A3" markets, some of the dividends are paid at the end of the trading horizon, and the rest are paid on a per-period basis. The results indicate that price bubbles are most likely in A2 markets, less likely in A3 markets, and least likely in A1 markets. Six distinct hypotheses are considered. The data suggest that the concentration of dividend value at a single point in time helps to create common expectations, and thus significantly reduce the incidence of bubbles. Also, the results underscore the difficulty facing econometric tests on field data where fundamental value has to be approximated.

Keywords and Phrases: Price bubbles, Asset market experiments, Expectations.

JEL Classification Numbers: C92, D84, G12.

1 Introduction

Economists have an ongoing interest in stock price "bubbles," or sustained price deviations from fundamental value[1]. Stock price volatility may affect wealth distribution or capital flow, and market crashes like October 1987 are often fol-

* We are grateful to Corinne Bronfman, Ron King, Jim Meehan, John Conlon and anonymous referees for comments. The views expressed in this paper do not necessarily represent those of the Federal Trade Commission, or of any individual Commissioner. The data and subject instructions are available at cost from the authors.

[1] Camerer (1989) surveys an extensive literature; also see the "Symposium on Bubbles" in the *Journal of Economic Perspectives*, vol. 4, no. 2, (1990).

lowed by regulatory discussions, *e.g.*, the Brady Commission. A variety of plausible explanations for bubbles have been offered, including "rational" (Delong et al., 1990), "intrinsic" (Froot and Obstfeld, 1991), and "contagion" bubbles (Topol, 1991). However, there is no widely accepted theory. Econometric tests for bubbles are useful, but there may be difficulties in estimating fundamental value or problems with specification error (Flood and Hodrick, 1990). Further, different techniques may yield conflicting results (*e.g.*, compare Dezhbakhsh and Demirguc-Kunt, 1990; West, 1987).

Even in controlled laboratory markets price bubbles are something of an enigma. Smith et al. (1988) observe "booms and crashes" in experimental stock markets where a long-lived asset is traded. Those bubbles are robust across a number of treatments, including short selling, margin buying and price change limits (King et al., 1993), but Ang and Schwarz (1992) find that short selling and margin buying dampen bubbles in markets with a short investment horizon. Although bubbles are attenuated when a futures market operates simultaneously with the spot market (Porter and Smith, 1995), they persist when a market for dividend information operates along with the spot market (King, 1991). Bubbles are observed in both a sealed bid-offer call market with no public display of bids and asks, and a continuous double auction where bids and asks are public information (Van Boening et al., 1993). When traders have a combination of knowledge about financial markets *and* design experience, bubbles are modest and dissipate quickly (Ackert and Church, 1998). But bubbles even occur in markets where reselling, and thus capital gains, are prohibited (Lei et al., 1999).

This study is motivated by evidence that bubbles in laboratory markets are a function of heterogeneous beliefs among traders. Smith et al. attribute their bubbles to endogenous capital gains expectations, which dissipate when traders participate in second and third sessions with the same group (also see Wellford, 1990). Camerer and Weigelt (1991) observe "information mirages" in some of their markets, where bubbles arise when traders are uncertain about the information set of other traders, even when all traders have the same information. Lei et al. conjecture that uncertainty about the task and/or asset structure contributes to confusion, which in turn manifests itself as a price bubble. These studies suggest that whatever their origin, diverse beliefs among traders contribute to, if not cause laboratory market bubbles.

Theory also emphasizes the role of heterogeneous beliefs in generating bubbles. Allen et al. (1993) present a model where such beliefs produce bubbles even though all agents are rational. Everyone knows that the stock is overpriced, but each agent thinks she can sell to someone else at a higher price before the bubble "bursts." While we do not test the Allen et al. theory per se, that model predicts price bubbles, and we observe them in our laboratory markets[2]. Daniel et al. (1998) model stock price over- and underreactions as a function of investor overconfidence and biased-self attribution. There, informed investors

[2] In their model, bubbles occur due to informational asymmetry where agents observe only their own trades, and thus can only speculate about other prices and the state of the world they imply. In our continuous double auction, bids, asks and contract prices are public information.

receive a noisy private signal, but they are overconfident and underestimate the error variance that signal. Further, their confidence reacts asymmetrically to public information: it rises when public information agrees with their information, but it does not fall by an equal amount when public information in contradictory. Again, we do not test this model *per se*, but it is another example where heterogeneous beliefs give rise to price deviations from intrinsic value.

Our research hypothesis is that expectations contribute to bubbles, and that those expectations are both heterogeneous and endogenous. Here, we investigate the dividend stream as a treatment variable, and we have three competing conjectures as to what will happen. First, more frequent dividend payments may focus a trader's attention myopically on the near term, fostering the heterogeneous beliefs across traders, and thus bubbles. Second, more frequent dividends could focus a trader's attention on the longer term income stream, fostering homogeneous expectations suppressing bubbles and yielding risk-adjusted expected value pricing. Finally, expectations may be independent of the dividend stream, in which case the frequency and/or magnitude of bubbles will be unaffected by changes in the dividend stream. Our results provide support for the first conjecture: concentration of dividend value in time helps create common endogenous expectations, and thus reduces bubbles.

Section 2 presents the experimental design that varies the dividend stream across twenty-six double auction markets for a long-lived asset. Six distinct hypotheses are outlined in Section 3. The results are presented in Section 4, and Section 5 concludes.

2 Experimental design and procedure

2.1 Design

Table 1 presents the 3x2 experimental design, which compares market parameters for twenty-six laboratory asset markets by asset type and experience level. All markets are fifteen periods in length, and each trading period lasts 240 seconds. In Table 1, markets are labeled by asset type, and by market replication number. In any given market, one of three alternative assets is traded. Asset "A1" pays a random dividend only after the final trading period. Asset "A2" pay a random *i.i.d.* dividend after each trading period. Finally, asset "A3" pays a random *i.i.d.* dividend after each period *and* pays an additional random dividend after the last period. The assets and their dividends are discussed in more detail below. Table 1 shows ten A1 markets, nine A2 markets and seven A3 markets[3]. An "x" after the

[3] These twenty-six markets are drawn from our ongoing research of laboratory asset markets, which accounts for the variation within asset types (*e.g.*, compare A1-1 and A1-7 in Table 1). We pool markets with different portfolios for three reasons. First, within a given market, all portfolios have identical period 1 expected values. Second, King (1991), King et al. (1993) and Van Boening et al. (1993) report A2 type markets where bubbles are robust to homogeneous initial portfolios. Third, in addition to the regressions reported below in Table 2, we also included variables to control for portfolio endowments, period 1 expected value and period 1 variance. None had any systematic effect on the results shown in Table 2, nor did they alter any of our interpretations stated in Section 4.

Table 1. Experimental design

Experiments	Endowments (U.S. \$, shares)	number	Dividend Distribution (U.S. \$)	Period 1 Expected Value and Variance
Asset 1 Markets				
A1-1, A1-2,	(3.50, 4)	3		
A1-3, A1-4,	(9.90, 2)	3	$D = (1.80, 2.40, 3.00)$	$E(D) = \$2.40$
A1-5, A1-6	(13.10, 1)	2	$p(D) = (1/4, 1/2, 1/4)$	$\sigma^2(D) = 0.180$
	(16.30, 0)	2		
A1-7, A1-8	(2.25, 3)	4	$D = (2.60, 3.60, 4.60)$	$E(D) = \$3.60$
	(5.85, 2)	4	$p(D) = (1/4, 1/2, 1/4)$	$\sigma^2(D) = 0.500$
	(9.45, 1)	4		
A1-9x, A1-10x	(2.25, 3)	3	$D = (1.00, 2.30, 3.60, 4.90, 6.20)$	$E(D) = \$3.60$
	(5.85, 2)	3	$p(D) = (1/9, 2/9, 3/9, 2/9, 1/9)$	$\sigma^2(D) = 2.253$
	(9.45, 1)	3		
Asset 2 Markets				
A2-1, A2-2,	(2.80, 4)	4	$d = (0.00, 0.08, 0.16, 0.40)$	$E(d) = \$0.16$
A2-3, A2-4,	(7.60, 2)	4	$p(d) = (1/4, 1/4, 1/4, 1/4)$	$E(D_1) = \$2.40$
A2-5, A2-6	(10.00, 1)	4		$\sigma^2(D_1) = 0.336$
A2-7x, A2-8x,	(2.25, 3)	3	$d = (0.00, 0.08, 0.28, 0.60)$	$E(d) = \$0.24$
A2-9x	(5.85, 2)	3	$p(d) = (1/4, 1/4, 1/4, 1/4)$	$E(D_1) = \$3.60$
	(9.45, 1)	3		$\sigma^2(D_1) = 0.804$
Asset 3 Markets				
A3-1, A3-2,	(2.80, 4)	4	$d = (0.00, 0.04, 0.08, 0.20)$	$E(d) = \$0.08$
A3-3	(7.60, 2)	4	$p(d) = (1/4, 1/4, 1/4, 1/4)$	$E(D_1) = \$2.40$*
	(10.00, 1)	4	buyout $B_t = \Sigma_{t=1}^{15} d_t \pm \$0.50$	$\sigma^2(D_1) = 0.586$*
			$p(\$0.50) = p(-\$0.50) = 1/2$	* *includes* B_t
A3-4, A3-5,	(2.25, 3)	3	$d = (0.00, 0.04, 0.14, 0.30)$	$E(d) = \$0.12$
A3-6x, A3-7x	(5.85, 2)	3	$p(d) = (1/4, 1/4, 1/4, 1/4)$	$E(D_1) = \$3.60$*
	(9.45, 1)	3	buyout $B_t = \Sigma_{t=1}^{15} d_t \pm \$1.00$	$\sigma^2(D_1) = 1.804$*
			$p(\$1.00) = p(-\$1.00) = 1/2$	* *includes* B_t

Note: Each market has 15 trading periods and each period lasts 240 seconds. "x" denotes once-experienced traders.

market name denotes experienced traders, *i.e.*, they were participating in their second market with the same asset type.

Each trader in a market is randomly assigned a portfolio consisting of cash and "shares" of an asset, and a trader number. A trader's endowment is private information, and at no time does any trader have information about other traders' portfolios. As shown in Table 1, there are ten subjects in market A1-1. Three traders have endowments of \$3.50 and 4 shares, three have endowments of \$9.90 and 2 shares, two have endowments of \$13.10 and one share, and two traders are endowed with \$16.30 and no shares.

Table 1 also shows the dividend distributions. Each share of A1 pays a common, random dividend D at the end of period 15. In any trading period $t < 15$, a share yields no income, other than capital gains or losses, and is a claim only on the future realization of the dividend. The probability distributions $p(D)$ are symmetric. A1 markets with inexperienced traders have a three-point discrete distribution, while those with experienced traders have a five-point distribution. For example, in market A1-1, the possible dividend draws are \$1.80, \$2.40 and \$3.00, with probability 1/4, 1/2 and 1/4, respectively. The fundamental (intrinsic) value of a single share in any trading period is the expected value of the dividend E(D).

Each A2 share pays a common, random *i.i.d.* dividend d at the end of each trading period $t = 1, \ldots, 15$. Thus in any period $t \leq 15$, ownership of a share is a claim to the dividend paid at the close of period t, as well as a claim on dividends paid in the remaining 16-t periods. After period 15, each share expires with no "scrap" or buyout value. The distributions of d are discrete four-point distributions where each draw is equally likely (p = 1/4). In A2 markets, the intrinsic value of a single share in period t equals the expected per-period dividend E(d) times the number of periods remaining, *i.e.*, $E(D_t) = E(d) \times (16 - t)$.

Each A3 share pays a common, random *i.i.d.* dividend d each period $t = 1, \ldots, 15$ and an additional random "buyout" dividend B at the end of period 15. As in the A2 markets, d has a discrete four-point distribution where each draw is equally likely (p = 1/4), but the individual dividends are half as large in the A3 markets, for the corresponding number of traders (*e.g.*, compare A2-1 and A3-1 distributions, or A2-7x and A3-6x). At the end of period 15, the buyout dividend B equals the sum the fifteen per-period dividends plus or minus a constant C, with p(+C) = p(−C) = 1/2. In markets with twelve traders, the constant C = \$0.50, while in markets with nine traders C = \$1.00. In period $t, B_t = \Sigma_{i=1}^{t} d_i + \Sigma_{j=t+1}^{15} d_j \pm C$, where d_i is the realized dividend in period i and d_j is the random dividend to be paid in period j. Thus, the intrinsic value of an A3 share fluctuates with each period with the realization d_i, *i.e.*, $E(D_t) = E(B_t) + E(d) = \Sigma_{i=1}^{t} d_i + [2 \times E(d) \times (16 - t)]$.

The three experimental assets do not perfectly parallel those found in naturally occurring markets, but there are some similarities[4]. Our objective is not to replicate naturally occurring markets but investigate systematic, controlled changes in dividend streams and their relation to the propensity of laboratory bubbles

2.2 Procedure

All our markets utilize the interactive PLATO system (Smith et al., 1988; Williams, 1980). Briefly, shares of an asset are traded via computer network over a series of real-time market periods. Trading rules are similar to those on the NYSE. Each trader, at a private computer terminal, enters bids to buy, or offers to sell single shares. The standing (highest) bid and the standing (lowest)

[4] The A1 asset is similar to an emerging market investment fund, with a buyout at the close of all trading on a pre-specified date, but it is not a growth stock, nor there is new information on the terminal value each period. Alternatively, the A1 asset can be viewed as a zero coupon bond. A2 and A3 assets are analogous to those traded in secondary markets, where dividends are regularly paid over the trading life of the asset. The A3 assets are also akin to secondary market assets that are bought out, as in a leveraged buyout, spin-off, or recapitalization. Of course, an important distinction between our markets and naturally occurring stock Exchanges is that in our markets, shares have a known, finite life, so that the expected asset price declines linearly with the age of the asset.

offer are displayed on each trader's screen. Other bids and offers are ordered by price priority in an electronic queue, but are not publicly displayed. Orders in the queue can be voided by the maker, but the standing bid or offer cannot be retracted. A trade occurs when the standing bid or standing offer is accepted by another trader. Upon electronic confirmation by the accepting party, the trade is recorded, both the standing bid and offer are cleared, and the market is open for new quotes (which may come from the queue).

A trader's profit for a market session is his working capital (or current cash position) after the close of the final trading period, plus final dividend payments. Working capital is displayed to each trader on their private computer screens as cash endowment + cumulative dividend earnings + cumulative sales revenue - cumulative expenditures. Current inventory, including the purchase price and time stamp, is also displayed on the screen. Any purchase or sale is immediately recorded in both the cash account and the inventory. Market information shown on each screen includes the period number, the time remaining in the period, the standing bid and offer, the nine most recent contract prices, and the current vote to end the period[5]. At the end of each period in the A2 and A3 markets, a dividend is drawn and displayed on each trader's screen (no dividend is drawn in the A1 markets until after the final period). Each trader's working capital is increased by the dividend per share times the number of shares held at the market close.

Between market periods, each trader receives a reminder as to the final period and the dividend distribution(s). The "average" (or expected), minimum and maximum values of his current portfolio over the remaining periods are also provided, as is an optional summary table with the last ten periods' mean, maximum and minimum prices, and the actual dividends. Also, the dividend distribution is explained to the traders during the instructions, and between periods they are reminded as to the possible draws and associated probabilities. All dividend parameters, and the final trading period are common information[6].

The market participants were recruited from business courses at the University of Arizona, Indiana University, and Washington University at St. Louis. After participating in a laboratory asset market, participants were invited back as "experienced" traders of a specific asset type (A1, A2, or A3). The typical market lasted about two to two and a half hours, including the computerized instructions. Each subject was paid in cash a $3.00 show-up fee plus any earnings from their participation in the laboratory market. Most subjects earned between $10 and $20 (U.S. currency), with a high around $30 and low of about $2, exclusive of the show-up fee.

---

[5] A period ends if all traders vote to end before time expires; voting to end does not affect one's ability to buy and sell until unanimity is achieved.

[6] In addition to the computerized instructions, a verbal announcement is made prior to period 1 reminding subjects that (a) all traders have identical dividend distributions, and receive identical dividend draws, (b) there are fifteen trading periods, and (c) "mode" errors should be avoided, such as entering a bid when one is actually in selling mode. Subjects are encouraged to ask questions.

3 Research hypotheses

Six hypotheses are evaluated with the experimental design. The first hypothesis is based on perfect foresight, under which Muthian-rational traders would not deviate from expected value. If traders value the asset by the expected value of the dividend stream, and use backward induction, then bubbles never occur (see Tirole, 1982). If traders form their expectations in the same manner across asset types, and those expectations are consistent with the intrinsic value, then deviations from intrinsic value should not be observed.

H1. *Perfect Foresight hypothesis.* Traders have perfect foresight and backward induct; therefore bubbles will not occur. On average, bubbles will not be observed in any market type: B(A1) = B(A2) = B(A3) = 0,

where B(.) functionally describes bubbles by asset type. Some specific measures of bubbles are discussed in the following section.

Liquidity, or non-information motivated trading may yield transitory price changes that deviate from intrinsic value (*e.g.*, Black, 1986). If the trading activity is due to "noise," then bubbles might be observed in our laboratory asset markets. As our experimental design does not control for noise trading, if bubbles arise according to this liquidity explanation, they should not vary with asset type.

H2. *Liquidity hypothesis.* Most traders have perfect foresight and backward induct, but some liquidity trading occurs, so bubbles are possible. Observed bubbles may suggest "noisy" rational expectations if on average bubbles are significant and the same across asset types: B(A1) = B(A2) = B(A3) > 0.

Trading dynamics may differ across asset types due to uncertainty about a share's value. Asset A1's intrinsic value is the same in all trading periods, while assets A2 and A3 have an intrinsic value that varies across trading periods. The A2 intrinsic value declines each period by the amount of the expected dividend E(d). Asset A3's intrinsic value is determined by both a per-period and a final dividend, and the final dividend depends in part on the realized per-period dividends, so the intrinsic value of an A3 share may increase or decrease from one period to the next. If greater uncertainty about future expected values contributes to bubbles, the Uncertainty hypothesis suggests that bubbles will be largest in A3 markets and smallest in A1 markets.

H3. *Uncertainty hypothesis.* Uncertainty about a share's value will increase the likelihood that bubbles will occur, and the severity of bubbles when they occur. This uncertainty is greatest in A3 markets and smallest in A1 markets. On average, bubbles will be largest in A3 markets and smallest in A1 markets: B(A1) < B(A2) < B(A3).

If trading strategies are also affected by dividend timing, so that more frequent dividends result in more myopic behavior, then A2 and A3 markets should be more prone to bubbles than A1 markets. The Dividend hypothesis predicts that bubbles will be smallest in A1 markets, and as both A2 and A3 markets pay dividends each period, those markets will be observationally equivalent.

H4. *Dividend hypothesis.* The trading horizon is affected in part by the frequency of dividends: the more frequently that dividends are paid, the more that traders will focus on myopically on the short-term, and less on long-term intrinsic value. On average, bubbles will be smallest in A1 markets, and equally severe in A2 and A3 markets: B(A1) < B(A2) = B(A3).

A fifth hypothesis is that the increased income each period (*i.e.*, the increase in each traders' working capital after dividends are paid) fuels bubbles. Under this scenario, as traders acquire more cash for a given inventory, they spend more, creating excess demand and thereby contributing to bubbles. As shown in Table 1, A2 shares have the largest *ex ante* per-period dividends, A3 shares the next largest, and A1 shares the smallest (or zero). The Income hypothesis predicts that bubbles will be the largest where the increase in income is greatest.

H5. *Income hypothesis.* Increased income over the course of a market fuels bubbles. On average, bubbles will be the greatest in markets with the largest injection of dividend income each period, and smallest in markets with no change in income: B(A1) < B(A3) < B(A2).

Hypotheses H3, H4 and H5 all predict that A1 markets will have the smallest bubbles, but each has a different prediction as to the relative ranking of A2 and A3 markets. Different effects may simultaneously be present. For example, if both the Uncertainty and Income effects are present, they could offset each other, and yield B(A2) = B(A3) on average. But this is precisely the prediction of the Dividend hypothesis. In the next section, we present a statistical test designed to help identify those effects that are present, and which are the most dominant.

A final hypothesis is based on an often-mentioned criticism of behavior of experimental subjects in asset markets. As this argument goes, deviations from intrinsic value are not speculative bubbles, but merely Bayesian learning. Subjects are unfamiliar with the artificial stocks traded in laboratory markets, and so they make errors during early trials (*e.g.*, Lei et al., 1999). Under this hypothesis, there should be few or no bubbles in markets with experienced subjects, and any observed bubbles will be independent of our dividend treatments:

H6. *Bayesian Learning hypothesis.* After gaining experience, traders have perfect foresight and backward induct; therefore bubbles will not occur. On average, bubbles will not be observed in any market type: B(A1) = B(A2) = B(A3) = 0, and subject experience will be the only significant treatment variable.

4 Results

4.1 Overview

Figure 1 charts mean normalized price deviations from intrinsic value per period for each of the three asset market types. Markets with inexperienced traders are shown on the left panel, and markets with experienced traders are shown on the right. Normalized deviations are $\Sigma_{i=1}^{q}$ [P_{ti} − E(D_t)]/NSO, where P_{ti} is the price of ith share traded in period t, q is the trading volume in period t, and NSO is the number of shares outstanding (see Table 1). To facilitate comparisons across the designs, the deviations are normalized by the number of shares outstanding. This measure may be thought of as the per share aggregate overvaluation (or undervaluation), relative to intrinsic value, in a period. The values shown on Figure 1 are means by period for the given asset type and experience level. For example, the value shown for period 1 of the A1 markets is the average of the eight normalized deviations from period 1 of markets A1-1 through A1-8.

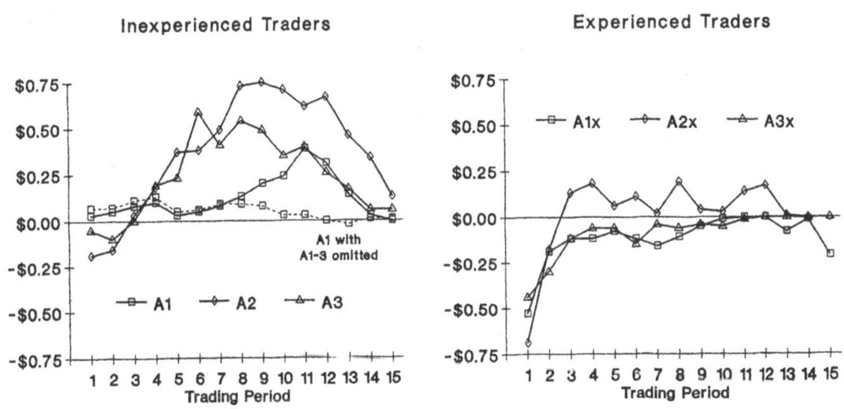

Figure 1. Mean normalized deviation from intrinsic value, by trading period and experience level

Figure 1 indicates that bubbles are most severe in A2 markets, and least severe in A1 markets. That is, the data suggest the relative ordering B(A1) < B(A3) < B(A2). These mean deviation plots are representative of the plots from the individual markets. Only one of the ten A1 markets exhibited a bubble (A1-3). The dashed line in the left panel of Figure 1 shows the mean deviation for A1 markets with A1-3 omitted; it further highlights the ordering B(A1) < B(A3) < B(A2). Bubbles occurred in nearly every A2 and A3 market, but were usually greater (*i.e.*, further from intrinsic value) in the A2 markets. When traders are experienced (right panel), the deviations from intrinsic value and the ordering by asset type are not as severe, although the A2 type (A2x) markets again exhibit the largest deviations. In both the A1x and A3x markets, deviations from intrinsic value tend to converge to zero from below[7].

[7] The period 15 divergence of A1x from zero is primarily due to a "mode error" where a trader inadvertently sold at a very low price. Several experiments had mode errors, which were documented

4.2 Measurement of bubbles

Regression analysis is reported in Table 2. For each of the twenty-six laboratory markets, six distinct bubble measures are calculated (and shown in Appendix Table 1), and then each measure is used as the dependent variable in a dummy-variable regression. Both parametric ANOVA regressions using the bubble measures, and nonparamteric regressions using ranks of the bubble measures are estimated[8]. Bubble measures are ranked in ascending order, so that a higher rank implies a larger measure.

Table 2. Regression analysis of bubble measures

Estimated Coefficient	Dependent Variable Y: Normalized I.V. Deviation	Price Variance	Price Amplitude	Share Turnover	a_{11}	a_{22}
ANOVA Regression: $Y = \beta_{A1}A1 + \beta_{A2}A2 + \beta_{A3}A3 + \beta_X X + \varepsilon$						
b_{A1}	2.07	0.68	0.70	5.40	0.01	–
(s.e.)	(1.07)	(0.32)	(0.18)	(0.36)	(0.01)	
b_{A2}	5.26	1.25	1.38	5.16	0.04	0.01
(s.e.)	(1.19)	(0.35)	(0.19)	(0.40)	(0.01)	(0.02)
b_{A3}	3.55	0.72	0.91	4.56	0.02	−0.03
(s.e.)	(1.30)	(0.38)	(0.21)	(0.44)	(0.01)	(0.02)
b_X	−4.59	−0.25	−0.44	−1.93	−0.005	0.03
(s.e.)	(1.45)	(0.43)	(0.24)	(0.49)	(0.01)	(0.02)
R^2	.37	.09	.32	.47	.31	.25
F	4.31	0.70	3.40	6.42	3.36	2.13
(p-value)	(.016)	(.569)	(.036)	(.003)	(.037)	(.158)
n	26	26	26	26	26	16
Nonparametric Regression: Rank of $Y = \beta_{A1}A1 + \beta_{A2}A2 + \beta_{A3}A3 + \beta_X X + \varepsilon$						
b_{A1}	12.9	10.6	9.3	17.6	9.6	–
(s.e.)	(1.94)	(2.42)	(1.90)	(2.01)	(2.07)	
b_{A2}	20.0	17.9	21.0	16.7	20.4	9.5
(s.e.)	(2.15)	(2.69)	(2.11)	(2.23)	(2.29)	(1.70)
b_{A3}	16.4	14.6	15.5	14.2	12.4	5.6
(s.e.)	(2.35)	(2.94)	(2.31)	(2.43)	(2.50)	(1.83)
b_X	−10.4	−2.8	−5.7	−10.5	−1.8	2.3
(s.e.)	(2.63)	(3.29)	(2.58)	(2.72)	(2.80)	(2.42)
R^2	.48	.18	.50	.44	.40	.23
F	6.68	1.62	7.23	5.70	4.88	1.95
(p-value)	(.002)	(.212)	(.001)	(.005)	(.010)	(.181)
n	26	26	26	26	26	16

Note: s.e. = standard error. See text for description of dependent and independent variables.

The four independent variables represent asset type (A1, A2 or A3) and experience level (X). If the observation of the bubble measure is from markets A1-1 through A1-10x, then A1 = 1, A2 = 0 and A3 = 0. Similarly, the nine

at the time of occurrence. We include mode errors in our data because (a) they appear to be randomly distributed across designs and (b) the increased price variance provides a stronger challenge to statistical tests, *i.e.*, make it more difficult to reject the null hypothesis.

[8] The nonparametric regression is only conditionally distribution-free, but it is less sensitive to outliers and nonnormal distributions than is the parametric regression. (see Conover, 1999).

A2 markets have A1 = 0, A2 = 1 and A3 = 0, while the seven A3 markets have A1 = 0, A2 = 0 and A3 = 1. If the observation is from one of the seven markets with experienced traders, then X = 1, but if it is from a market with inexperienced traders then X = 0. For example, A1-1 has X = 0 and A1-9x has X = 1. This regression design allows comparison across asset types while controlling for experience level.

In Table 2, Normalized I.V. Deviation is the sum of all prices in a given market, expressed as deviations from intrinsic value, divided by the number of shares outstanding. Price volatility is measured by the Price Variance relative to the mean price in a market. The Price Amplitude of a market's bubble is the maximum deviation (low to high) of mean period prices from the intrinsic value, normalized by $E(D_1)$ (see Table 1). Turnover is total trading volume in a market divided by the number of shares outstanding. Field studies typically use return variance, rather than a measure like Normalized I.V. Deviation, as intrinsic value is not observable for most naturally-occurring assets. We include Price Variance as an approximation of measures often applied in field studies of bubbles.

For the bubble measures Normalized I. V. Deviation, Price Variance, and Price Amplitude, larger values imply greater bubbles. However, there exist conflicting explanations of Turnover. If market volume is due to speculative activity, then greater Turnover can be associated with larger bubbles. But if a large number of trades occur around intrinsic value, traders may simultaneously infer that the market is highly competitive. In this case, bubbles would be unlikely and greater Turnover would be associated with smaller bubbles.

The bubble measures a_{11} and a_{22} are estimates of price dynamics. For each of the twenty-six laboratory markets, the following regression is estimated:

$$\overline{P}_t - \overline{P}_{t-1} = \alpha_{10} + \alpha_{11}(\text{Bids}_{t-1} - \text{Offers}_{t-1}) + \varepsilon_{1t} \qquad (1)$$

where $\overline{P}_t$ is the mean contract price in period t, Bids_{t-1} is the total number of bids to buy in period $t-1$, Offers_{t-1} is the total number of offers to sell in $t-1$, and ε_{1t} is a random error term. If lagged excess demand ($\text{Bids}_{t-1} - \text{Offers}_{t-1}$) is a reliable proxy for endogenous expectations[9], then $\alpha_{11} > 0$ and $\alpha_{10} = -E(d)$. That is, the change in mean price from period $t-1$ to t should be comprised of three terms: (i) a decline equal to the decline in the intrinsic value of a share[10], (ii) an adjustment proportional to changing expectations, and (iii) random variation. From each regression, the estimated coefficient a_{11} is used as an observation on the bubble measure in Table 2[11].

[9] Smith et al. first reported that equation (1) characterizes price dynamics in these types of laboratory asset markets, and the result is replicated elsewhere. But there is some disagreement in the literature as to what α_{11} measures. We interpret α_{11} as proxy for (*i.e.*, indirect measure of) endogenous heterogeneous expectations. See the conclusion for further discussion.

[10] The price will decline by less than $E(d)$ if on average traders are risk averse in dividends (Smith et al., p. 1141).

[11] The Durbin-Watson statistics associated with the twenty-six regressions of equations (1), and the sixteen regressions of equation (2) below, indicate that autocorrelation may be present (about 20% reject H_0: $\rho = 0$, about 30% are indeterminant), and the standard errors may be biased. Thus, we do not make statistical inference based on these regressions. However, the coefficient estimates are still unbiased, and we use the estimated coefficients as observations from each market.

For the sixteen A2 and A3 markets, the lagged income change is added to the equation to test whether the price dynamics can be explained by the increase in aggregate cash holdings due to dividend payments[12]:

$$\overline{P}_t - \overline{P}_{t-1} = \alpha_{20} + \alpha_{21}(\text{Bids}_{t-1} - \text{Offers}_{t-1}) + \alpha_{22}(\Delta\text{Income}_{t-1}) + \varepsilon_{2t} \qquad (2)$$

where each variable is defined as in equation (1), and $\Delta\text{Income}_{t-1}$ is the change in income due to the dividend paid to each share after period $t-1$ (and prior to period t). The a_{22} estimates from the twenty-six regressions of equation (2) are used as the observations on the bubble measure. As the A1 markets have no interperiod change in income, equation (2) is not estimated those ten markets, and the variable A1 is omitted from the Table 2 regressions where a_{22} is the dependent variable.

Table 2 presents the results. The parametric and nonparametric models have similar explanatory power, although the nonparameteric regressions tend to have a better fit (higher R^2, lower p-value). Four of the six bubble measures have significant F-statistics (p-value $< .04$); the exceptions are Price Variance and a_{22}. The lack of explanatory power for the Price Variance regressions underscores the difficulty in identifying bubbles in field markets when average prices are used in place of intrinsic values. Normalized I.V. Deviation, Price Variance, Price Amplitude, and a_{11} all estimate β_{A2} as the largest coefficient and β_{A1} as the smallest. Share Turnover estimates β_{A1} as the largest coefficient and β_{A3} as the smallest; see the discussion above regarding turnover and bubbles. (The a_{22} results are discussed below). The estimates are consistent with the ordering B(A1) < B(A3) < B(A2) observed in Figure 1.

4.3 Analysis of research hypotheses

Table 3 reports statistical tests by asset type, and for experience effects, based on the estimated regressions (Table 2). The statistics shown are F-statistics, except for the Bayesian Learning hypothesis where t-statistics are shown. For the nonparametric regressions, the F and t statistics are equivalent, respectively, to Kruskal-Wallis and Mann-Whitney statistics based on ranks (see Conover, 1999). For simplicity, all tests are evaluated at the 5% significance level. The parametric and nonparametric methods again yield similar results, although the nonparametric statistics typically have a larger magnitude, and thus lower p-value than the corresponding parametric values. As the two methods only twice contradict one another (*i.e.*, one rejects H_0 and the other does not), our summary and interpretation will not systematically distinguish between them.

The pairwise tests and the estimated coefficients in Table 2 reveal three regularities in the comparisons of A1, A2, and A3. First, tests for measures Normalized I.V. Deviations, Price Amplitude and a_{11} reject H_0: $\beta_{A1} = \beta_{A2}$ (the nonparametric regression adds Price Variance), and in Table 2 these measures estimate

[12] We also considered total dividend income, which produced similar results, as did normalizing the price change by an index based on the change in dividend income relative to the previous period, and relative to the initial (period 1) aggregate cash holdings.

Table 3. Statistical tests based on regressions

Null Hypothesis	Normalized I.V. Deviation	Price Variance	Price Amplitude	Share Turnover	a_{11}	a_{22}
ANOVA Regression: $Y = \beta_{A1}A1 + \beta_{A2}A2 + \beta_{A3}A3 + \beta_X X + \varepsilon$						
Pairwise tests						
$H_0 : \beta_{A1} = \beta_{A2}$	4.45**	1.61	7.63**	0.21	8.98**	–
$H_0 : \beta_{A1} = \beta_{A3}$	0.84	0.01	0.65	2.43	0.17	–
$H_0 : \beta_{A2} = \beta_{A3}$	1.07	1.19	3.06	1.22	5.49**	4.70**
Perfect Foresight hypothesis						
$H_0 : \beta_{A1} = \beta_{A2} = \beta_{A3} = 0$	8.64**	5.81**	23.40**	139.3**	12.45**	–
Liquidity hypothesis						
$H_0 : \beta_{A1} = \beta_{A2} = \beta_{A3}$	2.22	0.96	3.93**	1.25	5.01**	–
Bayesian Learning hypothesis						
$H_0 : \beta_X \geq 0$	−3.16**	−0.59	−1.86**	−3.97**	−0.54	1.13
Nonparametric Regression: Rank of $Y = \beta_{A1}A1 + \beta_{A2}A2 + \beta_{A3}A3 + \beta_X X + \varepsilon$						
Pairwise tests						
$H_0 : \beta_{A1} = \beta_{A2}$	19.35**	4.52**	18.94**	0.10	13.87**	–
$H_0 : \beta_{A1} = \beta_{A3}$	1.46	1.20	4.61**	1.31	0.83	–
$H_0 : \beta_{A2} = \beta_{A3}$	1.48	0.77	3.58	0.70	6.40**	4.84**
Perfect Foresight hypothesis						
$H_0 : \beta_{A1} = \beta_{A2} = \beta_{A3} = 0$	49.19**	24.32**	46.83**	46.31**	35.21**	–
Liquidity hypothesis						
$H_0 : \beta_{A1} = \beta_{A2} = \beta_{A3}$	3.41	2.27	9.50**	0.68	7.29**	–
Bayesian Learning hypothesis						
$H_0 : \beta_X \geq 0$	−3.94**	−0.84	−2.19**	−3.87**	−0.67	0.93

Note: ** indicates reject H_0 at the 5% level. F statistics are reported for the *Pairwise Tests*, the *Perfect Foresight Hypothesis*, and the *Liquidity Hypothesis*; t statistics are reported for the*Bayesian Learning Hypothesis*. For the nonparametric regressions, the F and t statistics are equivalent to Kruskal-Wallis and Mann-Whitney statistics, respectively (see Conover, 1999).

$\beta_{A1} < \beta_{A2}$. Thus on average, these bubble measures are significantly larger in A2 markets than in A1 markets. Second, $H_0 : \beta_{A1} = \beta_{A3}$ is not rejected for any bubble measure (except nonparametric Amplitude), and for every measure but Turnover, the estimated coefficient in Table 2 always indicates $\beta_{A1} < \beta_{A3}$. While bubbles tend to be larger in A3 markets relative to A1 markets, the difference is not great. Third, tests for measures a_{11} and a_{22} reject $H_0 : \beta_{A2} = \beta_{A3}$, and the Price Amplitude test rejects H_0 at the 10% level (not shown in Table 3). Without exception the estimated coefficients in Table 2 indicate $\beta_{A3} < \beta_{A2}$. So bubbles tend to be larger in A2 markets relative to A3 markets, but the degree varies by bubble measure. Our interpretation of these regularities is that they support the ordering $B(A1) \leq B(A3) < B(A2)$. That is, we observe a systematic difference between A1 and A2, with A3 in between. Below is analysis of our research hypotheses.

H1: *Perfect Foresight hypothesis.* H1 predicts $B(A1) = B(A2) = B(A3) = 0$, which is inconsistent with the data in Figure 1, and the pairwise tests in Table

3, as bubbles are observed in the laboratory asset markets. Furthermore, Table 3 reports F-statistics for the Perfect Foresight hypothesis $H_0 : \beta_{A1} = \beta_{A2} = \beta_{A3} = 0$. The null hypothesis is unilaterally rejected.

H2: *Liquidity hypothesis.* H2 predicts B(A1) = B(A2) = B(A3), which is inconsistent with the ordering B(A1) $\leq$ B(A3) $<$ B(A2) which is observed in Figure 1 and supported by the pairwise tests in Table 3. In addition, Table 3 shows F-statistics for the Liquidity hypothesis $H_0 : \beta_{A1} = \beta_{A2} = \beta_{A3}$. The hypothesis is rejected for Price Amplitude and a_{11}. Furthermore, although it is not shown in Table 3, Normalized I.V. Deviation rejects the null at the 10% level (nonparametric) and 15% level (parametric). By at least two bubble measures, and possibly three, bubbles are not the same across asset types.

H3: *Uncertainty hypothesis.* H3 predicts B(A1) $<$ B(A2) $<$ B(A3), but the data contradict B(A2) $<$ B(A3). As mentioned above, all twelve regressions estimate $\beta_{A2} > \beta_{A3}$, and the difference is greatest for Price Amplitude and a_{11}. Thus, the severity of bubbles in the observed markets cannot be attributed to increased uncertainty about the value of an asset.

H4: *Dividend hypothesis.* H4 appears to be consistent with the data, at least in a modified form. H4 predicts B(A1) $<$ B(A2) = B(A3). The data suggest that B(A1) $\leq$ B(A3) $<$ B(A2), as the difference between A1 and A3 is typically small. In particular, the pairwise tests for a_{11} indicate that B(A1) = B(A3) $<$ B(A2). Recall that both A2 and A3 pay a large, final dividend while A3 does not. Perhaps the correct, or modified, interpretation of the Dividend Effect is not that more frequent dividends tend to induce myopic behavior, but rather that the presence of a large, final dividend helps traders focus on longer term strategies, and away from myopic short term strategies.

H5: *Income hypothesis.* H5 predicts B(A1) $<$ B(A3) $<$ B(A2), which is consistent with the data. But the failure to observe systematic differences between A1, where there is no income change, and A3, where dividends are paid each period, casts doubt on H5 (see pairwise tests for $H_0 : \beta_{A1} = \beta_{A3}$ in Table 3; only nonparametric Amplitude rejects H_0). In fact, the ANOVA regression with bubble measure a_{11} estimates $\beta_{A3} < 0$ (see Table 2), implying that on average, the income effect is *negative* in A3 markets. The rejection of H_0: $\beta_{A2} = \beta_{A3}$ under a_{22} is apparently due to the fact that the regression estimates $\beta_{A3} < 0$ and $\beta_{A2} > 0$; at test for H_0: $\beta_{A2} \geq 0$ cannot be rejected (see estimated coefficient and standard error in Table 2; $t = 0.50$). Thus, if there is a positive income effect in A2 markets, it is very weak. Furthermore, the a_{22} regressions in Table 2 have very little explanatory power. The influx of dividend income does not appear to contribute to the formation of bubbles in our markets.

H6: *Bayesian Learning hypothesis.* The expected sign of the coefficient β_X is negative, *i.e.*, experienced subjects are less prone to bubbles. In general, experience reduces bubbles in our markets: ten of the twelve estimates of β_X are negative in Table 2, and the two exceptions are the a_{22} estimates. But while experience reduces Normalized I.V. Deviation, Price Amplitude and Turnover (*i.e.*,

H_0: $\beta_X \geq 0$ is rejected in Table 3), it does not significantly reduce Price Variance or a_{11} (H_0 is not rejected). Thus, we do find evidence of Bayesian learning, but, surprisingly, the price adjustment process described by equation (1) is not sensitive to experience level. Additionally, the positive sign of the β_X estimate in the a_{22} regressions can be interpreted as marginal evidence that experience *increases* income effects, which contradicts the Bayesian hypothesis. Finally, we note that the analysis of hypotheses H1 through H5 was made *after* controlling for experience level, *i.e.*, Bayesian learning.

5 Conclusion

This study examines price bubbles in twenty-six laboratory asset markets with long-lived assets. In each market, one of three assets is traded, where assets are differentiated by the timing of the dividend payments. Six distinct hypotheses are considered, and the results suggest that bubbles are affected by the dividend stream, even when controlling for experience level. The strongest evidence for the dividend-timing hypothesis is the "lagged excess demand" price adjustment effect. The a_{11} data clearly reject three hypotheses, and the absence of an income effect rules out a fourth. Subject experience has a negligible effect on the a_{11} bubble measure, while it has significant effect on others, ruling out a fifth hypothesis. Also, we find price variance to be a very uninformative bubble measure. This underscores the difficulty facing econometric tests on field data where fundamental value has to be approximated.

Smith et al. interpret lagged excess demand as a proxy for endogenous capital gains expectations, but there are other plausible interpretations. Van Boening (1991) finds that value-weighted version of *rejected* excess bids explains price changes in sealed-bid uniform price auctions. One interpretation is that subjects respond to their own individual acceptance or rejection rates. Lei et al. find an excess demand effect in double auctions where there *no* possibility of capital gains. This astonishing result implies that endogenous capital expectations are not a necessary condition for bubbles. As an alternative, Lei et al. offer an "active learning" hypothesis and a "confusion" hypothesis[13]. We also identify an excess demand effect, and find it sensitive to the timing of dividend stream but not to the experience level. If the Lei et al. explanations are applicable here, then we should observe evidence of a liquidity effect (noise or confusion), an income effect (as subjects use added income to trade more according to their preconditioning), and an experience effect (which would mitigate both the confusion and the preconditioning). But we observe no such evidence, particularly with regards to the lagged excess demand effect. Currently, we view lagged excess demand as a proxy for expectations that are both heterogeneous and endogenous.

[13] Surprisingly, Lei et al. do not examine the effect of subject experience. It would be interesting to see if they observed bubbles and an excess demand effect in their NoSpec markets with experienced subjects.

The diversity of data within and across studies suggests that lagged excess demand is a robust proxy for *some* endogenous effect. Smith et al. distinguish between rationality *in the sense of Muth* and rationality *in the sense of Nash.* In the former, expectations are supported by outcomes that in turn support an underlying theory; here it is risk- adjusted intrinsic value. By that metric, laboratory bubbles are undeniably "irrational" or represent "systematic errors." In the latter sense, expectations are rational as long as they are supported or reinforced by outcomes. Thus the class of potential explanations is clearly large, and includes those mentioned above. But the robustness of both bubbles and the excess demand effect would seem to suggest that some consistent, reasonable explanation exists. Extant theory addresses heterogeneous expectations, but endogenous beliefs are relatively new (*e.g.,* Camerer and Ho, 1999). Our conclusion here is that the concentration of dividend value in time helps create common endogenous expectations, and thus reduces bubbles. But we acknowledge that a formal theoretical explanation is an area of future work. While our markets are not exact replicas of naturally occurring markets, they augment theory and field research by providing controlled data on asset pricing.

Appendix Table 1. Bubble measures by market

	Bubble Measure					
Normalized Market	Price I.V. Deviation	Price Variance	Share Amplitude	Turnover	a_{11}	a_{22}
A1-1	1.604	0.003	0.060	5.250	−0.005	–
A1-2	2.491	0.354	0.687	6.450	0.016	–
A1-3	2.124	0.401	0.656	3.100	−0.001	–
A1-4	9.291	3.960	2.583	6.550	0.078	–
A1-5	4.513	0.241	0.541	6.350	0.008	–
A1-6	3.765	0.046	0.129	5.850	0.003	–
A1-7	−6.458	0.148	0.289	5.541	0.017	–
A1-8	−2.445	0.454	0.528	4.583	0.010	–
A1-9x	−1.948	0.242	0.287	3.056	0.003	–
A1-10x	−1.413	0.459	0.323	3.389	0.001	–
A2-1	5.743	1.447	1.708	6.250	0.039	−0.045
A2-2	9.560	2.056	2.065	5.000	0.049	−0.011
A2-3	4.835	0.584	1.077	6.071	0.026	0.010
A2-4	5.324	0.288	0.924	3.788	0.028	0.025
A2-5	5.165	0.267	0.851	6.607	0.025	0.028
A2-6	2.360	2.149	1.702	3.357	0.054	0.008
A2-7x	1.608	0.139	0.771	3.278	0.033	−0.008
A2-8x	0.951	2.229	1.047	3.722	0.063	0.086
A2-9x	−2.013	1.300	0.962	2.611	0.041	0.061
A3-1	7.158	0.646	1.022	4.250	0.014	0.077
A3-2	4.710	0.745	0.396	3.607	−0.013	−0.036
A3-3	4.574	0.582	1.113	3.571	0.035	−0.037
A3-4	2.893	1.910	1.269	6.333	0.038	−0.032
A3-5	−1.357	0.248	0.824	4.444	0.029	−0.094
A3-6x	−0.073	0.310	0.297	2.111	0.014	0.014
A3-7x	−2.226	0.078	0.558	3.722	−0.0003	−0.051

References

Ackert, L.F., Church, B.K.: The effect of subject pool and design experience on rationality in experimental asset markets. Working paper 98-18, Federal Reserve Bank of Atlanta, Atlanta, GA (1998)

Allen, F., Morris, S., Postelwaite, A.: Finite bubbles with short sale constraints and asymmetric information. Journal of Economic Theory **61**, 206–229 (1993)

Ang, J.S., Schwarz, T.V.: The formation and control of asset bubbles: an experimental study. Tms, Florida State University (1992)

Black, F.: Noise. Journal of Finance **41**(3), 529–543 (1986)

Camerer, C.: Bubbles and fads in asset prices. Journal of Economic Surveys **3**(1), 3–41 (1989)

Camerer, C., Ho, T.-H.: Experience-weighted attraction learning in normal-form games. Econometrica **67**(4), 827–874 (1999)

Camerer, C., Weigelt, K.: (1991) Information mirages in experimental asset markets. Journal of Business **64**(4), 463–493

Conover, W.J.: Practical nonparametric statistics, 3rd edn. New York, NY: Wiley 1999

Daniel, K., Hirschleifer, D., Subrahmanyam, A.: Investor psychology and security market under and overreations. Journal of Finance **53**(6), 1839–1885 (1998)

Delong, Br.J., Schleifer, A., Summers, L.H., Waldmann, R.J.: Positive feedback investment strategies and destabilizing rational speculation. Journal of Finance **45**(2), 379–396 (1990)

Dezhbakhsh, H., Demirguc-Kunt, A.: On the presence of speculative bubbles in stock prices. Journal of Financial and Quantitative Analysis **25**(1), 101–112 (1990)

Flood, R.P., Hodrick, R.J.: On testing for speculative bubbles. Journal of Economic Perspectives **4**(2), 85–101 (1990)

Froot, K.A., Obstfeld, M.: Intrinsic bubbles: the case of stock prices. American Economic Review **81**(5), 1189–1214 (1991)

King, R.R.: Private information acquisition in experimental markets prone to bubble and crash. The Journal of Financial Research **14**(3), 197–206 (1991)

King, R.R., Smith, V.L., Williams, A.W., Van Boening, M.V.: The robustness of bubbles and crashes in experimental spot markets. In: Day, R.H., Chen, P. (eds.) Nonlinear dynamics and evolutionary economics. New York: Oxford Press 1993

Lei, V., Noussair, C.N., Plott, C.R.: Non-speculative bubbles in experimental asset markets: lack of common knowledge of rationality vs. actual irrationality, Tms, Purdue University (1999)

Miller, R.G., Jr.: Simultaneous statistical inference, 2nd edn. New York: Springer 1980

Porter, D., Smith, V.L.: Futures contracts and dividend uncertainty in experimental asset markets. Journal of Business **68**(4), 509–541 (1995),

Smith, V.L.: Microeconomic systems as an experimental science. American Economic Review **72**(5), 923–955 (1982)

Smith, V.L., Suchanek, G.L. Williams, A.W.: Bubbles, crashes, and endogenous expectations in experimental spot asset markets. Econometrica **56**(5), 1119–1151 (1988)

Tirole, J.: On the possibility of speculation under rational expectations. Econometrica **50**(5), 1163–1187 (1982)

Topol, R.: Bubbles and volatility of stock prices: effect of mimetic contagion. The Economic Journal **101**, 786–800 (1991)

Van Boening, M.V.: Call versus continuous auctions: an experimental study of market organization. University of Arizona Ph.D. dissertation (1991)

Van Boening, M.V., Williams, A.W. LaMaster, S.: Price bubbles and crashes in experimental call markets. Economic Letters **41**, 179–185 (1993)

Wellford, C.P.: Tender offer versus market acquisition methods: an experimental analysis, takeovers and horizontal mergers: policy and performance. University of Arizona Ph.D. Dissertation (1990)

West, K.D.: A specification test for speculative bubbles. Quarterly Journal of Economics **102**, 553–580 (1987)

Williams, A.W.: Computerized double auction markets: some initial experimental results. Journal of Business **53**(3), 235–258 (1980)

Bidding up, buying out and cooling-off: an examination of auctions with withdrawal rights*

J. Asker

Department of Economics, Harvard University, Littauer Center, 1875 Cambridge St, Cambridge, MA 02138, USA (e-mail: asker@fas.harvard.edu)

Summary. This paper considers a model in which bidders in an auction are faced with uncertainty as to their final valuation of the auctioned object. This uncertainty is resolved after the auction has taken place. It is argued that the inclusion of a cooling-off right raises the expected revenue to the seller when bidders face a risk of the object being a strict 'bad', in that owning the object incurs negative utility to the winner of the auction. The model is then tested in a laboratory setting. The evidence from this experiment supports the predictions of the theory.

Keywords and Phrases: Auctions, Withdrawal rights, Auction experiments, Cooling-off rights.

JEL Classification Numbers: C92, D44, K11.

1 Auctions and cooling-off rights

A fairly common assumption in economics is that contracts are binding. In many markets, however, one party to a contract has a right to withdraw from the contract after it has been entered into. This right, when it applies to buyers, is often called a cooling-off right. Such a right may arise from the terms of the contract or may be imposed by statute.

---

* This paper is a revised version of my honours thesis written at the Australian National University. I am grateful to Flavio Menezes, Simon Grant and Andrew Wait for their helpful comments. I would like to thank two anonymous referees for their thoughtful suggestions. Cathy O'Neil, Nick Carroll, Marnie Griffith, Arvid Streimann, Toby Buscombe and Steve Roberts lent invaluable assistance in conducting and refining the experiment. Financial assistance from the Department of Economics at the Australian National University is gratefully acknowledged.

When a cooling-off right was introduced to the residential housing market in Victoria, Australia, the Hon. W.A. Landeryou, MLC, commented in the second reading speech of the Sale of Land (Amendment) Bill [11/24/1982] that

> "The cooling-off period will provide purchasers of land with some protection against impetuous buying or persuasive sales techniques and will enable a purchaser to obtain further advice in respect of the transaction."

In other words, if an object has different valuations in different states of the world, cooling-off rights exist to enable the buyers to ascertain which state of the world they are in.

There are many examples of markets where cooling-off rights exist. Examples of markets in Australia where statutory cooling-off rights exist are: residential housing markets; motor vehicle markets; and door-to-door sales (also in the USA). The details of the rights differ from jurisdiction to jurisdiction. However, there are several trends: the cooling-off right tends to last for 3 to 7 business days after the contract has been entered into; a nominal fee tends to be attached to the exercise of the right (often 1% of the buying price); and the right does not apply to sales by auction.

Von Ungern-Sternberg [13] reports instances of non-statutory cooling-off rights. He notes that it is relatively common for the winner of a tendering process in the Swiss and German construction industries to be given a similar option to withdraw from a contract. The idea of the right in this industry is to allow winning firms to evaluate their position with regard to other commitments before being bound to a new job. It is difficult to measure how prevalent such non-statutory cooling-off rights are as they depend on the express and implied terms of individual contracts. However, it is noteworthy that instances have arisen when parties have litigated over such terms, for instance in *Darley v John Valentine Health Group Pty Ltd (In Liq) (1987) 21 IR 441*, an Australian case where a cooling-off right was contained in a health club franchise agreement.

There seems sufficient evidence to suppose that cooling-off rights operate in several different types of markets around the world. This raises the interesting question of how such rights affect behavior in markets. This paper attempts to address this question in the context of a first price sealed bid auction, a stylized representation of markets that operate by tendering or bidding processes. It takes the independent private valuations (IPV) model developed by Vickrey [12] and incorporates the notion of a cooling-off right. The model is then tested empirically. A laboratory experiment was conducted in which undergraduate economics students at the Australian National University were faced with the exact problem faced by the bidders in the theoretical model. The data from the experiment is used to verify the equilibrium strategies predicted by the theory. Thus the contribution of this paper is two-fold: it examines the effect of cooling-off rights within the most comprehensively understood auction framework; and it offers an empirical evaluation of the resulting model.

A key innovation in the paper is the modeling of valuations. In the model values, while privately known, are state dependent. Bidders know their values in each state but are uncertain about which state they will find themselves in after the auction. This state is revealed during the cooling-off period. To give this structure some intuition, consider the following example.

A residential home is put up for sale. Prospective buyers may have the opportunity to inspect the home once or twice before they have to enter a bid. These inspections give prospective buyers an accurate impression of their value of the house, if what they see is what they get. However, in the back of their minds, they may acknowledge the possibility of other states of the world; perhaps they fear the house is infected with termites (a 'disaster' state of the world) or perhaps they suspect the council plans to build a school nearby (a possibly good 'surprise' state of the world). On the day of the house auction it is hard to verify either of these suspicions; however after the auction both may be easily verified in a few days (by a building inspection or a call to the local council, respectively).

In both of these scenarios, termites or school, the prospective buyers of the house have different valuations depending on the state that they find themselves in. As compared with the state where what you see is what you get, the presence of termites would dramatically lower the value of the house (possibly making it negative if the problem was severe enough and the purchasers attach negative utility to the inconvenience it creates). In the second case, having a new school nearby may raise the value of the house.

Conceivably, the presence of the cooling-off right will dramatically affect bidding behavior in each example. In the case of suspected termites, bidders can ignore this possibility because they can withdraw from the contract if their fears are realized. This would make bidders more willing to bid aggressively for the house. Similarly, the possibility of having a new school nearby is more likely to induce higher bids if bidders have the option of withdrawing if they find they have bid too high and a school is not planned. Without a cooling-off right a bidder would want to hedge their bids in both instances, protecting themselves against the risk of being in the least attractive state of the world.

The question of how a cooling-off right affects behavior in auctions has not been widely explored in the literature. One paper that examines cooling-off rights and auctions explicitly is von Ungern-Sternberg [13]. This paper examines a multi-object auction where the bidder can withdraw their bid after winning the auction. The model considers a procurement auction where capacity-constrained firms bid for two jobs simultaneously. The firm will withdraw from a contract if both contracts are won and the firm is too constrained to perform both profitably. Von Ungern-Sternberg demonstrates that, in his setting, given sufficiently capacity-constrained firms, the cooling-off right is desirable for the employer of the tendering firms. This is due to the ability of firms to bid more aggressively without having to hedge against the possibility of taking on more than they can handle.

In von Ungern-Sternberg's framework types are distributed over a known distribution with an unknown mean, with each bidder knowing their precise valuation. This allows considerable simplification of the strategic elements in the

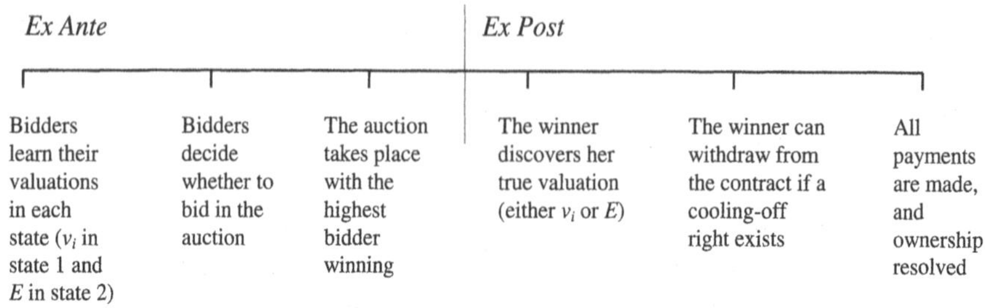

Figure 1. The timing of the model

auction he considers; specifically, bidders are unable to infer their probability of winning from their valuations. This makes it desirable to reconsider the effect of cooling-off rights in the standard IPV structure adopted in this paper.

As in von Ungern-Sternberg's paper, it is found that the hedging behavior of bidders lies at the heart of the problem. When faced with the chance of disaster, the expected revenue of the seller in the auction rises with the introduction of a cooling-off right. This is because bidders with a cooling-off right are able to ignore the disaster case and bid aggressively, while bidders with no such right have to hedge against it. The negative effect this hedging has on revenue is greater than the loss the seller experiences when the cooling-off right is exercised. The experimental data lends support to these theoretical conclusions.

This paper is organized as follows: Part 2 presents a model of an auction with a cooling-off right; Part 3 presents the empirical research; while Part 4 offers some closing remarks.

2 A theoretical model of an auction with a cooling-off right

2.1 The structure of the model

Figure 1 shows the timing of the model used to analyze the effect of a cooling-off period on a first price sealed bid auction.

When a potential bidder first inspects the object that is to be auctioned she forms a set of valuations. Each valuation corresponds to a state of the world. There are two potential states of the world. The value corresponding to state 1 is denoted v_i and is an independent draw from a commonly known distribution, F. However, the value of v_i is private information to bidder i.

In state 2 the value is denoted E. E is common to all bidders and known to be common. It corresponds to the value of the object in the "disaster" or "surprise" cases that were used earlier in discussing cooling-off rights.

The bidder also knows the probability of each state arising. The probability of state 2 arising is γ, with the probability of state 1 being $(1 - \gamma)$, where $\gamma \in (0, 1)$. After observing v_i, E and γ the bidder decides whether to participate in the auction. The highest bid wins the auction.

The seller is assumed to value the object at zero in both states of the world and have the same information as bidders about the relative likelihood of each state occurring.

If the bidder wins, the bidder is now in the *ex post* region of Figure 1. The state of the world (either state 1 or state 2) is realised and the bidder observes their true (or *ex post*) valuation (that is, either E or v_i). At this stage, if the cooling-off right exists, the bidder may choose to withdraw from the contract.

Once these steps are complete the payoffs to the parties are realized.

As in Vickrey [12], I assume risk neutral bidders, with bidder's state one valuations drawn from a uniform distribution on [0,1]. This has the advantage of making the exposition simple and easily applicable to an experimental setting.

In the following discussion the pure symmetric bidding strategies are derived for the benchmark case, where no cooling-off right exists. Then the equilibrium strategies are considered for the case where bidders have a costless cooling-off right. These two cases are compared on the basis of the bidding strategies and the expected revenue for the seller.

2.2 The benchmark case: no cooling-off right

In the benchmark case the bidder does not have the right to cool-off after the contract to buy has been entered into. This means the outcome of the auction is binding. In this setting the expected value of the object to the bidder is $(1-\gamma)\,v_i + \gamma E$. Thus, each bidder faces an individual rationality (or participation) constraint that must be satisfied if the bidder is to participate in the auction. Given that the only constraint on bids is that they be non-negative, this constraint is

$$(1-\gamma)\,v_i + \gamma E \geq 0\,. \tag{1}$$

When E is non-negative this constraint is satisfied for all types of bidders. When E is strictly negative this constraint will not be satisfied for some types of bidder and, hence, some types of bidder will not participate in the auction. Thus when E is negative the shape of the auction resembles an auction with a reserve price, where the reserve price is equal to $v^* = -\frac{\gamma E}{1-\gamma}$. When E is non-negative $v^* = 0$. That is, no bidder with an expected valuation less than v^* will enter a bid.

Assume that the equilibrium strategy is strictly monotonic and symmetric. Let $V(b_i)$ be the inverse of the bidding function to be followed by all bidders. It follows that the profit of bidder i is

$$\pi = [(1-\gamma)\,v_i + \gamma E - b_i]\,[V\,(b_i)]^{n-1} \tag{2}$$

where the last part of the expression, $[V\,(b_i)]^{n-1}$, is the probability that bidder i wins the auction (or has the highest valuation, v_i). The first order condition, with respect to b_i, is

$$0 = (n-1)\,[v_i]^{n-2}\,v_i'\,(b_i)\,[(1-\gamma)\,v_i + \gamma E - b_i] - [v_i]^{n-1}\,.$$

The assumption of symmetry allows $V(b_i)$ to be replaced by v_i. We can now proceed by either deriving $b_i(v_i)$ directly from this differential equation, or, leaving it defined implicitly, use the envelope theorem to derive $b_i(v_i)$ indirectly. This derivation follows the later route. From Equation 2, making b_i the maximization variable and v_i the parameter being varied, the envelope theorem yields $(1-\gamma)\left[V\left(b_i\right)\right]^{n-1} = \pi'\left(v_i\right)$. This means that

$$\pi\left(v_i\right) = \pi\left(v^*\right) + \int_{v^*}^{v_i} (1-\gamma) x^{n-1} dx \,. \tag{3}$$

Noting that $\pi\left(v^*\right) = 0$, as this 'marginal' bidder is indifferent between winning and losing the auction, we can use (2) and (3) to obtain

$$b_i = (1-\gamma) v_i + \gamma E - \frac{1}{v_i^{n-1}} \int_{v^*}^{v_i} (1-\gamma) x^{n-1} dx \,. \tag{4}$$

From the participation constraint (Equation (1)), we have the initial conditions that when $E \geq 0$, $\pi_i = 0$ if $v_i = 0$ and when $E < 0$, $\pi_i = 0$ if $v_i = -\frac{\gamma E}{1-\gamma}$. Equation (4) then gives the following symmetric equilibrium strategy for a risk neutral bidder in an auction with no cooling-off right:

$$b_i\left(v_i\right) = \begin{cases} (1-\gamma)\frac{n-1}{n} v_i + \gamma \mathrm{E} & \text{if } E \geq 0 \\ (1-\gamma)\left[\frac{(n-1)}{n} v_i + \frac{1}{n v_i^{n-1}}\left(\frac{-\gamma E}{1-\gamma}\right)^n\right] + \gamma E & \text{if } E < 0 \text{ and } v_i \geq -\frac{\gamma E}{1-\gamma} \\ no & \text{if } E < 0 \text{ and } v_i < -\frac{\gamma E}{1-\gamma} \end{cases} . \tag{5}$$

This bid function has an intuitive explanation. When $E \geq 0$ the equilibrium bid is the weighted average of the optimal bid in each possible state of the world. If the first state of the world were certain to occur the optimal strategy is to bid $\frac{(n-1)}{n} v_i$. However the chance of this happening is $(1-\gamma)$, so the bid $\frac{(n-1)}{n} v_i$ has a weight $(1-\gamma)$. In the second state of the world (where the true valuation is E) the best thing to do is bid E. Since E is common to all bidders, a form of (inverted) Bertrand competition emerges in this state of the world. Thus bids must converge on E. The bidder weights this strategy by γ, the chance of state 2 eventuating.

When $E < 0$ the bid function is again a weighted sum of the bid functions in the two states of the world. In state 1, the bid function is that of an IPV auction with reserve price $-\frac{\gamma E}{1-\gamma}$, since that reserve price is the value of v_i at which a risk neutral bidder is indifferent between winning and losing the auction. In state 2, each bidder wishes to bid the negative amount E, which would not be accepted by the auctioneer on its own, but can enter the overall bid function when combined with the positive bid function from state 1.

This case, where no cooling-right exists, is the benchmark against which to compare an auction in which bidders have the option to withdraw from the contract after they discover the state of the world.

2.3 Bidding in an auction with a costless cooling-off right

The introduction of a costless cooing-off right changes the payoff function, conditional on winning, from

$$\pi\Big|_{\text{win}} = \begin{cases} v_i - b_i & \text{in state 1} \\ E - b_i & \text{in state 2} \end{cases}$$

to

$$\pi\Big|_{\text{win}} = \begin{cases} v_i - b_i & \text{in state 1 and } v_i - b_i \geq 0 \\ 0 & \text{in state 1 and } v_i - b_i < 0 \\ E - b_i & \text{in state 2 and } E - b_i \geq 0 \\ 0 & \text{in state 2 and } E - b_i < 0 \end{cases}.$$

Now the winner can observe her profit after the state of the world is realized and decide whether, on the basis of this new information, she wishes to purchase the item. The new payoff function is made complicated by the fact that the bidder can now withdraw from the contract without cost. This means that whenever the winner of the auction will realize a loss, this loss can be reduced to zero.

To make the problem tractable it is helpful to divide bidders into two groups, those with $v_i \leq E$ and those with $v_i > E$.

Consider the problem facing bidders with $v_i \leq E$. If this type places a bid such that $b_i < E$ the best response from any competing bidder is to bid $b_i + \varepsilon$; that is, to bid marginally higher than b_i. By doing so the competing bidder gets, at least, the profit $E - (b_i + \varepsilon)$ should state two arise. In state one the competing bidder earns a profit of at least zero, due to the cooling-off right. Thus all bidders have an incentive to bid higher in response to a bid of less than E, regardless of the value of v_i or γ. Thus a form of Bertrand competition arises which drives all bids up to at least the level of E. Any bid less than E may be bettered by any bidder, and all bidders have an incentive to better such a bid.

This establishes the fact that in a pure strategy equilibrium E must be the absolute lower bound of any bidding strategy. However when faced with a bid of E the best response of a bidder in the group characterised by $v_i \leq E$ is to bid anything in their action set as, regardless of whether they win or lose the auction, their profit will be zero. Any bid greater than E will be withdrawn if it wins, while any bid less than E will not win. A bid of E will be withdrawn if it wins and state one arises, and earns zero profit in state two.

Any bidder with $v_i \leq E$ is indifferent between any possible bid in equilibrium. This creates multiple equilibria in the game. For example, $b_i = E + 10$ and $b_i = E + v_i$ are both equilibrium bidding strategies for bidders in this group. The symmetric pure strategy equilibrium bid function for $v_i \leq E$ can be thought of as $b_i = A(v_i)$ where $A(v_i)$ maps from the type space to $[E, \infty)$.

All but one of these possible equilibria has bidders entering spurious bids. Bids may be considered spurious if the cooling-off right is certain to be exercised. For example, if bidders are bidding $b_i = E+10$ when $v_i \leq E$ and this bid wins the auction, the winning bidder is certain to exercise the cooling-off right regardless of the state of the world that eventuates. The reason that such bids are included in the set of best responses is that the cooling-off right is costless. If $v_i \leq E$ a zero payoff is incurred by bidding E or spurious bids of anything higher than E.

There are several arguments in support of rejecting the spurious equilibria in favor of equilibria where the ownership of the object to be auctioned is transferred in at least one state of the world. First, when bidders enter spurious bids they know that they will never get possession of the object. Once this observation is made it becomes difficult to see why a bidder would enter an auction if they have no intention of acquiring possession of the object. Second, non-spurious equilibria are attractive as they might form the limit of the case where a fee is attached to the exercise of the cooling-off right. While this is a conjecture, it makes intuitive sense. If we consider an epsilon (small) fee attached to the cooling-off right and then reduce this fee toward zero, the hedging behavior of the bidder against this fee will become insignificant and strategies will approach the non-spurious equilibria. However, such a fee will always remove any spurious bids from the best response set. Lastly, spurious bids require more actions by the bidder than non-spurious bids. The bidder has to make a positive action by electing to cool-off, whereas if a non-spurious bid is entered the bidder may not have to act at this stage in the game. However, the payoff for both types of bid is the same. It seems counter-intuitive to expect a bidder to choose to engage in more actions for the same payoff when doing so incurs no strategic advantage.

On this basis, the equilibrium bids of bidders with $v_i \leq E$ is restricted to $b_i = E$. This is the only non-spurious bid function that constitutes a symmetric pure strategy equilibrium.

It remains to consider those bidders with $v_i > E$.

If $E < 0$ all types will elect to cool-off if state 2 eventuates. Since this can be done costlessly E will be ignored and all types will bid according to their normal strategy for a first price seal bid auction, as first proposed by Vickrey [12], so that $b_i(v_i) = \frac{n-1}{n} v_i$.

If $E \geq 0$ then types with $v_i \leq E$ bid $b_i = E$. It was argued earlier, in the context of types with $v_i \leq E$, that E is the lower bound of any equilibrium bidding strategy. This argument is directly applicable to types with $v_i > E$. This implies that any bid less than E is pointless in equilibrium as it has a certain profit of zero. This makes the problem of bidders with $v_i > E$ isomorphic with the problem of bidders in an auction with a reserve price of E – in both cases bids in equilibrium are restricted to be above E. Thus for $v_i > E$, $b_i = \frac{n-1}{n} v_i + \frac{E^n}{n v_i^{n-1}}$ when $E \geq 0$.

Hence, refining equilibria such that spurious bidding is eliminated, the equilibrium bidding function in the presence of a costless cooling-off right is

$$b_i = \begin{cases} E & \text{if } v_i \leq E \\ \frac{n-1}{n} v_i + \frac{E^n}{n v_i^{n-1}} & \text{if } v_i > E \text{ and } E \geq 0 \\ \frac{n-1}{n} v_i & \text{if } E < 0 \end{cases} . \tag{6}$$

2.4 Comparing equilibria on the basis of bidding strategies

Equation (6) gives the non-spurious equilibrium bidding strategy in the presence of a costless cooling-off right. Equation (5) gives the equilibrium bidding strategy in the absence of a cooling-off right. Comparing the bidding strategies under the two regimes establishes the following proposition:

Proposition 1. *In the symmetric, pure strategy, non-spurious equilibria derived above, the bidding strategy under cooling-off always results in a higher bid, for a given type, than the bidding strategy used when a cooling-off right does not exist.*

Proof. See Appendix.

This is because the existence of a cooling-off right promotes more aggressive bidding as it allows bidders to limit their downside.

From Proposition 1 one can derive the following corollary:

Corollary 1. *In the symmetric, pure strategy, non-spurious equilibria derived above, the expected winning bid with a costless cooling-off right is always greater than that in an auction without a cooling-off right.*

If every type enters a higher bid under costless cooling-off, then the expected winning bid must also be higher as the entire range of bids is shifted upwards with the introduction of the cooling-off right.

2.5 Comparing the equilibria on the basis of expected revenue

Table 1 reports the expected revenue of auctions for both types of cooling-off regime and various ranges of E. Expected revenues are calculated by taking the sum of the bids of each type, weighted by the probability that that type wins. Table 1 also compares the expected revenue of auctions with and without cooling-off; the regime with the larger expected value is indicated using an inequality. Propositions 2 through 4 establish these relative magnitudes.

Proposition 2. *When E is strictly positive, bidders play symmetric, pure strategies, and spurious equilibria are eliminated, the expected revenue from an auction without a cooling-off right is strictly greater than the expected revenue from an auction with a costless cooling-off right.*

Proof. See Appendix.

When a costless cooling-off right is present, bidders will focus their bidding on the state of the world in which they have the highest valuation, and expect to cool-off in the other state. This leads to more aggressive bidding. From the seller's perspective this aggressive bidding is off-set by the fact that in one state of the world the winner of the auction will withdraw from the contract, resulting in zero revenue for the seller. This ability to withdraw costlessly means that bidders do not bother to hedge against their least preferred state of the world

Table 1. Expected revenue across auctions

Value of E	With cooling-off	Relative magnitude	Without cooling-off[a]
$E < 0$	$(1-\gamma)\frac{n-1}{n+1}$	$>$	$\gamma E\left[1-\left(\frac{-\gamma E}{1-\gamma}\right)^n\right] + (1-\gamma)\frac{n-1}{n+1}\left[1-\left(\frac{-\gamma E}{1-\gamma}\right)^{n+1}\right] + (1-\gamma)\left(\frac{-\gamma E}{1-\gamma}\right)^n\left(1+\frac{\gamma E}{1-\gamma}\right)$
$E = 0$	$(1-\gamma)\frac{n-1}{n+1}$	$=$	$(1-\gamma)\frac{n-1}{n+1}$
$0 < E \leq 1$	$\gamma E + (1-\gamma)\left(\frac{n-1}{n+1} + E^n\right)(1-E)$	$<$	$(1-\gamma)\frac{n-1}{n+1} + \gamma E$
$E > 1$	γE	$<$	$(1-\gamma)\frac{n-1}{n+1} + \gamma E$

[a] Note that $0 \leq \frac{-\gamma E}{1-\gamma} \leq 1$

occurring. Where E is greater than zero, those bidders who have v_i greater than E will behave as if they are in an auction with a reserve price of E. However, the amount that this pushes up bids can never be equal to the size of the reserve price itself. Hence, it does not fully compensate the seller for the revenue lost from the bidder being able to withdraw in state 2. When compared with an auction without cooling-off, the ability of bidders to disregard one state of the world results in 'under-hedging' - their bids do not incorporate the hedging needed in auctions without cooling-off. From the seller's point of view this results in a situation analogous to an auction without cooling-off, where one of v_i (if E is greater than v_i) or E is set to zero. When both states of the world result in the object for sale being a good (i.e. $E > 0$), this means that the seller is unable to extract value in both states of the world and hence loses revenue.

Proposition 3. *When E equals zero, bidders play symmetric, pure strategies, and spurious equilibria are eliminated, auctions with a costless cooling-off right and auctions with no cooling-off right are revenue equivalent.*

This follows straight from the observation that, when E equals 0, auctions under both regimes mirror the first price auction considered by Vickrey [12].

Proposition 4. *When E is strictly negative, bidders play symmetric, pure strategies, and spurious equilibria are eliminated, the expected revenue from an auction without a cooling-off right is strictly less than the expected revenue from an auction with a costless cooling-off right.*

Proof. See Appendix.

The cooling-off right creates a tendency for the bidder to under-hedge when faced with risk. In the case where E $>$ 0 this lead to a loss in revenue as it inhibited the seller's ability to extract expected value. When E $<$ 0 the opposite is observed, because the bidder does not have to hedge against the 'bad' state of the world, the seller is not made to accept bids that incorporate compensation for the possibility of state 2 arising. The seller benefits from the under-hedging, as, in this instance, it more than compensates for the revenue lost when the cooling-off right is exercised. As E tends to large negative amounts, and when no cooling-off right exists, this effect becomes starker as large numbers of bidders choose not to participate in the auction. These bidders are kept in the auction when the cooling-off right is introduced.

These propositions have been established in the context of a first price sealed bid auction. This corresponds to the empirical testing of the model in part 3. It is conjectured that the revenue equivalence result for risk neutral bidders (Milgrom and Weber [8] for its most general statement) may allow us to extend the results across other auction types.

It is also interesting to note that the expected surplus of bidders is increased by a cooling-off right when $E < 0$ and decreased when $E > 0$ (calculated using the bid functions in Equations (5) and (6)). This coincides with the effects on expected revenue set out in Propositions 2 through 4. Thus, when it is more efficient to have a cooling-off right *ex ante*, in the sense of increasing expected joint surplus, it is also revenue increasing. When $E < 0$ a cooling-off right is also *ex post* efficient in that the agent with the highest value will always get possession of the object. This is not the case without a cooling-off right as E may be so low that no bidder bids even though, *ex post*, exchange would have been welfare enhancing. These efficiency results may go some of the way toward explaining why we observe parties agreeing to include cooling-off periods in some contracts.

3 Testing the model in the laboratory

This section seeks to verify the bidding strategies that were derived for auctions with, and without, a cooling-off right (contained in Equations (5) and (6)). Those aspects of the predicted strategies tested experimentally are:

1. The point predictions of the bid functions: Do bids by participants in the experiment deviate significantly from those predicted by theory?
2. The proposed refinement excluding spurious bids: How often do spurious bids occur?
3. The explanatory power of the theory: What proportion of the observed variation in the bids does the theory explain?
4. Coefficients in linear bid functions: When bid functions are linear in the variables, do the estimated coefficients differ significantly from those predicted?
5. Changes in the bid functions: As E changes, do the bids exhibit structural breaks when, and only when, the form of the predicted bid function changes?

If these strategies are supported by the data, then Propositions 1 through 4 are similarly supported.

3.1 The design of the experiment

This experiment was conducted on six groups of six subjects. 10 first price sealed bid auctions were conducted on each group. In each auction the subject was given two values corresponding to v and E in the preceding discussion. The subject was then invited to enter a bid of either *no*, indicating that they did not wish to participate in the auction, or any non-negative amount expressed as dollars and cents. The *ex post* valuation was decided by a dice roll. There was a one-in-six chance that E was the *ex post* valuation, and a five-in-six chance that v was the *ex post* valuation. Once the *ex post* valuation was resolved, the profit of the winning bidder was calculated by subtracting the winning bid from the *ex post* valuation.

The values of E were common to all bidders. This was common knowledge. The values of v that were assigned to each subject were independent draws from a uniform distribution over the integers in the set $\{0, 1, 2, ..., 199, 200\}$, expressed in cents. Thus v could range from \$0.00 to \$2.00 Australian.

In the first five auctions there was no cooling-off right. In the last five auctions there was a costless cooling-off right. An automatic decision rule was imposed on the auctions with a cooling-off right so that if the winning bidder was going to realize a loss this loss was reduced to zero. Thus the auctions with cooling-off rights were explained in the instructions as "you cannot make a loss in an auction if you win".

The subjects were given \$5.00 for arriving at the experiment on time. This \$5.00 then served as a float, so that, if a winning bid turned out to be higher than the *ex post* valuation in an auction without cooling-off, a negative profit could be realised. The final payoff to a subject was calculated as \$5.00 plus the net profit of that subject over the ten auctions. The highest payoff paid was \$6.93 and the lowest was \$3.38. The mean payoff was \$5.21.

This up-front lump-sum payment of \$5 may have affected behavior in the auction by dampening the anticipation of negative profits. One of the main channels through which cooling-off affects bidding is via the possibility of making a negative profit. If participants perceived this up-front payment as substantially removing this possibility, then the payment would distort bidding behavior away from the predicted bids. This raises a problem, common to experimental work, in distinguishing rejection due to the model not capturing behavior, from rejection due to a design problem in the experiment. On the other hand, if the participants had incorporated the payment into the reference point from which they judge gains and losses, then the problem will not exist. The potential impact on the interpretation of results needs to be weighed against the practical difficulty of inviting participants to engage in an experiment that costs them money. Since, as argued later, the data from this experiment tend to support the theory, it seems unlikely that bids were significantly distorted by the up-front payment of \$5.

The progressive profit of each subject was not reported during the course of the experiment. The primary reason for this was that keeping 36 subjects up to date as to their profit would have taken too long with the resources at hand. It is unclear whether not reporting profit would have changed the common problem of wealth effects influencing behavior as profit changes during the experiment. Generally, there is little that can be done to address this problem in experiments of this kind.

The 36 subjects were gathered together in one room and split into groups of six. The subjects in each group bid against each other in the subsequent auctions. It was decided to have six subjects in each group as experiments in the past have indicated that this was a sufficient size to effectively deter any collusive behavior (Cox, Roberson and Smith [2]).

After being divided into groups the subjects were given a set of written instructions. The same booklet also contained the forms on which participants entered their bids in each auction. It was made clear to the subjects, both verbally and in the instructions, that any form of communication between subjects would render the experiment a failure and was prohibited.

Before the experiment began, subjects were given two practice runs. In these practice runs, and in the ten real auctions, subjects had two minutes to make their bid. This was felt to be more than adequate time.

It was useful to hold all the auctions for all the groups at the same time in the same room. This meant that any information given to one group in response to a question was given to all groups. It also meant that the timing of the experiment was the same for all groups. It follows that any other environmental factors that may affect results are constant across the groups.

Six laboratory assistants were assigned to oversee the auction process, one for each group. These assistants were familiar with the experiment structure, having been the subjects of a test run of the experiment. The assistants were instructed to submit any substantive questions from the participants to the author, who oversaw the experiment.

The subjects were drawn from undergraduate economics students at the Australian National University, in August 1998, with the author presenting a short address to each of several economics classes. From the 50 students who responded, the 36 subjects were randomly selected.

3.2 A description of the data

The experiment yielded 360 separate data points (bids), taken from 60 different auctions. The sixty auctions were divided into 10 sets of 6, with each set characterised by different values of E. These parameter values are shown in Table 2. The values of E, in Australian dollars, ranged from -0.4 to 3. These values were chosen to test all of the variations in the bid functions predicted by the theory. The theoretical predictions for each value of E were tested using 36 observations. Each auction type is referred to using the following mnemonic: the letter, N or C, refers to no cooling-off and cooling-off, respectively; the number, 1–5, refers to the level of E, in ascending order. Thus, N5 is the auction with no cooling-off right and the highest level of E, \$3.

Table 2. Parameter values for each auction

Auction type	Value of E	Costless cooling-off?
N1	−0.40	N
N2	0.00	N
N3	0.50	N
N4	1.20	N
N5	3.00	N
C1	−0.40	Y
C2	0.00	Y
C3	0.50	Y
C4	1.20	Y
C5	3.00	Y

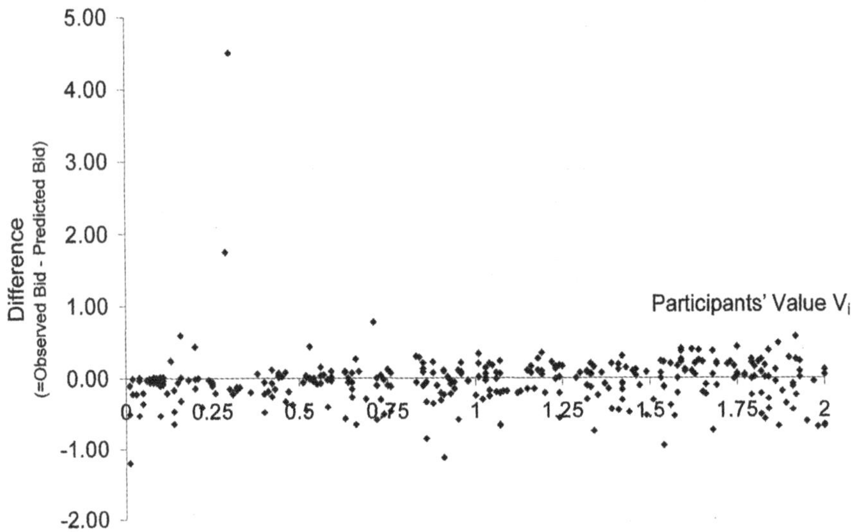

Figure 2. All auctions excluding C5 (mean difference: −0.05)

Each observed bid was compared with the predicted bid generated from the theory. The difference between the prediction and the observed bid was calculated, with a negative difference indicating that the observed bid was lower than the predicted bid. These differences are plotted for all auctions except C5 in Figure 2. Auction C5, which is somewhat unique, is shown separately in Figure 3.

In most of the auction types there is a very slight upward trend in the data points. This is evident in Figure 2. As the value of v_i increases, the difference between the predicted value of the bid and the actual bid tends to increase. For small values of v_i there is a slight tendency for the predictions to overestimate bids, and for high values of v_i there is a slight tendency for the predictions to underestimate bids. Generally, however, the differences tend to be scattered around zero.

Auction C5, however, is an exception to this generalization. All the bids in auction C5 were at or below the predicted bid. This suggests that something

Table 3. Spurious bidding

Participant no.	Value of v	Value of E	Observed bid
c2	0.29	−0.40	1.99
c2	0.30	0.50	5.00
d1	0.71	0.50	1.38
d1	1.16	1.20	1.27
e2	1.16	1.20	1.36

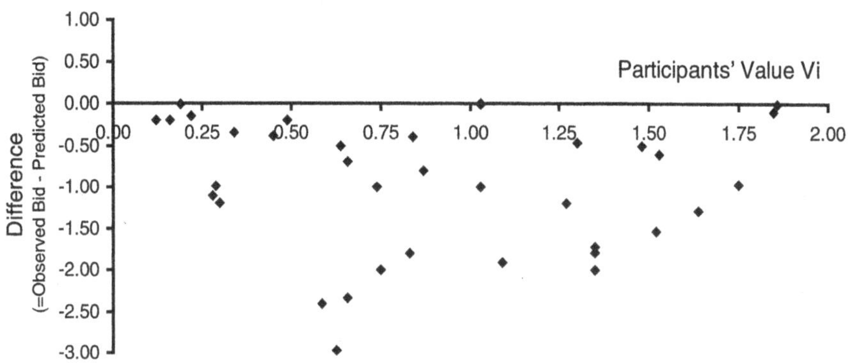

Figure 3. Auction 5 (mean difference: −0.967)

about auction C5 makes it significantly different from the rest of the auctions. The possible reasons for this will be explored later.

Before engaging in econometric analysis it is useful to explore some of the patterns that emerge from a cursory examination of the data.

Spurious bidding behavior was observed in some of the auctions with cooling-off. Table 3 records the parameter values for these bids and the bids themselves. Table 3 shows that, out of 180 bids collected from auctions with cooling-off, only five were spurious in the sense of inviting automatic cooling-off regardless of the state of the world that arose. All of the 180 observations from auctions with cooling-off rights invited spurious bids – to see this note that in any of these auctions it is a Nash equilibrium for all types to bid, say, \$100, which is much more than either v_i or E.

These spurious bids can be seen as positive signals that people were aware of the structure of the bidding problem and had a feel for the equilibrium strategies and consequent payoffs. It is also comforting to note that the theory acknowledges the possibility of this behavior, while the fact only 5 spurious bids were observed suggests that the proposed equilibrium refinement was perhaps justified as a first approximation. These spurious bids raise the interesting question of how to generalize the theory to accommodate occasional deviations.

A significant proportion of the subjects (at least 7) exhibited a reluctance to bid higher than their value of v in the auctions with no cooling-off right. The theory predicted that bids should have been higher than v in 105 of the

Table 4. Loss averse bids (in auctions with no cooling-off)

Participant no.	Value of E	Value of v	Observed bid	Predicted bid
d3	1.20	0.51	0.50	0.55
d4	1.20	0.33	0.30	0.43
d6	1.20	0.32	0.30	0.42
f5	1.20	0.14	0.14	0.30
d1	3.00	0.10	0.05	0.57
e1	3.00	0.48	0.47	0.83
f4	3.00	0.91	0.90	1.13
f5	3.00	1.17	1.17	1.31

observations. In 36 of these observations this failed to occur. Table 4 shows examples of this behavior.

This behavior seems to be due to something more than 'normal' errors in bidding. The mean differences between observed and predicted bids for these 36 observations is -0.36 compared with -0.14 for the entire experiment. A simple t-test rejects the null hypothesis that these means are the same (p= 0.005). The behavior may be explained by fairly strong aversion to the chance of making a loss. In each of the instances shown in Table 4 the subject's predicted bid was above their value of v. However, the subjects placed their bid at, or just below, their value of v. This suggests that subjects had no wish to incur the possibility of a making a loss should they win in the auction and v turned out to be the true valuation. Interestingly, there were no clear cases of people being averse to bidding above E, even from those in the above sample. This suggests that the relatively high likelihood of having v_i as the true valuation was an important factor in the decision of when to remove any chance of making a loss.

Overall the auction procedure was a fairly efficient allocation mechanism. 74% of all the auctions conducted (not counting the auctions of type C5) resulted in the predicted winner winning the auction. Each of the two cooling-off regimes incurred seven inefficient auctions. Of the 14 auctions that did not have the predicted winner winning, 10 were won by the bidder with the second highest expected value. It is not useful to consider auction C5 in respect to efficiency as the parameter values were such that, in equilibrium, it was expected that all subjects had an equal chance of winning.

Since the key results of the theoretical model presented in this paper concern expected revenue it is of some interest to examine the actual revenue received in this experiment. Table 5 reports the average revenue of the six auctions of each auction type. It also reports the paired difference of auctions with, and without, cooling-off for common values of E. The null hypothesis that each difference is zero is tested in each case (the test statistic is reported).

The results in this table must be treated with considerable caution. The number of observations is small because there are only six auctions to pool data from in each case. The sensitivity of the comparisons to differences in v_i and whether cooling-off occurs must also be kept in mind. Lastly, with such a low number

Table 5. Comparison of average revenue

Value of E	No cooling-off	Cooling-off	Paired difference	t-statistic[a]
−0.40	1.34	1.19	−0.152	−2.206*
0	1.47	0.93	−0.537	−4.765***
0.5	1.47	1.19	−0.272	−3.347***
1.2	1.41	1.26	−0.147	−1.762
3	1.74	0.42	−1.325	−7.450***

[a] The null hypothesis is that the paired difference is equal to zero. The test statistic is $t = \frac{\bar{x}_1 - \bar{x}_2}{\sqrt{s^2\left(\frac{1}{n_1} + \frac{1}{n_2}\right)}}$ which has a t distribution with $n_1 + n_2 - 2$ degrees of freedom. The pooled estimate of the population variance is s^2.

* indicates that the null hypothesis can be rejected with only a 0.10 chance of committing a type one error.
** indicates that the null hypothesis can be rejected with only a 0.05 chance of committing a type one error.
*** indicates that the null hypothesis can be rejected with only a 0.01 chance of committing a type one error.

of observations it is necessary to assume that the population variances are equal when performing the t-test. It is hard to justify this assumption from the theory. That said, it is nice to observe that as E increases there is an increased tendency for no cooling-off to dominate cooling-off in terms of revenue.

3.3 Analysis of the data

The mean difference between the observed bids and the theoretical predictions was calculated for various combinations of auctions. A hypothesis test was then conducted with a null hypothesis that the mean difference was zero. The results of these tests are reported in Table 6.

For the overall sample of 60 auctions and 360 bids the mean difference was −0.138. The corresponding test statistic was −5.020, strongly rejecting the null hypothesis that the mean difference is zero. This suggests that the theoretical model is not a good predictor of the bidding in these auctions. However, much of this error in prediction can be attributed to certain parameter settings. In particular, the bidding in auction type C5 was poorly predicted. The influence of auction C5 on the results is shown by comparing the test statistic for all the auctions (auctions N1–5 and C1–5) with that for auctions N1–5 and C1–4; the removal of auction C5 more than halves the value of the test statistic. Although making this adjustment still means that the null hypothesis can be rejected at any significance level greater than 0.025, it illustrates the degree to which auction C5 skews the hypothesis test.

Examining the tests closely reveals that the theory performs well as a predictor of bids when the bulk of the probability mass of the random variable v lies above the value of E. That is, for cases with, and without, cooling-off the theory

Table 6. Tests of mean differences

Auction nos.	All	N1–5, C1–4	N1–5	N1–4	C1–5	C1–4
Mean difference	−0.138	−0.046	−0.062	−0.021	−0.214	−0.026
Standard deviation	0.522	0.386	0.294	0.274	0.670	0.478
Test statistic[a]	−5.020***	−2.150**	−2.822***	−0.929	−4.300***	−0.666
Auction nos.	N1	N2	N3	N4	N5	C1
Mean difference	0.001	−0.009	−0.007	−0.070	−0.225	0.023
Standard deviation	0.352	0.260	0.277	0.184	0.321	0.357
Test statistic[a]	0.030	−0.212	−0.162	−2.273***	−4.208***	0.382
Auction nos.	C2	C3	C4	C5	N1–2	C1–2
Mean difference	−0.068	0.089	−0.150	−0.967	−0.004	−0.023
Standard deviation	0.222	0.795	0.299	0.799	0.307	0.299
Test statistic[a]	−1.830*	0.671	−3.007***	−7.265***	−0.102	−0.641

[a] The null hypothesis is that the mean difference is equal to zero. The test statistic is $t = \frac{\bar{x} \cdot \sqrt{n}}{s}$ which has a t distribution with $n - 1$ degrees of freedom.

* indicates that the null hypothesis can be rejected with only a 0.10 chance of committing a type one error.

** indicates that the null hypothesis can be rejected with only a 0.05 chance of committing a type one error.

*** indicates that the null hypothesis can be rejected with only a 0.01 chance of committing a type one error.

performs well when $E = 0.50$, 0.00 or −0.40. In auctions N1, N2, N3, C1, and C3 the null hypothesis cannot be rejected at any meaningful level of significance. Auction C2 is somewhat borderline in that it has a p-value of 0.06. This can be compared with the results of auctions N4, N5, C4 and C5, all of which reject the null hypothesis at a significance level of 0.01 or more. These results illustrate a stark division in the subjects approach to evaluating choice over risk. When the subjects were faced with a small chance of making a relatively significant loss, subjects tended to develop strategies more or less in line with those predicted. However, when subjects were faced with the prospect of a small chance of making a relatively significant gain they did not adopt strategies similar to those predicted. Instead, they tended to entered bids consistently lower than predicted. This suggests that the subjects were more risk averse when faced with a small chance of a gain, as opposed to a loss.

This is an interesting result as it bears some similarity to the predictions of the prospect and cumulative prospect theories of choice under risk and uncertainty (Kahneman and Tversky [5] and Tversky and Kahneman [11], respectively). In both theories the value function is concave for gains and convex for losses, implying that we should frequently observe risk aversion in gambles over gains and risk seeking in gambles over losses. This is consistent with the behavior presented here. However, in both theories the decision makers true risk preferences are determined jointly by their value function and their weighting function.

When probabilities over gains and losses are sufficiently low the shape of the weighting function may reverse the above characterization, so that risk seeking is observed in gambles over gains and risk aversion in gambles over losses. So a key issue in deciding how these experimental results relate to the various versions of prospect theory is whether the probability over gains and losses, $0.1\overline{6}$, is 'low'. In an experiment designed to explore this aspect of the theory Tversky and Kahneman [11] considered a probability of 0.1 as low and a probability of 0.5 as high (in the sense of inviting risk aversion over gains). $0.1\overline{6}$ lies between these ranges. Hence, the behavior observed in this experiment, while perhaps explained by prospect theory, cannot be easily characterized as confirming or contradicting it.

Comparing the predictions of the model on the basis of the different cooling-off regimes does not lead to any conclusion that the model performs better under one regime than the other. If the result for auctions N1–5 is compared to that for auctions C1–5, the respective test statistics are −2.8 and −4.3. However, both these test statistics are sufficient to reject the null hypothesis at the 0.01 significance level, thus comparing predictive power is, in a sense, pointless. If we exclude the auctions where $E = 3.00$ (i.e. auctions N5 and C5), the predictions for the case with cooling-off do slightly better, on the basis of the test statistics, than the case without cooling-off. However this is a very rough yardstick with which to compare predictive power. The most that can be said is that, in both cases, the predictions seem to perform quite strongly as in neither case is the null hypothesis rejected.

Auction C5 deserves special mention. It stands out as being the auction in which the theory did not perform well. The theory has posited that the symmetric pure strategy equilibrium in this auction is to bid $b_i = E$. However, there also exist asymmetric pure strategy Nash equilibria, of the form

$$\begin{aligned} b_i &\in [E, \infty) \\ b_j &\in [E, \infty) \text{ and} \\ b_k &\in no \cup [0, \infty) \end{aligned}$$

where b_k is the bid entered by those individuals in the set of bidders not including bidders i and j. This result is apparent once we observe that, if two bidders (i and j) enter bids greater than E, in equilibrium the payoffs of all other bidders (k) are zero regardless of what they do. Hence, these other bidders (k) can choose any bid. In the experiment this translates into a prediction that at least two bidders in auction C5 would bid \$2.99 or more (the predictions have to adjusted for the fact that the bidding space is discrete in cents). Once these predictions are adjusted in this manner, and compared to the observed bids, we see that one of the auctions of type C5 conforms exactly to an asymmetric Bayesian Nash equilibrium in pure strategies. Two of the others come close. It is some comfort that this tendency for strategies in Bertrand games to diverge from the standard Nash equilibrium has been observed in other experimental studies (see Plott [9], for instance). Baye and Morgan [1] investigate possible reasons for this behavior

Table 7. Percentage of the variation in bids explained by the theory

Model	% of variation in bids explained by model
All Auction Types	
Theory	0.41
$bid = \beta_0 + \beta_1 v_i + \beta_2 E$	0.43
Auction Types N1–5, C1–4	
Theory	0.55
$bid = \beta_0 + \beta_1 v_i + \beta_2 E$	0.54
Auctions N1–2, C1–2	
Theory	0.67
$bid = \beta_0 + \beta_1 v_i + \beta_2 E$	0.66

in the data. They find that two equilibrium concepts that incorporate notions of bounded rationality, Radner's [10] epsilon equilibrium and McKelvey and Palfrey's [6] quantal response equilibrium, are significantly more successful in organizing their data than the Nash equilibrium concept. This suggests that some notion of bounded rationality may explain the unique bidding behavior in auction C5.

The preceding theoretical discussion brought out some important normative prescriptions for when a cooling-off regime might be desirable for a seller. It was argued that a cooling-off regime benefits the seller when the value of E is less than zero. With this in mind particular attention was paid to the quality of the predictions in auctions N1, N2, C1 and C2. When auctions N1 and N2 were analyzed together the test statistic had a p-value of 0.94, while for auctions 6 and 7 the combined test statistic had a p-value of 0.54. Both these results suggest that that the null hypothesis was fairly robust for both cooling-off regimes when E was less or equal to zero.

To get a more precise impression of the quality of the predictions the percentage of the variation in the observed data explained by the theory was calculated. This is a calibration exercise where the calibration parameters are given by the equilibrium bid functions (Equations (5) and (6)). The r-squared statistic from an OLS linear regression was calculated to provide a point of reference. Table 7 reports the results.

When all the auctions are considered together the theory is able to explain 41% of the observed variation in the bids. This compares to 43% for the estimated linear model. When auction C5 is omitted from the data set, the theory outperforms the linear models, explaining 55% of the variation in bids. The nonlinearity of the theory allows it to outperform the linear model, despite the fact that the linear model has more degrees of freedom. The linear model explains 54% of the variation. Auctions N1, N2, C1 and C2 are particularly important from a theoretical viewpoint, as in these auctions it is advantageous for a seller

Table 8. Coefficient estimates for linear strategies

Regression model:	$bid = \beta_0 + \beta_1 v_i + \beta_2 E$		
Parameter	Prediction	Estimate[a]	t-statistic[b]
Auctions N2–5			
β_0	0	−0.065 (0.045)	−1.446
β_1	0.694	0.779 (0.035)	2.383**
β_2	0.166	0.085 (0.019)	−4.258***
Auctions C1–2			
β_0	0	0.021 (0.073)	0.291
β_1	0.833	0.739 (0.057)	−1.642
β_2	0	−0.220 (0.173)	−1.272

[a] The standard errors are in parentheses.
[b] The null hypothesis is that the estimate is equal to the prediction. The test statistic is $t = \frac{\beta - B}{s(\beta)}$ which has a t distribution with $n - (k + 1)$ degrees of freedom.

* indicates that the null hypothesis can be rejected with only a 0.10 chance of committing a type one error.
** indicates that the null hypothesis can be rejected with only a 0.05 chance of committing a type one error.
*** indicates that the null hypothesis can be rejected with only a 0.01 chance of committing a type one error.

to have a cooling-off period. Overall, the theory is successful in explaining 67% of the variation in bids in auctions N1, N2, C1 and C2. This compares to 66% by the linear model. This suggests that the theoretical model is a good explanator of bidding behavior when $E \leq 0$.

Where bidding strategies are linear it is possible to test the coefficients predicted by the theory using linear OLS regressions. Auctions N2–5 and C1–2 have linear bid functions. Table 8 compares the estimated parameters from the regression to those predicted by the theory. In both sets of auctions it is not possible to reject the null hypothesis that the intercept is zero. In fact it is not possible to reject any of the theoretical predictions in auctions C1 and C2. In auctions N2–5 the predictions of the coefficients of v_i and E are both rejected. This suggests that the coefficients predicted by the theory are most accurate in auctions with cooling-off right where E is less than or equal to zero.

The last testable hypothesis from the theory is that as E changes, so do the strategies played by bidders. Equations (5) and (6) predict that auctions N1, C3, C4 and C5 have unique biding strategies, while auctions N2, N3, N4, and N5 should have the same strategies. Auctions C1 and C2 should also share the same strategies. This should manifest itself in coefficient estimates being structurally

Table 9. Chow test statistics[a] for instances where structural breaks were expected

Auction nos.	N1	N2–5	C1–2	C3	C4	C5
N1	–					
N2–5	1.498	–				
C1–2	5.096***	3.414**	–			
C3	na	8.398***	5.550***	–		
C4	na	14.451***	9.573***	na	–	
C5	na	82.242***	18.278***	na	na	–

[a] The null hypothesis is that the coefficient estimates are the same. The structure of the linear regression model used to test this was $Bid = \beta_1 v_i + \beta_2 E$. The test statistic is $\mu = \frac{(T-2k)(SSE_R - \Sigma SSE_U)}{k \Sigma SSE_U}$ which is distributed over $F(k, T-2k)$.

* indicates that the null hypothesis can be rejected with only a 0.10 chance of committing a type one error.
** indicates that the null hypothesis can be rejected with only a 0.05 chance of committing a type one error.
*** indicates that the null hypothesis can be rejected with only a 0.01 chance of committing a type one error.

stable in those auctions in which strategies are the same. It is possible to test for this using structural break tests. This was done using the Chow test.

The Chow test uses linear regression modeling to test for significant changes in coefficient estimates, which indicate an underlying change in the modeled behavior. Because the Chow test uses linear regression techniques it is not appropriate to use it to test for structural changes between auctions N1, C3, C4 and C5. This is because the theory predicts that, in auctions N1, C3, and C4, the coefficient of v_i is a step function (see Equations (5) and (6)). Since the appropriate equilibrium model for auction C5 is uncertain (after looking at the data) it is also included in this group. However, it is appropriate to use the Chow test where at least one of the equilibrium strategies is linear. Table 9 reports the results of structural break tests done where a structural break was expected. Table 10 reports the results of structural break tests done where a structural break was not expected.

Table 9 shows that, with the exception of auction N1 versus auctions N2–5, structural breaks exist where they are expected. This is strong evidence in support of the proposition that strategies change as the value of E changes. In particular, it is good to see that in all instances that were amenable to comparison, strategies under cooling-off were significantly different from those used in auctions where no cooling-off right existed.

Table 10 shows that in no instance, where strategies were expected to be the same, was there a structural change in coefficient estimates. This supports the proposition that the strategies being played in auctions N2, N3, N4 and N5 were the same. The data supports the same conclusion for auctions C1 and C2.

In totality, these structural break tests provided strong evidence in support of the prediction that as parameter values change so do the bidding strategies.

Table 10. Chow test statistics[a] for instances where structural breaks were not expected

Auctions	Test statistic
N2 vs. N3-5	0.192
N3 vs. N2, N4 andN 5	0.100
N4 vs. N2, N3 and N5	0.285
N5 vs. N2-4	0.337
C1 vs. C2	0.070

[a] The null hypothesis is that the coefficient estimates are the same. The structure of the linear regression model used to test this was $Bid = \beta_1 v_i + \beta_2 E$. The test statistic is $\mu = \frac{(T-2k)(SSE_R - \Sigma SSE_U)}{k\,\Sigma SSE_U}$ which is distributed over $F(k, T-2k)$.

* indicates that the null hypothesis can be rejected with only a 0.10 chance of committing a type one error.

Moreover, the tests also support the proposition that these changes occur as predicted by the theory.

3.4 Commentary on the empirical results

The experimental data lends support to the theoretical predictions in part 2, however in auction C5 the bids were consistently below the predicted level. In this instance $E = 3$ and a cooling-off right exists. Such a systematic deviation from predicted behavior seems to suggest some other strategy was being played. The pattern may also be consistent with the notions of bounded rationality explored by Baye and Morgan [1]. Although this behavior contradicts the theoretical predictions, it tends to strengthen the conclusions regarding the comparison of expected revenues. When a cooling-off right exists, if bidders exhibit a tendency to bid below E when E is greater than the maximum value of v_i, the expected winning bid will be lower than predicted. This implies that the revenue to the seller will be even less than predicted. This supports the conclusion that when $E > 0$ the sellers expected revenue drops when a cooling-off right is added.

On the whole, the experimental evidence lends support to the theoretical conclusions. Changes in strategies occur as predicted, and for most values of E the point predictions of the strategies are fairly close to those observed in the data. However, as in most experimental work, it may be overly ambitious to expect the data to fit the specialized parameters and assumptions of the theory perfectly. It is also encouraging to see the data support the refinement of the equilibrium under cooling-off that ruled out spurious bidding (Equation (6)). Where the predictions can be rejected, the rejection is in favor of behavior that makes the revenue conclusions in part 2 stronger.

Hence the data supports the conclusion that cooling-off is a desirable element when $E < 0$ and undesirable when $E > 0$.

4 Concluding remarks

This paper has addressed the question of when a cooling-off right is a desirable element in an auction from the seller's point of view. It has been argued that a seller will be made better off, from an expected revenue point of view, by introducing a cooling-off right when there exists a state of the world in which the object being auctioned gives dis-utility to the bidders. However, when the object gives positive utility in all states of the world it is not in the seller's best interests to introduce a cooling-off right.

From experimental testing we learn that, when a state of the world exists where the good may yield negative utility to a buyer ($E < 0$), the theory predicts behavior particularly well. The point predictions of the theory are unable to be rejected, over 65% of the variation in bidding is explained and the predicted coefficients of the linear bid functions in auctions with cooling-off are not rejected. This is encouraging as the implications of the model are strongest for this case – when $E < 0$ a cooling-off right raises revenue.

More generally, the experiment confirms that strategies change with parameters and cooling-off rights. What is more, the Chow tests suggest that the theory accurately predicts when these changes take place. The experiment also confirmed that the model is justified in ruling out spurious bidding in auctions with cooling-off. Only 5 out of a possible 180 spurious bids were observed.

The experiment illustrated behavior that was not predicted by the theory. In particular, the pervasive under-bidding in auction C5 was unforeseen. This behavior seems consistent with other, similar, Bertrand game experiments. It suggests that the point predictions of the model may be poor when E lies above the support of the privately held values v_i. However, the flavor of the revenue conclusions is strengthened by this experimental observation.

The experiment also suggested some tendency for subjects to be more risk averse when faced with the small chance of a gain, as opposed to a loss. This was discussed with reference to the prospect theories of choice under risk and uncertainty. While no firm conclusions could be drawn, the tendency was suggestive of behavior predicted by prospect theory.

Overall, the experiment seems to support the theoretical model presented in part 2. However, it also suggests at least two areas where the model could be improved; in capturing bidding behavior when E is very high and allowing for different responses to different types of risks. It has also been noted that, in practice, a nominal fee is often attached to the exercise of the cooling-off right. Bidding behavior in the presence of such a fee remains an avenue for further research.

Appendix

Proof of Proposition 1

When $E \geq 1$ this proposition is established by inspection. When $1 > E \geq 0$ the bidding strategy with cooling-off is

$$b_i = \begin{cases} E & \text{if } v_i \leq E \\ \dfrac{n-1}{n} v_i + \dfrac{E^n}{n v_i^{n-1}} & \text{if } v_i > E \end{cases}$$

and without cooling-off is $b_i(v_i) = (1-\gamma)\frac{n-1}{n}v_i + \gamma E$. Clearly $(1-\gamma)\frac{n-1}{n}v_i + \gamma E < E$ and by contradiction $\frac{n-1}{n}v_i + \frac{E^n}{nv_i^{n-1}} > (1-\gamma)\frac{n-1}{n}v_i + \gamma \text{E}$.

That is, if $\frac{n-1}{n}v_i + \frac{E^n}{nv_i^{n-1}} \leq (1-\gamma)\frac{n-1}{n}v_i + \gamma\text{E}$ then $\frac{E^n}{nv_i^{n-1}} \leq -\gamma\frac{n-1}{n}v_i$ which is a contradiction as it implies that a negative number is greater than a positive number.

When $E < 0$ it has to be established that

$$(1-\gamma)\frac{(n-1)}{n}v_i + \gamma E + \frac{1-\gamma}{nv_i^{n-1}}\left(\frac{-\gamma E}{1-\gamma}\right)^n < \frac{n-1}{n}v_i \tag{A1}$$

in the region $\frac{-v_i(1-\gamma)}{\gamma} \leq E < 0$. It is sufficient to establish that in this region $\theta = \gamma E + \frac{1-\gamma}{nv_i^{n-1}}\left(\frac{-\gamma E}{1-\gamma}\right)^n \leq 0$. Now, noting that

$$\begin{aligned} \frac{d\theta}{dE} &= \gamma\left(1 - \left(\frac{1}{v_i^{n-1}}\right)\left(\frac{-\gamma E}{1-\gamma}\right)^{n-1}\right) \\ &= 0 \quad \text{when} \quad E = -\frac{(1-\gamma)v_i}{\gamma} \end{aligned}$$

and that when $E = 0$, $\frac{d\theta}{dE} = \gamma > 0$, it must be the case that Equation [A1] holds as we have a function that is monotonic in the relevant region with a positive slope, so that as E decreases that function also decreases in value, with the condition that at the upper bound of the relevant region the function equals zero. Hence we can support Proposition 1. □

Proof of Proposition 2

By observation, from Table 1, when $E > 1$ the expected revenue from an auction without a cooling-off right will exceed that from an auction with costless cooling-off. When $0 < E \leq 1$ the entries in the third row of Table 1 must be compared. Consider the case where $(1-\gamma)(1-E)\left(\frac{n-1}{n+1} + E^n\right) + \gamma E \geq (1-\gamma)\frac{n-1}{n+1} + \gamma E \quad \forall E \in (0,1]$. This is equivalent to

$$E^{n-1} - E^n \geq \frac{n-1}{n+1}. \tag{A2}$$

Now when $E = 0$, $E^{n-1} - E^n = 0$ and when $E = 1$, $E^{n-1} - E^n = 0$. If we let $f(E) = E^{n-1} - E^n$ then, $\frac{df(E)}{dE} = E^{n-2}[(n-1) - nE] = 0$.

Hence stationary points exist at $E = 0$ and $E = \frac{n-1}{n}$. From the second order sufficient condition, when $E = \frac{n-1}{n}$, $f(E)$ is at a maximum. Hence for a given number of bidders the maximum value of $f(E)$ is $\left(\frac{n-1}{n}\right)^{n-1}\left(\frac{1}{n}\right)$. Since $\left(\frac{n-1}{n}\right)^{n-1}$ has to be less than one, it must be true that $\left(\frac{n-1}{n}\right)^{n-1}\left(\frac{1}{n}\right) < \frac{1}{n}$. Thus it follows that $\left(\frac{n-1}{n}\right)^{n-1}\left(\frac{1}{n}\right) < \frac{1}{2}$. Since the minimum value of $\frac{n-1}{n+1}$ is one half, it follows that $E^{n-1} - E^n < \frac{n-1}{n+1} \quad \forall E \in (0,1]$.

Hence Equation [A2] is contradicted. Thus it is established by contradiction that

$$(1-\gamma)(1-E)\left(\frac{n-1}{n+1} + E^n\right) + \gamma E < (1-\gamma)\frac{n-1}{n+1} + \gamma E \qquad \forall E \in (0,1] \ .$$

Thus for all strictly positive values of E the expected revenue from an auction with no cooling-off rights exceeds the expected revenue from an auction with a costless cooling-off right, establishing Proposition 2. □

Proof of Proposition 4

When E is strictly negative, it is necessary to compare the entries in the first row of Table 1. If we let

$$\begin{aligned} g(\cdot) &= \gamma E\left[1 - \left(\frac{-\gamma E}{1-\gamma}\right)^n\right] + (1-\gamma)\frac{n-1}{n+1}\left[1 - \left(\frac{-\gamma E}{1-\gamma}\right)^{n+1}\right] \\ &\quad + (1-\gamma)\left(\frac{-\gamma E}{1-\gamma}\right)^n\left(1 + \frac{\gamma E}{1-\gamma}\right) \end{aligned}$$

where $E \in \left[\frac{-(1-\gamma)}{\gamma}, 0\right)$, then

$$\begin{aligned} \frac{\partial g(\cdot)}{\partial E} &= \gamma\left[1 - \left(\frac{-\gamma E}{1-\gamma}\right)^n\right] + \gamma E\left[\frac{\gamma n}{1-\gamma}\left(\frac{-\gamma E}{1-\gamma}\right)^{n-1}\right] \\ &\quad + \gamma(n-1)\left(\frac{-\gamma E}{1-\gamma}\right)^n - \gamma n\left(\frac{-\gamma E}{1-\gamma}\right)^{n-1}\left(1 + \frac{\gamma E}{1-\gamma}\right) + \gamma\left(\frac{-\gamma E}{1-\gamma}\right)^n \\ &= \gamma + \gamma(n-1)\left(\frac{-\gamma E}{1-\gamma}\right)^n - \gamma n\left(\frac{-\gamma E}{1-\gamma}\right)^{n-1} . \end{aligned}$$

It is difficult to sign this derivative. However, it can be seen that when $E = 0$, $\frac{\partial g(\cdot)}{\partial E} = \gamma$ and when $E = \frac{-(1-\gamma)}{\gamma}$, $\frac{\partial g(\cdot)}{\partial E} = 0$. If there are no concavity changes in the interval $\left[0, \frac{-(1-\gamma)}{\gamma}\right]$, other than $E = \frac{-(1-\gamma)}{\gamma}$ or 0, then it can be concluded that, in this interval, $\frac{\partial g(\cdot)}{\partial E} \geq 0$. Taking the second derivative of $g(\cdot)$ yields;

$$\begin{aligned}\frac{\partial^2 g(\cdot)}{\partial E^2} &= \left(\frac{-\gamma E}{1-\gamma}\right)^{n-2}\left(\frac{-\gamma E}{1-\gamma}-1\right)\\ &= 0 \quad \text{when} \quad E=0 \quad \text{and} \quad \frac{-(1-\gamma)}{\gamma}.\end{aligned}$$

Hence $\frac{\partial g(\cdot)}{\partial E} \geq 0$. This means that, without cooling-off, expected revenue decreases as E decreases in this interval. As the expected revenue, with a cooling-off right, is constant as E falls, and when E = 0 the revenue under both regime is the same, it must be true that $(1-\gamma)\frac{n-1}{n+1} > g(\cdot) \quad \forall E \in \left[\frac{-(1-\gamma)}{\gamma}, 0\right)$. That is, when E is strictly negative the expected revenue from an auction held under a costless cooling-off regime will be higher than that with no cooling-off, establishing Proposition 4. □

References

1. Baye, M.R., Morgan, J.: Bounded rationality in homogenous product pricing games. Working Paper, Princeton University (1999)
2. Cox, J.C., Roberson, B., Smith, V.L.: Theory and behaviour of single object auctions. In: Smith, V.L. (ed.) Research in experimental economics. Greenwich: JAI Press 1982
3. Davis, D.D., Holt, C.A.: Experimental economics. Princeton: Princeton University Press 1993
4. Harrison, G.W.: Theory and misbehaviour of first price auctions. American Economic Review **79**, 749–762 (1989)
5. Kahneman, D., Tversky, A.: Prospect theory: an analysis of decision under risk. Econometrica **47**, 263–291 (1979)
6. McKelvey, R., Palfrey, T.: Quantal response equilibria for normal form games. Games and Economic Behavior **10**, 6–38 (1995)
7. Matthews, S.A.: A technical primer on auction theory I: Independent private values. Northwestern University Center for Mathematical Studies in Economics and Management Science Discussion Paper 1096 (1995)
8. Milgrom, P.R., Weber, R.J.: A theory of auctions and competitive bidding. Econometrica **50**, 1089–1122 (1982)
9. Plott, C.: Industrial organization theory and experimental economics. Journal of Economic Literature **20**, 1485–1527 (1982)
10. Radner, R.: Collusive behavior in oligopolies with long but finite lives. Journal of Economic Theory **22**, 136–156 (1980)
11. Tversky, A., Kahneman, D.: Advances in prospect theory: Cumulative representation of uncertainty. Journal of Risk and Uncertainty **5**, 297–323 (1992)
12. Vickrey, W.: Counterspeculation, auctions and competitive sealed tenders. Journal of Finance **16**, 8–37 (1961)
13. von Ungern-Sternberg, T.: Swiss auctions. Economica **58**, 341–357 (1991)

Markets for contracts: experiments exploring the compatibility of games and markets for games*

Charles R. Plott[1] **and Dean V. Williamson**[2]

[1] Division of Humanities and Social Sciences, California Institute of Technology, Pasadena, CA 91125, USA (e-mail: cplott@hss.caltech.edu)
[2] Antitrust Division, US Department of Justice, 600 E Street, NW #10000, Washington, DC 20530, USA (e-mail: dean.williamson@usdoj.gov)

Summary. The research explores the relationship between games and the economic environment in which the games might be embedded. The focus is on a market institution in which agents buy and sell rights to participate in a follow-on stage of strategic interaction. The central question posed concerns how two different types of processes, the game and the market, interact. The market converges to a competitive equilibrium that is consistent with the Nash equilibrium that obtains in the game, and the convergence of the market to a competitive equilibrium lags the convergence of behaviors in the game to a Nash equilibrium.

Keywords and Phrases: Compatibility of markets and games, Simultaneous equilibration, Dynamic adjustment processes, Institutions.

JEL Classification Numbers: C72, C90.

1 Introduction

The research presented here provides new evidence about how agents resolve strategic uncertainty. The research involves an experimental market through which agents acquire the right to participate in a follow-on stage of strategic interaction (a game). The results indicate that, in the particular environment examined here, agents' expectations of behaviors in the game determine prices in the market process. Agents anticipate outcomes, and they factor their expectations into their pricing decisions. Meanwhile, experimental results in other studies

* We thank Tim Cason and an anonymous referee for thoughtful comments and suggestions. We conducted the research with support from the National Science Foundation and the Laboratory for Experimental Economics and Political Science.

indicate that prices in selected types of market processes may permit agents to align expectations on outcomes in games. In these studies, prices often dictate outcomes rather than the reverse. How can the new results and the previous results be rationalized? Although the research presented here does not provide a definitive answer, the research goes some way toward focusing the types of questions succeeding research might explore. In particular, the whole body of experimental results suggests that agents grope for some means of coordinating behaviors on outcomes in game processes. Institutional environments provide instruments – or, perhaps, impediments – to coordination. Some instruments appear to lend themselves to decentralized dynamic processes that are consistent with such game-theoretic concepts as forward induction whereas other instruments seem more amenable to processes that can be characterized by backward induction. A question of how agents might identify and sort through institutional instruments that are suggestive of one type of dynamic process or another awaits a more general theory than is currently available.

The research presented here takes a step toward motivating a more general theory by examining the compatibility of game theoretical models with the classical model of market equilibrium. The experimental approach is "exploratory" in that it is motivated by questions of economic behavior in the context of institutions even though there is neither good theory nor a clear line of previous experiments that point to what might be expected. Yet, the experiments seem to be central to both the thrust of theory and applications of theory. Thus, we report on the outcomes of data generated in a particular institutional setting. While the models we apply are very suggestive, we leave open to speculation and further theory a more fundamental explanation of what we report.

The experimental design links a market process to a contract process. A contract is modeled as a game, and the purchase and sale of contracts (or games) is modeled as a market. The market involves the purchase and sale of rights to participate in a follow-on stage of strategic interaction (the game). Compatibility concerns how the two different processes – games and markets – interact when they exist side by side as subsystems in economic environments. The questions posed concern 1) the convergence to equilibrium in each process, 2) the joint convergence of the entire system to an equilibrium, and 3) the selection of equilibria in each process. For example, are equilibria in the two processes reinforcing, each promoting the convergence of the other process, or do behaviors in one process complicate convergence of the other process? Is the selection of an equilibrium in one process systematically linked to the selection of an equilibrium in the other? Of course, the focus on institutions produces a deeper question about which we can only speculate: what dynamic processes drive the *joint* convergence of equilibria in the two processes.

Exploring the compatibility of markets and games is, of course, distinct from a mechanism design exercise out of which a joint game and market process might emerge. A design exercise involves crafting purposeful mechanisms, the constituent parts of each of which perforce constitute an organic whole. As dictated by some solution concept, a design exercise would impose the compatibility of a market mechanism and an attending game process. Instead, we distinguish two ways by which one might explore the compatibility of market processes and game processes – processes that may or may not have emerged from a purposeful design exercise. One way to explore such processes would be to ask what types of market institutions would motivate selection of particular equilibria in very particular games. The research might involve, for example, tailoring games and market mechanisms that lend themselves to tests of particular solution concepts. The games themselves, such as coordination games, may be of interest in themselves in that they provide a means of characterizing, at least in a stylized way, strategic contexts that commonly arise in applications. Theory might suggest that particular market mechanisms would motivate selection of efficient equilibria in a coordination game whereas others might motivate selection of Pareto dominated equilibria. The research presented here takes up different questions. Rather than judiciously choose a particular game and a particular market mechanism, the research pairs a game process with a market process, both which may be of interest of themselves, to explore how the processes interact. The style of research approaches what one might do in exploring real world applications that implicate market processes and follow-on game processes. The kinds of questions that emerge from the research, then, do not involve explicit tests of selected theories but rather compel us to indicate what we can say about such joint processes. Interestingly, we find that, in the context examined here, we are able to identify and partially characterize features of the observed behaviors, and the research suggests directions by which further exploratory research may proceed.

Clearly there are many alternative environments in which a study such as this could be initiated. The setting we chose reflects many arbitrary components, because, in part, there seemed to be no obviously unique place to begin. Contracting problems and even markets for contracts emerge naturally and abundantly in the context of industrial organization. Procurement contracts, for example, engage buyers and suppliers in strategic interactions that can be modeled as principal-agent games, and such contracts are often awarded to suppliers via market mechanisms such as auctions. Equity and debt certificates ("stocks and bonds") constitute contracts that confer various rights over the control of corporations and can thus be modeled as games. Obviously these contracts are exchanged in markets. Futures contracts engage buyers and suppliers in a game that involves follow-on delivery of goods and services. Insurance contracts are clearly a case of markets for games. The insurance contract involves the insurer and the insured in classic relationships of moral hazard. Wage contracts frequently can be viewed as games involving principal-agent components, and of course the labor market constitutes a market for contracts. Guarantees and warranties that accompany sales are in fact simply games. Thus the purchase and sale of games is common and arise naturally in the course of commerce. In all such cases two equilibrating processes exist side by side – the market process and the game. The problems are pervasive.[1]

[1] For example, the original research questions were motivated by contracting problems that pertain to the financing of Mediterranean trade in the late Middle Ages. The prominent features of the

One way, of course, to begin to make analysis of markets for games tractable is to distinguish stages of interaction, i.e., to specify an extensive form game, and to apply some game theoretic solution concept. Very often the structure of an extensive form game is simple to describe, but solving the same, simple game may very well require complex analysis. The game may yield a multiplicity of equilibria under given solution concepts, and identifying a compelling equilibrium may strain the available theory. Two complications are encountered here. First, although a market process itself, such as the price mechanism employed in this study, may be easy to describe, it may not be amenable to a characterization as an extensive form game. Auction theorists continue to extract clever results that pertain, for the most part, to price mechanisms by which a party, the "auctioneer," representing one side of a market exploits highly structured competition between parties on the other side of a market, but mechanisms that involve rather unstructured competition on both sides of a market such as a "double auction" remain beyond the reach of any tractable, game-theoretic model. (See, for example, Easley and Ledyard [12] on modeling the double-auction.) Second, the items traded in the market process themselves are rather complex objects. In this case, the objects are games that are, in fact, amenable to the most accessible and orthodox mode of game-theoretic analysis, but linking them to a market process raises a number of interesting prospects. Among other things, conduct in the market process might confound convergence in the game. At the same time previous realizations of conduct in the game might setup price dynamics in the market that in turn might disrupt agents' expectations of successive conduct in the game. Alternatively, in contexts involving multiple equilibria such as the one examined here, expectations might fail to become aligned on any particular equilibria.

The results reported here are striking. In the environment examined here, equilibria in the game are found to be systematically linked to competitive equilibria in the market. The convergence of the market to equilibrium lags the convergence of behaviors in the game to equilibrium, and the games move quickly to a solution of the game even though the market remained in substantial disequilibrium. The markets then settle into the equilibrium that would exist if rational expectations prevail, i.e., if agents are able to predict perfectly what the outcome of the game would be. Thus, the research demonstrates that equilibria in games and in the institutional contexts in which they are embedded can be jointly determined, and, of course, that fact suggests that institutional context might influence selection.

The remainder paper proceeds in 5 parts. The next segment, Part II, discusses other (mostly experimental) research that relates to equilibrium selection. Part III

---

organization of the Mediterranean trade, it turns out, emerge in a natural class of problems. The Mediterranean trade problem involves the joint equilibration of markets for traded commodities, of markets for agency contracts, and of principal-agent games. Specifically, the Mediterranean trade implicates the simultaneous equilibration of (geographically dispersed) markets. Coordination between these markets involves inputs from dispersed agents, and the contracting of these inputs generates a sequence of strategic interactions (games). Multiple equilibria may emerge in the games that obtain between the various functionaries of trade.

details the experimental design, and Part IV presents the models and predictions. Results are presented in Part V, and Part VI concludes.

2 Background and experimental research

The research was motivated chiefly by the two observations: 1) markets for contracts emerge in a broad range of applied contexts, and 2) the success with which certain types of solution concepts are able to characterize play in a game seems to depend on the institutional context in which the game is embedded. In some contexts, for example, traditional solution concepts help characterize play in a particular game whereas in other contexts they fail to characterize play in that same game. These contexts generally involve multiplicity of equilibria in games and the attendant problem of equilibrium selection. Much experimental research on games has approached problems such as the selection of equilibria by examining behaviors in increasingly simple environments, and yet other experimental research suggests that institutional context helps motivate equilibration in games as well as resolve the selection of equilibria. In exploring these and related issues, we designed an experiment that would be distinguished not by the starkness of the strategic environment but rather by the fact that it features some degree of institutional complexity.

Three branches of research suggest that the behavior in games can be influenced by placing games in particular institutional environments. First, such research as Cooper, DeJong, Forsythe, and Ross [6], [7] and Van Huyck, Battalio, and Beil [22], [23], and [24] study special ways of allocating rights to participate in games. Among other things, the research indicates that various mechanisms can motivate behaviors that are often consistent with forward induction. Van Huyck, Battalio, and Beil [22], for example, feature experiments in which rights to participate in a "median-action" coordination game are distributed by an English Clock auction. The auction mechanism provided agents with a means of coordinating behaviors on efficient outcomes in the game. While the results of Van Huyck, Battalio and Beil are suggestive of forward induction, Cachon and Camerer [2] articulate a principle of "loss-aversion" that also goes some way toward characterizing equilibrium selection in an experiment featuring median action games. Indeed, their equally thoughtful experimental design permits them to distinguish market mechanisms that are differentially conducive to forward induction and loss-avoidance, and they find roles for both types of processes. A second body of research on "cheap talk" suggests that embedding particular games in contexts that permit specially tailored pre-game communication can promote equilibration as well as the selection of equilibria. This research also points to a concept of forward induction. (See the survey of Crawford [9]. Examples include Cooper, DeJong, Forsythe, and Ross [7] and [8].)

A third body of research points to the possibility of very different institutional influences. It suggests that in properly structured environments variables exist that are unanticipated by game theory and may dominate or even confound convergence. Examples abound of games in which behaviors fail to line up with predictions in even starkly simple strategic contexts. (See Camerer and Thaler [4] for examples.) The resulting paradoxes have motivated exciting lines of research. Implicit in some of this research is that the shortcomings of theory results from 1) the failure of theory to incorporate psychology that routinely emerges in strategic interactions, or 2) the failure of human actors to satisfy the cognitive competencies required of traditional conceptualizations of "rationality." Behavioral game theory, for example, has involved efforts to innovate new equilibrium solution concepts, such as Mathew Rabin's "Fairness Equilibrium," that are informed by research in psychology. (See Rabin [17] and Camerer [3].) Related research has responded by operationalizing concepts such as "altruism" (See, for example, McKelvey and Palfrey [16], Eckel and Grossman [13], Cooper, DeJong, Forsythe, and Ross [5]), "spitefulness" (Levine [15]), "manners" (Camerer and Thaler [4]), "fairness norms" (Kagel, Kim, and Moser [14]), and "trust" (Berg, Dickhaut, and McCabe [1]). Other research more closely linked to the evolutionary game theory literature has involved efforts to operationalize concepts of bounded rationality (Stahl [18] and Stahl and Wilson [19]) by articulating dynamic processes by which "conventions" or "stable" outcomes emerge. (See, for example, Crawford [10], [11], Van Huyck, Battalio, and Beil [20], and Van Huyk, Battalio, Mathur, and Van Huyck [21].)

The research presented here takes up a tack that is parallel to the research on special theories of equilibrium selection in games and the role of institutions in facilitating equilibration and equilibrium selection. Rather than focus on increasingly simple environments, we focus on games that take place in a richer institutional context – a context that incorporates many of the features of situations for which applications of theory are intended. In particular, we examine an environment in which agents assume diverse roles and payoffs are private information, and then the research asks of the experimental design whether or not any kind of equilibration can occur. Previous research on the role of institutions in facilitating equilibration and equilibrium selection involved environments in which agents were symmetric and in which payoffs could be credibly modeled as common knowledge. Such environments were explicitly designed to support processes suggestive of such concepts as forward induction. In the research presented here, the structure of strategic interactions does not accommodate behaviors that can be characterized by forward induction in an obvious way, and, indeed, testing forward induction is not the objective. Prices do not provide an unambiguous means of signaling behaviors in the follow-on game. Instead, equilibria of the game map into equilibria of the market process in an intuitively accessible way that can also be characterized by backward induction. As will become obvious, the compatibility of the concept of a market and a game theoretic equilibrium will be established. Whether or not the vehicle was forward induction or backward induction is left open to speculation.

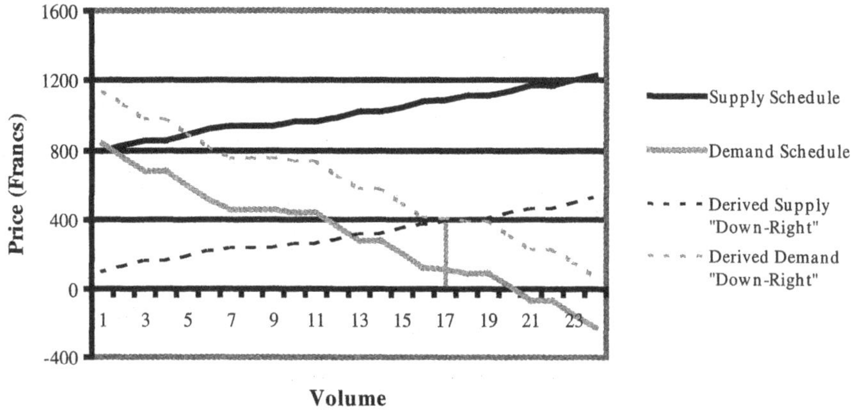

Figure 1. Supply and demand schedules

3 Experimental design and procedures

Four experimental sessions were conducted. The sessions were conducted in 1996 on February 14, February 22, November 13, and November 14. Hereafter each experimental session will be identified as Feb 14, Feb 22, Nov 13, and Nov 14. Respectively, 8, 12, 10, and 14 subjects participated in these four sessions. All subjects were students drawn from the Caltech student body. None had experience with this particular set of experiments, but some may have had experience in either markets or games in other experiments.

In each of four experimental sessions, an even number of agents participated in repeated rounds of a two-stage framework. In the first stage of each round agents bought or sold rights to participate in a second stage of strategic interaction. Exchange was organized via a double auction operationalized with standard software on a computer network. Agents were provided with sets of marginal cost schedules *or* with marginal redemption value schedules. Agents used a new schedule in each round. Costs and redemption values were denominated in an experimental currency called "francs." These marginal cost schedules determined the marginal cost of supplying to the market as many as 10 units. These units were indicated to agents as "assets." Similarly, marginal redemption value schedules determined positive marginal payoffs for as many as 10 units that a buyer could acquire in the market. The experimenter provided half of the agents with marginal cost schedules and the other half with marginal redemption value schedules. Those agents with cost schedules were designated "sellers," and the other agents were designated "buyers." One can aggregate the buyers' marginal benefit schedules into a market demand schedule and can aggregate sellers' marginal cost schedules into a market supply schedule. The demand and supply schedules (solid lines) presented in Figure 1 are representative of those that would obtain in a market with 6 sellers and 6 buyers.

In the first stage of each round, agents purchased and sold assets in a double auction that lasted 5 minutes. Buyers and sellers were permitted to speculate or "trade" during the process of price formation. That is, buyers could both buy

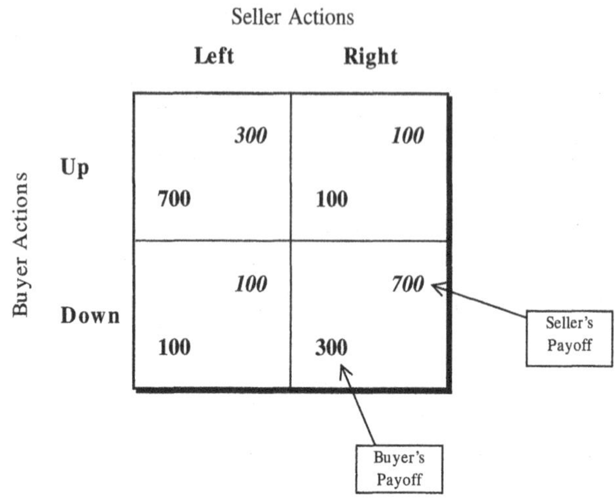

Figure 2

units and resell units and sellers could do the same. Total costs and total benefits were determined by the net inventories buyers and sellers maintained at the end of each double auction. These costs and benefits were recorded on record sheets, and in each round agents' inventories were restored to zero.

After each double auction closed, buyers assumed the right to choose the actions "Up" or "Down" in a binary choice process, the Battle-of-the-Sexes game. (The corresponding 2×2 payoff matrix is presented in Figure 2.) Sellers assumed the right to choose "Left" or "Right." Agents chose one action for each of the units they sold or purchased. These action choices were made without knowledge of other agents' choices. Agents recorded each of their choices on a separate piece of paper. The pieces of paper or "tickets" were color-coded and were labeled with the index number of the current round of the experiment and with the an index number assigned to each of the agents. Sellers' tickets were printed on green paper whereas buyers' tickets where printed on pink paper. Agents were supplied with ten tickets for each of as many as 22 rounds of interaction.

After all agents had chosen their actions, the experimenter shuffled the sellers' tickets (the green tickets) and paired each of a seller's tickets with a buyer's ticket. In this way sellers' and buyers' tickets were randomly and anonymously matched. The resulting pairs dictated one of four possible pairs of payoffs. For example, a pair of actions "Left," "Right" determined payoffs of 700 francs to the buyer of that unit and 300 francs to the seller of that unit. The experimenter indicated payoffs of 100, 300, or 700, as appropriate on each of the buyers' and sellers' tickets. Tickets were re-distributed to the agents, and agents were permitted to account their payoffs on account sheets.

The aggregate frequencies of choices were public information. After payoffs were distributed, the experimenter announced and posted aggregate results from the stage of strategic interaction. Posting the results entailed indicating the frequencies with which pairs occurred in each of the four cells of the 2×2 matrix. The four marginal frequencies were also posted. Thus both buyers and seller could determine the relative frequencies with which actions were chosen, although they could not identify the actions of any particular person or persons.

For each agent, accounting for payoffs entailed 1) summing payoffs indicated on each of the returned tickets, 2) summing, as appropriate, total costs or total benefits, and 3) identifying revenues or expenditures from the double auction. In turn, summing these three quantities generated each agent's total payoff from participation in the entire two-stage process. Agents accounted payoffs at the end of each round. Payoffs from the entire experiment were generated by summing the payoffs from each round.

Notation

In the succeeding sections we articulate models, predictions, and results. The models operationalize mechanical adjustment processes. It is these processes that permit examination of the data. Before we proceed, however, we need some notation.

An equilibrium is a 4-tuple (P^*, Q^*, p^*, q^*) where
P = price in the double auction,
Q = volume in the double auction,
$(p, 1-p)$ = row player's strategy over the strategy set {Up, Down},
$(q, 1-q)$ = column player's strategy over the strategy set {Left, Right},
(p^*, q^*) constitutes a static equilibrium, under some solution concept, of the second stage, and
(P^*, Q^*) conforms to competitive equilibrium price and quantity in the market.

The models of market equilibration share a common structure. *Given* unit payoffs (b, s) to buyers and sellers in the second stage, the models of market equilibration satisfy the following:

$$P \in [D(Q)+b] \cup [S(Q)-s]$$

where	$D(Q)$	=	inverse demand correspondence,
	$S(Q)$	=	inverse supply correspondence,
	b	=	unit payoff to buyers, under some solution concept, in the second stage,
	s	=	unit payoff to sellers in the second stage,
and	$D(Q)+b$	=	derived demand,
	$S(Q)-s$	=	derived supply.

Consider, for example, an equilibrium that corresponds to $(p^*, q^*) = (0, 0)$, i.e., the "Down-Right" pure-strategy Nash equilibrium of the second stage. The attending payoffs for each unit to the buyers and sellers in the second stage of interaction correspond to $(b, s) = (300\,\text{francs}, 700\,\text{francs})$. In turn, the derived demand schedule $D(Q)+b$ corresponds to the demand curve $D(Q)$ shifted upward by 300 francs. Similarly, the derived supply schedule $S(Q) - s$ corresponds to the supply curve $S(Q)$ shifted downward by 700 francs. The derived demand and derived supply schedules, represented by the dashed lines in Figure 1, characterize a competitive equilibrium that clears at a price of 400 francs. The competitive equilibrium volume of 17 units corresponds to a market with 6 sellers and 6 buyers.

4 Models and predictions

A particular advantage of the experimental method is that it permits examination of dynamic processes by which equilibria (if any) emerge. Accordingly, analysis of the experimental data is organized around both static predictions and dynamic adjustment processes. The predictions pertain to the identification of static equilibria (P^*, Q^*, p^*, q^*) of the two-stage process, to the selection of equilibria, and to the evolution of prices.

Models

The models pertain to price formation, the determination of volume in the market and of action choices in the second stage of the two-stage apparatus, and to the relation (if any) of prices to action choices. Specifically, the models include two sets of partial equilibrium models: those that characterize equilibration in the second stage process, and those that characterize equilibration in the market process. The real power of the models, however, derives from the characterization of the simultaneous equilibration of the market process and second stage process. Such models link equilibria in the second stage with equilibria in the market and raise deep issues about the dynamic processes through which coordination between the market and second stage process is achieved.

The data are examined via three different classes of models. These models can be distinguished from each other by the nature of belief formation and the nature of the individual decision process. Two types of belief formation, labeled "Cournot expectations" and "perfect foresight," are distinguished. The notion of Cournot expectations involves dynamic adjustment in the second stage of interaction to some subset of the history of the entire two stage process whereas the notion of perfect foresight does not implicate any kind of dynamic adjustment. To operationalize Cournot expectations, we assume that agents generate beliefs with reference to the immediate history of interaction. In particular, agents determine how to behave in the second stage of interaction by referring to the last realization of second stage interaction. Perfect foresight, on the other hand, is a static concept. Agents forecast (correctly) outcomes in the second stage of interaction. In a general equilibrium of the entire two-stage process, beliefs implicate the pricing of units.

Two types of decision-making processes are distinguished: "Myopic" processes by which agents perceive the entire two-stage process as a game against nature, and "Self-aware" processes by which agents perceive the strategic interdependence of payoffs. These processes link outcomes in the market to outcomes in the second stage of interaction. Under the model of self-aware decision-making, agents factor other agents actions into their own choices of actions in the second stage and into pricing decisions. Myopic behaviors, however, are less sophisticated. Under this behavioral model, agents fail to factor their own responses into their pricing decisions. Rather, agents act as if they were playing a game against nature rather than against other forward-looking agents.

The two models about belief formation and the two models about individual decision-making identify four general equilibrium models, three of which we operationalize. We identify and label these models in the following table:

		Belief formation	
		Cournot expectations	Perfect foresight
Individual decision-making	Myopic	Myopia	Not operationalized
	Self-aware	Cournot	Rational Expectations (RE)

The Myopia model incorporates the Cournot expectations hypothesis with the myopic behavioral hypothesis, whereas the Cournot model incorporates Cournot expectations with the self-aware behavioral hypothesis. Finally, the "Rational Expectations" model (hereafter "RE"), incorporates perfect foresight with the self-aware behavioral hypothesis.

These models determine 1) unit payoffs (b, s), 2) market outcomes (P, Q), and 3) the nature of coordination between market outcomes and unit payoffs. Consider, for example, a case in which agents observed a realization of the game process that generated relative frequencies $(p, q) = (1/2, 1/2)$. A buyer operating under Myopia anticipates frequencies $(p, q) = (1/2, 1/2)$ in the succeeding realization of the game process and best-replies to $q = 1/2$ with $p = 1$, i.e., the buyer plays "Up" on every unit he acquires in the market. This buyer is myopic in that he does not factor his best-reply into the expected payoff of the game but rather determines the expected payoff from a game as a function of $(p, q) = (1/2, 1/2)$. Frequencies $(p, q) = (1/2, 1/2)$ yield an expected payoff of 300 francs. Accordingly, the buyer's derived demand schedule corresponds to his marginal benefit schedule shifted up by 300 francs. Similar calculations indicate that a seller operating under Myopia expects a payoff from each game of 300 francs, so his derived demand correspond to his marginal cost schedule shifted down by 300 francs. The Cournot buyer also best-replies to $q = 1/2$ with $p = 1$, but the Cournot buyer factors his best-reply into the expected payoff. He takes $q = 1/2$ and his best-reply of $p = 1$ as given and determines an expected payoff of $(1/2 \times 700) + (1/2 \times 300) = 500$. The derived demand schedule of the Cournot buyer corresponds to his marginal benefit schedule shifted up by 500 francs. The derived supply schedule of the Cournot seller would correspond to his marginal cost schedule shifted down by 500 francs. The buyer or seller operating under RE, however, anticipates correctly the realization (p, q) that will emerge in play of the game, determines his best-reply, and factors his best-reply and the realization (p, q) into his expected payoff. He derives his demand schedule or cost schedule accordingly. In equilibrium, behaviors in the second stage of interaction conform to the Nash hypothesis, Nash outcomes of the second stage imply payoffs to buyers and sellers, and agents factor these payoffs into their pricing decisions. In this way RE links Nash equilibria of the second stage to competitive equilibria in the market.

a) A static Rational Expectations model (RE)

Under RE agents know the structure of the model and coordinate behaviors on outcomes of the two-stage process. Agents have perfect foresight: they anticipate behaviors will converge in the second stage of each round, and they anticipate *which* behaviors will obtain. More specifically, agents anticipate a Nash equilibrium (p^*, q^*) of the second stage, and the behaviors (p^*, q^*) imply expected payoffs $b(p^*, q^*)$ and $s(p^*, q^*)$ to buyers and sellers. Agents factor these payoffs into their pricing decision in the preceding double auction. The usual analysis of competitive markets provides an obvious way of factoring payoffs into the double auction: buyers and sellers derive their demand and supply schedules in the preceding double auction by factoring their game payoffs b and s into their marginal surplus computations. Specifically, buyers buy their kth units if they can secure prices which do not exceed the sum of the unit game payoff b and the marginal benefit of the kth unit. Sellers sell their kth units if they can secure prices which exceed the difference between the marginal cost of the kth unit and the unit game payoff s. These derived demands and supplies in turn imply competitive equilibria (P^*, Q^*) in the double auction. We label the predictions (P^*, Q^*, p^*, q^*) generated in this way as "RE-consistent."

The supply and demand schedules were constructed such that the payoffs that would obtain to the pure-strategy Nash Equilibrium corresponding to "Up," "Right" would yield a competitive equilibrium price of 800 (or, rather, an RE-consistent interval 790–810). At the same time, the supply and demand apparatus yields an equilibrium price of 400 (or 390–410) in the event the pure-strategy Nash equilibrium corresponding to "Down-Left" emerges.

The static predictions (P^*, Q^*, p^*, q^*) are presented in the following table:

Price (P^*)	Quantity (Q^*)[a]	Strategies (p^*, q^*)	Surplus[a]
790–810	17	(1, 1)	8,540
690–710	9	(3/4, 1/4)	1,980
390–410	17	(0, 0)	8,540

[a]Values correspond to the double auction conducted with 6 buyers and 6 sellers.

Whereas buyers and sellers oppositely rank the pure-strategy Nash equilibria (1, 1) and (0, 0) of the second stage, the surpluses available to each agent under each of the pure strategy equilibria (800, Q^*, 1, 1) and (400, Q^*, 0, 0) of the entire two-stage process are identical. All players rank the pure strategy equilibria above the mixed-strategy equilibrium (700, Q^*_{mixed}, 0.75, 0.25).

b) Cournot

The Cournot model assigns to agents a lesser degree of rationality than RE. Agents assume outcomes of the two-stage process result from the active strategizing of the other agents in the market, and agents best-reply to some subset of past realizations of second stage play, which, in this case, is restricted to the latest realization (p, q). Agents factor into their pricing decision the last realization (p, q) and their own best-replies to that realization. Under this model, however, agents do not factor other agents' best-replies into their own best-replies. Rather, they assume that the last realization characterizes other agents' next-period choices.

Second-stage interaction converges on one of the two pure strategy equilibria of the two stage process or collapses into an infinite, non-convergent sequence of out-of-equilibrium play. The frequencies (p_t, q_t) in a round t of interaction may not correspond to any Nash equilibrium of the second stage but may generate a non-convergent sequence of mis-coordinated play. For example, a realization in a round t $(p_t, q_t) = (1, 0)$ generates a pair of best-replies in the succeeding round $t + 1$ $(p_{t+1}, q_{t+1}) = (0, 1)$ which in turn generates $(p_{t+2}, q_{t+2}) = (1, 0)$, and so on. On the other hand, a realization in some round t $(p_t, q_t) = (1, 1)$ or (0, 0) generates a forward invariant sequence $(p_t, q_t) = (p_{t+k}, q_{t+k})$ for any $k > 0$.

c) Myopia

Myopia assigns to agents a lesser degree of rationality than Cournot, and Myopia admits the Cournot-consistent as well as the RE-consistent predictions. Under the RE model, price and volume anticipate the frequencies (p, q). Under Myopia, prices and quantities respond to previous realizations of second stage interaction. Again, agents assume that observed frequencies are representative of the forthcoming frequencies, and they price their units accordingly. The distinction from the Cournot model is that agents do not factor their best-replies into their pricing decisions.

Myopia does not pin down a countable set of predictions. Note however that the Cournot model can be distinguished from Myopia in the data in that Cournot predicts a larger period-on-period volume than Myopia, because under Cournot agents factor there own best-replies into their own decision to buy or sell units.

Table 1 indicates the price intervals and volumes that the three models predict round-on-round in the experimental session Feb 22. The RE price and volume predictions correspond to the price-intervals and volumes that were RE-consistent with the frequencies (p, q) that actually obtained in each round. In contrast, the Cournot and Myopia predictions are functions of the last realizations of the frequencies (p, q).

Table 1. Feb 22 predictions

	Price predictions			Volume predictions		
Round	RE	Cournot	Myopia	RE	Cournot	Myopia
1	654–718			11		
2	628–639	633–674	654–718	11	12	11
3	518–527	620–727	628–639	14	11	11
4	502–589	484–562	518–527	12	14	14
5	396–485	532–580	502–589	15	14	12
6	390–410	396–500	396–485	17	15	15
7	390–487	390–410	390–410	15	17	17
8	390–487	390–500	390–487	15	15	15
9	390–410	390–500	390–487	17	15	15
10	390–410	390–410	390–410	17	17	17
11	390–410	390–410	390–410	17	17	17
12	390–410	390–410	390–410	17	17	17
13	390–410	390–410	390–410	17	17	17
14	390–410	390–410	390–410	17	17	17

5 Results

The first general result is that prices, volumes, and behaviors systematically converge *and* that they converge to an RE-consistent state. The second general result is that the convergence of prices lags the convergence of realizations in the second stage. The suggestion is that prices respond to the convergence of realizations in the second stage and not that prices signal realizations. Specifically, the results suggest that 1) agents factor their own action choices as well as the second-stage realizations they anticipate into their pricing decisions in the double-auction, and 2) their action choices are sensitive to previous second-stage realizations. In this way agents link previous realizations of the game to yet unrealized outcomes of the second-stage, and, in turn, they link anticipated realizations to their pricing decisions.

Result 1: In each of the experimental sessions, behaviors in the second stage of strategic interaction converge to one of the two pure-strategy Nash equilibria.

Support: In 3 of 4 sessions behaviors converge to the pure-strategy Nash equilibrium that corresponds to "Down-Right" or $(p, q) = (0, 0)$. In the one other session, behaviors converge on the other pure-strategy Nash equilibrium that corresponds to "Up-Left" or $(p, q) = (1, 1)$.

In session Feb 22, behaviors in the Battle-of-the-Sexes converged to the pure-strategy Nash equilibrium conforming to "Down-Right" by round 6. A single buyer deviated with a single unit in both of the succeeding rounds, but in the remaining 6 rounds behaviors conformed to $(p, q) = (0, 0)$ (see "Observed Frequencies" in Table 3).

In session Feb 14, behaviors converged to (0, 0) by round 5. (See "Observed Frequencies" in Table 2.) A single buyer deviated from (0, 0) in round 6, but in the remaining 4 rounds behaviors conformed to (0, 0). Behaviors converged to

Table 2. Feb 14 results

Round	Observed frequencies (p, q)	Avg game payoffs Buyers	Sellers	Observed Avg price	Volume	System efficiency
1	(1.00, 0.00)	100.00	100.00	836	5	−13.48%
2	(0.50, 0.33)	266.67	333.33	842	6	26.06%
3	(0.63, 0.25)	325.00	375.00	833	6	19.28%
4	(0.50, 0.00)	200.00	400.00	780	6	58.16%
5	(0.00, 0.00)	300.00	700.00	734	9	73.67%
6	(0.33, 0.00)	233.33	500.00	644	9	45.04%
7	(0.00, 0.00)	300.00	700.00	574	9	92.55%
8	(0.00, 0.00)	300.00	700.00	480	9	85.82%
9	(0.00, 0.00)	300.00	700.00	459	9	94.68%
10	(0.00, 0.00)	300.00	700.00	411	11	98.58%

Table 3. Feb 22 results

Round	Observed frequencies (p, q)	Avg game payoffs Buyers	Sellers	Observed Avg price	Volume	System efficiency
1	(0.57, 0.36)	285.71	314.29	742	14	−6.44%
2	(0.60, 0.07)	166.67	300.00	725	15	−18.65%
3	(0.27, 0.09)	300.00	554.55	692	11	53.49%
4	(0.35, 0.00)	229.41	488.24	631	17	19.20%
5	(0.08, 0.00)	284.62	653.85	535	13	80.21%
6	(0.00, 0.00)	300.00	700.00	425	11	77.28%
7	(0.07, 0.00)	286.67	660.00	369	15	82.55%
8	(0.07, 0.00)	286.67	660.00	344	15	79.63%
9	(0.00, 0.00)	300.00	700.00	355	17	92.04%
10	(0.00, 0.00)	300.00	700.00	361	17	97.42%
11	(0.00, 0.00)	300.00	700.00	386	15	98.95%
12	(0.00, 0.00)	300.00	700.00	398	16	99.77%
13	(0.00, 0.00)	300.00	700.00	402	16	98.36%
14	(0.00, 0.00)	300.00	700.00	402	16	99.06%

the other equilibrium (1, 1) by round 4 of session Nov 13. In rounds 6 through 11 some sellers deviated from the equilibrium, but behaviors conformed to (1, 1) in the closing 8 rounds of the session. Lastly, behaviors converged on (0, 0) by round 10 of session Nov 14. A single agent deviated with 8 units in round 13, and the same agent deviated with 3 units in round 14, but in the remaining 4 rounds behaviors conformed to (0, 0) (see "Observed Frequencies" in Tables 4 and 5).

Result 2: In each of the experimental sessions, prices converge to RE-consistent levels.

Support: In sessions Feb 14, Feb 22 and Nov 14, prices converged to 400, a price RE-consistent with the pure-strategy Nash equilibrium (0, 0) to which behaviors converged in each of those three sessions. In session Nov 13, prices converged to 800, a price RE-consistent with the pure-strategy Nash equilibrium (1, 1).

Table 4. Nov 13 results

Round	Observed frequencies (p, q)	Avg game payoffs Buyers	Avg game payoffs Sellers	Observed Avg price	Observed Volume	System efficiency
1	(0.43, 0.86)	271.43	157.14	838	7	14.27%
2	(0.90, 0.80)	600.00	320.00	728	10	56.05%
3	(0.78, 1.00)	566.67	255.56	720	9	45.97%
4	(1.00, 1.00)	700.00	300.00	724	11	72.91%
5	(1.00, 1.00)	700.00	300.00	730	17	71.76%
6	(0.73, 1.00)	536.36	245.45	746	11	43.52%
7	(0.71, 1.00)	528.57	242.86	739	14	45.97%
8	(0.64, 1.00)	481.82	227.27	751	11	25.58%
9	(0.73, 1.00)	536.36	245.45	755	11	43.23%
10	(0.75, 1.00)	550.00	250.00	751	12	49.57%
11	(0.77, 1.00)	561.54	253.85	752	13	46.54%
12	(1.00, 1.00)	700.00	300.00	751	15	93.52%
13	(1.00, 1.00)	700.00	300.00	753	15	94.24%
14	(1.00, 1.00)	700.00	300.00	765	14	95.10%
15	(1.00, 1.00)	700.00	300.00	770	12	92.36%
16	(1.00, 1.00)	700.00	300.00	782	12	96.83%
17	(1.00, 1.00)	700.00	300.00	786	16	91.79%
18	(1.00, 1.00)	700.00	300.00	790	15	95.53%
19	(1.00, 1.00)	700.00	300.00	805	15	95.53%

Table 5. Nov 14 results

Round	Observed frequencies (p, q)	Avg game payoffs Buyers	Avg game payoffs Sellers	Observed Avg price	Observed Volume	System efficiency
1	(0.63, 0.19)	287.50	362.50	555	16	24.75%
2	(0.33, 0.40)	206.67	313.33	588	15	16.70%
3	(0.63, 0.13)	200.00	300.00	576	16	5.53%
4	(0.44, 0.00)	212.50	437.50	582	16	30.08%
5	(0.29, 0.00)	241.18	523.53	546	17	45.98%
6	(0.24, 0.06)	241.18	523.53	545	17	54.73%
7	(0.06, 0.00)	288.89	666.67	533	18	83.60%
8	(0.20, 0.00)	260.00	580.00	510	20	58.25%
9	(0.15, 0.00)	270.00	610.00	465	20	70.32%
10	(0.00, 0.00)	300.00	700.00	453	18	87.83%
11	(0.00, 0.00)	300.00	700.00	440	20	93.16%
12	(0.00, 0.00)	300.00	700.00	438	21	91.35%
13	(0.38, 0.00)	223.81	471.43	441	21	28.27%
14	(0.12, 0.00)	276.47	629.41	442	17	76.46%
15	(0.00, 0.00)	300.00	700.00	443	18	94.37%
16	(0.00, 0.00)	300.00	700.00	436	20	93.46%
17	(0.00, 0.00)	300.00	700.00	412	22	98.89%
18	(0.00, 0.00)	300.00	700.00	431	19	95.27%

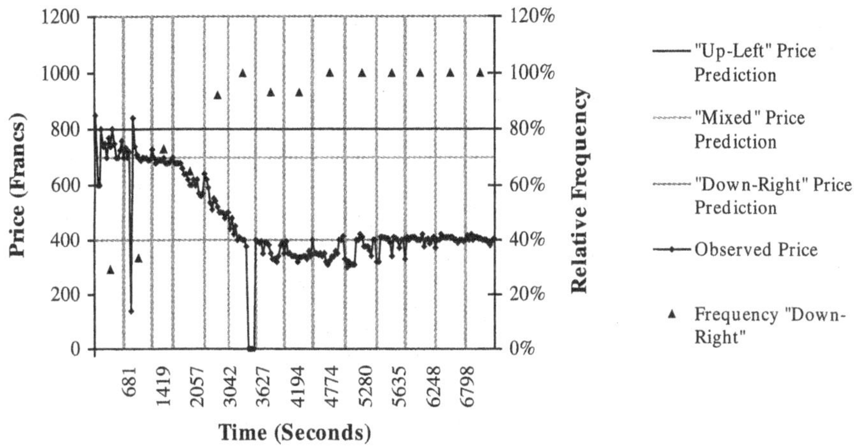

Figure 3. Prices and behaviors converge to the RE-consistent values (400, "Down-Right") on Feb 22

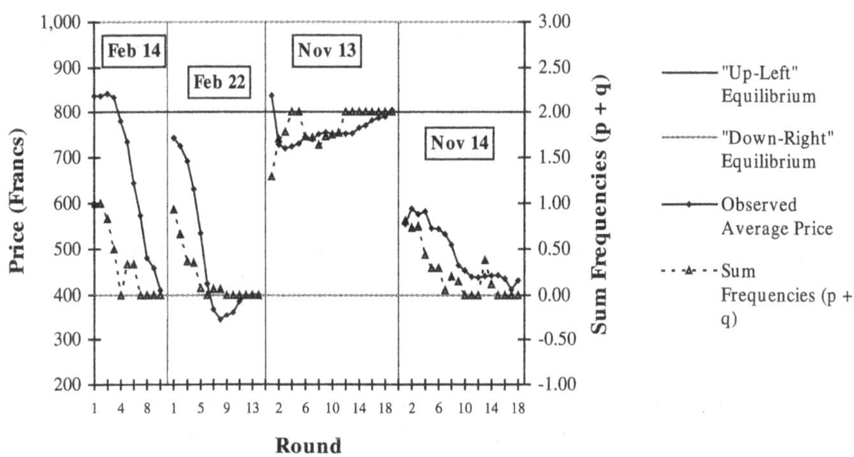

Figure 4. Prices and the sum of frequencies $(p+q)$ converge to joint rational-expectations equilibria

Figure 3 exhibits the prices at which units were transacted over the course of fourteen market periods in experimental session Feb 22. After each market closed, test subjects participated in a round of strategic interaction. The round-on-round relative frequency with which the outcome "Down-Right" emerged is mapped against the right axis of Figure 3. By round 6, behaviors in the second stage of strategic interaction conformed to "Down-Right," and prices subsequently converged to the RE-consistent price of 400.

Figure 4 presents prices and behaviors from all four experimental sessions. Average round-on-round prices are mapped against the left axis, and the statistic $(p+q)$, the round-on-round sum of observed frequencies, is mapped against the right axis. The condition $(p+q) = 2.00$ indicates that behaviors in the second stage conformed to "Up-Left," and $(p+q) = 0.00$ indicates that behaviors conformed to "Down-Left." In the three sessions in which the statistic $(p+q)$ converged to the value 0.00, prices converged to the RE-consistent price 400. In session Nov

Table 6. Equations: $\text{Volume}_t = \alpha_1 d_1 + \alpha_2 d_2 + \alpha_3 d_3 + \alpha_4 d_4 + \beta\ \text{Volume}_{t-1} + u_t, U_t = \rho_1 u_{t-1} + \rho_2 u_{t-2} + \varepsilon_t, \varepsilon_t \sim N(0, \sigma^2)$

Regressor	ML coefficent Estimates $\hat{\alpha}, \hat{\beta}, \hat{\rho}_i$	Std. error	Estimated Session volume $\hat{V}_i = \frac{\hat{\alpha}_i}{1-\hat{\beta}}$	Predicted V_i	t-ratio: $\left(\frac{\hat{V}_i - V_i}{\sigma_v}\right)$
Feb 14 dummy: d_1	3.16	0.7737	9.43	12	−1.41
Feb 22 dummy: d_2	5.16	1.3363	15.38	17	−0.89
Nov 13 dummy: d_3	4.60	1.1416	13.72	14	−0.15
Nov 14 dummy: d_4	6.41	1.6541	19.09	21	−1.04
Volume_{t-1}	0.66	0.0889			
U_{t-1}	−0.46	0.1273			
U_{t-2}	−0.28	0.1273			

Std. err. of volume_t, σ_V	1.83
DW-statistic	2.09

13 $(p+q)$ converged to 2.00, and average prices converged on the corresponding price of 800.

Result 3: In each of the experimental sessions, volumes converge to RE-consistent levels.

Support: We apply a descriptive statistical criterion that indicates that volumes in all four sessions converge to RE-consistent levels. We apply an ARMA(1, 2) of the form $Q_t = \alpha_1 d_1 + \alpha_2 d_2 + \alpha_3 d_3 + \alpha_4 d_4 + \beta Q_{t-1} + u_t$ where $u_t = \rho_1 u_{t-1} + \rho_2 u_{t-2} + \varepsilon_t, \varepsilon_t$ is assumed to conform to a "white noise" process, and the terms d_i are dummy variables identifying each of the four experimental sessions. The ARMA generates steady-state estimates of volumes in each of the four sessions that are in each instance lower than the volumes predicted by RE, but these estimates cannot be statistically distinguished from the RE-consistent predictions. (See regression results in Table 6.)

The ARMA(1,2) employs 57 observations of volumes of 61 of the observations generated from all four experimental sessions. Four observations are omitted in order to accommodate the single lag on volume Q_{t-1}. The model pools the data to generate an estimate of the effect of the lag, but the four dummy variables permit estimation of an "intercept" α_i for each of the four experimental sessions.

The next result characterizes the evolution of prices in the system. Principally, the convergence of prices lags the convergence of behaviors in the game. The lag of prices itself suggests that prices respond to the equilibrium realizations that emerge in the second stage.

Result 4: In all four experimental sessions, the convergence of prices to RE-consistent values lags the convergence of behaviors to pure-strategy Nash equilibria.

Support: Figure 4 indicates the convergence of behaviors in the second stage and the follow-on convergence of prices. Within the first four rounds of all four sessions, behaviors on the side of the market receiving 700 in the equilibrium of the second stage game make initial contact with the equilibrium that eventually obtains. In sessions Feb 14, Feb 22, and Nov 14, sellers actions all conform to "Right" ($q = 0$) in the first round, the fourth round, and the fourth round, respectively. In each of the three sessions behaviors in the second stage subsequently converge to the equilibrium conforming to "Down-Right". By the fourth stage, prices have yet to converge. Average prices in the fourth round of each of these sessions were 780, 631, and 582. Prices in each of these sessions eventually converged to 400. In session Nov 13 the equilibrium conforming to "Up-Left" eventually obtained. In this case, buyers actions first made contact with the equilibrium behaviors ($p = 1$) by round 4. Average prices were 724, well short of the equilibrium price of 800.

Result 5 provides a clue about the strategic sophistication agents exercise, and it suggests that their behaviors can be partially characterized by backward induction. First, the result suggest that agents use the latest realizations of the second stage game to form expectations about the next stage. More importantly, however, the result suggests that agents are sophisticated enough 1) to factor expected payoffs into their pricing decisions – they apply backward induction – and 2) to factor there own best-replies to expected behaviors into their calculations of expected payoffs.

Expressed in the language of the models, Result 5 indicates that the Cournot model captures the convergence of prices with more success than RE or Myopia. The idea is that, round-on-round, each model indicates a price interval that is consistent with the realizations in the previous round of second stage interaction, or, in the case of RE, with the yet unrealized outcome of the succeeding round of second stage interaction. The predicted price intervals are functions of the strategic adjustment processes that each model implies. In turn, the differential success with which these price intervals capture realized prices suggests that one or another of the underlying strategic adjustment processes more closely characterizes the actual strategic adjustment processes driving realizations in the second stage of interaction and in the market process.

The result is expressed in terms of the prices at which the *last* unit was transacted in each market period, although the result could have been articulated in terms of average prices or some other price statistic.

Result 5: The Cournot model characterizes the evolution of prices with markedly more success than the RE and Myopia models.

Support: Round-on-round, the adjustment process underlying the Cournot model predicts price intervals that are consistent with the price of the last unit transacted in 82% of the trading periods. The Myopia and RE models generate predictions that are consistent with 35% and 25% of the same price data, respectively.

In each round the various models predict frequencies (p, q). These frequencies imply payoffs to buyers and sellers, and demand and supply may be derived from these payoffs. Because demand and supply can be characterized as step functions, they indicate competitive equilibria that can each be identified by a price interval. Under the Cournot model, for example, agents best-reply in a round t to previous frequencies (p_{t-1}, q_{t-1}), and they determine price and volume by factoring the last period's frequencies *and* their best-replies to those frequencies into their calculations. In 47 of the 57 rounds (82%), the last price at which a unit was traded is contained in the Cournot-consistent price interval. In contrast, the price intervals that correspond to the competitive equilibria implied by the RE model are consistent with the price of the last unit traded in 15 of 61 rounds (25%), and the price intervals that correspond to the competitive equilibria implied by the Myopia model are consistent with the price of the last unit traded in 20 of 57 rounds (35%).

6 Conclusion

How do agents resolve strategic uncertainty in games? There is a body of experimental evidence, including the research presented here, that suggests that what happens depends systematically on features of the larger institutional context in which games are embedded. Some research examines games in stark institutional settings with the purpose of articulating and exploring concepts of complex behaviors such a "altruism," "fairness," and "spite." Other research examines games in richer institutional settings – settings in which agents may be unable to operationalize behavioral concepts such as "fairness" – with the purpose of exploring how context might facilitate equilibration and even the selection of equilibria in games. Some research, for example, indicates that embedding games in auction mechanisms can permit agents to use prices as a means of coordinating behaviors, whereas the research presented here suggests that pairing a game with a double auction can generate complex behaviors by which prices become coordinated with equilibria in the game. The whole body of research suggests that there exist both behavioral and institutional devices by which agents seem to coordinate behaviors.

Different types of devices seem to dominate in different environments. Quite plainly, competitive behaviors dominate altruism in an expansive range of environments, yet altruistic behaviors seem to emerge (if not trump) competition in other environments. Behaviors that are consistent with competition seem to dominate in more institutionally complex environments. Within some of these institutionally complex environments, price coordination mechanisms go some way toward characterizing equilibration and the selection of equilibria in games. In environments featuring yet more complex types of price competition, such as the one examined here, price coordination gives way to a dynamic process by which prices become coordinated after convergence in the game.

Price coordination mechanisms have been the focus of attention of several studies. In the experiments of Van Huyck, Battalio and Beil, agents participate in a market through which the right to participate in a second stage of interaction is distributed. In their experiments market volume is predetermined, and the structure of payoffs is common knowledge. In Cooper, Dejong, Forsythe, and

Ross [7], 18 agents bid for 9 units where each unit entitles the owner one action choice in a 9-player coordination game. In this framework, a forward-induction model entails mapping prices into Nash equilibria in an obvious way.

The framework we present is less accommodating to price coordination, and, indeed, it is not our purpose to craft a test of such mechanisms. In the framework presented here agents do not know the entire structure of the two-stage apparatus. They know their own marginal cost or marginal redemption value schedules, and they commonly know the structure of the second stage process, but they do not know the demand and supply schedules. Accordingly, it is not obvious how agents might use prices to signal behaviors in the second stage. How would, for example, an agent know that a price of 800 is consistent (in the RE model) with the outcome ("Up-Left")?

The first three results presented here indicate that behaviors converge to RE-consistent equilibria, and the remaining results begin to illuminate the dynamic processes by which behaviors converge. From Result 4 one sees that agents do not coordinate behaviors in the game via equilibrium prices. Rather, the convergence of prices in the double auction lags convergence of behaviors in the second stage. Result 5 indicates that prices respond to realizations in the game. Specifically, the data indicate that agents factor their own actions in the second stage into their pricing decisions in the first stage. What the results do not indicate is how agents select an equilibrium from among the two pure-strategy Nash equilibria in the game.

We speculate that both learning dynamics and solution concepts, whether new or traditional, can be systematically linked to attributes of the institutional context in which games are embedded. That is, we suggest that the issues of equilibration and equilibrium selection can be naturally embedded within the larger issue of the institutional sensitivity of equilibria. In this context the research makes progress in characterizing a mapping from institutions into learning dynamics and into equilibria. First and foremost, the research demonstrates that agents' behaviors converge on static outcomes that are consistent not merely with a Nash prediction of behaviors in a static game but are consistent with a joint application of supply-demand analysis and Nash predictions. While none of the models completely characterize the dynamics exhibited in the experimental data, one model outperforms all of the others. That model is one in which agents best-reply to some subset of past realizations of strategic interaction and in which agents factor their own actions into future realizations. This model outperforms those in which agents fail to factor their own actions into their calculus and those in which agents forecast (perfectly) other agents' actions.

References

1. Berg, J.E., Dickhaut, J.W., McCabe, K.: Trust, reciprocity, and social history. Games and Economic Behavior **10**, 122–142 (1995)
2. Cachon, G.P., Camerer, C.F.: Loss-avoidance and forward induction in experimental coordination games. Quarterly Journal of Economics **111**, 165–194 (1996)

3. Camerer, C.F.: Progress in behavioral game theory. Journal of Economic Perspectives **11**, 167–188 (1997)
4. Camerer, C.F., Thaler, R.H.: Anomalies: ultimatums, dictators and manners. Journal of Economic Perspectives **9**, 209–219 (1995)
5. Cooper, R., DeJong, D.V., Forsythe, R., Ross, T.W.: Cooperation without reputation: experimental evidence from prisoner's dilemma games. Games and Economic Behavior **12**, 187–218 (1996)
6. Cooper, R., DeJong, D.V., Forsythe, R., Ross, T.W.: Alternative institutions for resolving coordination problems: experimental evidence on forward induction and preplay communication, pp. 129–146. In: Friedman, J.W. (ed.) Problems of coordination in economic activity. Boston: Kluwer 1994
7. Cooper, R., DeJong, D.V., Forsythe, R., Ross, T.W.: Forward induction in the battle-of-the-sexes game. American Economic Review **83**, 1303–1316 (1993)
8. Cooper, R., DeJong, D.V., Forsythe, R., Ross, T.W.: Communication in the battle of the sexes game: some experimental results. Rand Journal of Economics **20**, 568–587 (1989)
9. Crawford, V.: A survey of experiments on communication via cheap talk. Journal of Economic Theory **78**, 286–298 (1998)
10. Crawford, V.: Adaptive dynamics in coordination games. Econometrica **63**, 103–144 (1995)
11. Crawford, V.: An 'evolutionary' interpretation of Van Huyck, Battalio, and Beil's experimental results on coordination. Games and Economic Behavior **3**, 25–59 (1991)
12. Easley, D., Ledyard, J.O.: Three theories of price formation and exchange I double oral auctions, pp. 63–97. In Friedman, D., Rust, J. (eds.) The double auction market: institutions, theories, and evidence. New York: Addison-Wesley 1993
13. Eckel, C.C., Grossman, P.J.: Altruism in anonymous dictator games. Games and Economic Behavior **16**, 181–191 (1996)
14. Kagel, J.H., Kim, C., Moser, D.: Fairness in ultimatum games with asymmetric information and asymmetric payoffs. Games and Economic Behavior **13**, 100–110 (1996)
15. Levine, D.: Modeling altruism and spitefulness in experiments. Unpublished manuscript (1997)
16. McKelvey, R.D., Palfrey, T.R.: An experimental study of the centipede game. Econometrica **60**, 803–836 (1992)
17. Rabin, M.: Incorporating fairness into game theory and economics. American Economic Review **83**, 1281–1302 (1993)
18. Stahl, D.O.: Boundedly rational rule learning in a guessing game. Games and Economic Behavior **16**, 303–330 (1996)
19. Stahl, D.O., Wilson, P.W.: On players' models of other players: theory and experimental evidence. Games and Economic Behavior **10**, 218–254 (1995)
20. Van Huyck, J.B., Cook, J.P., Battalio, R.C.: Adaptive behavior and coordination failure. Journal of Economic Behavior & Organization **32**, 483–504 (1997)
21. Van Huyck, J.B., Battalio, R.C., Mathur, S., Van Huyck, P.: On the origin of convention: evidence from symmetric bargaining games. International Journal of Game Theory **24**, 187–212 (1995)
22. Van Huyck, J.B., Battalio, R.C., Beil, R.O.: Asset markets as an equilibrium selection mechanism: coordination failure, game form auctions, and tacit communication. Games and Economic Behavior **5**, 485–504 (1993)
23. Van Huyck, J.B., Battalio, R.C., Beil, R.O.: Strategic Uncertainty, Equilibrium Selection, and Coordination Failure in Average Opinion Games. Quarterly Journal of Economics **106**, 885–910 (1991)
24. Van Huyck, J.B., Battalio, R.C., Beil, R.O.: Tacit coordination games, strategic uncertainty, and coordination failure. American Economic Review **80**, 234–248 (1990)

An experimental study of coordination and learning in iterated two-market entry games*

Amnon Rapoport[1], **Darryl A. Seale**[2], **and Eyal Winter**[3]

[1] Department of Management and Policy, University of Arizona, 405 McClelland Hall, Tucson, AZ 85721, USA (e-mail: amnon@u.arizona.edu)
[2] University of Nevada, Las Vegas, AZ 89154, USA
[3] The Hebrew University of Jerusalem, Jerusalem, ISRAEL

Summary. Tacit coordination in large groups is studied in an iterated market entry game with complete information and multiple market capacities that are varied randomly from period to period. On each period, each player must decide independently whether to enter any of the markets, and if entering, which of the two markets to enter. Across symmetric and asymmetric markets, we find remarkable coordination on the aggregate level, which is accounted for by the Nash equilibrium, together with considerable individual differences in frequency of entry and decision rules. With experience, the decisions of most players converge to decision rules with cutoff values on the combined market capacity that determine whether or not to enter but not which of the two markets to enter. This latter decision is determined probabilistically by the differential market capacities. The aggregate and individual results are accounted for quite well by a reinforcement-based learning model that combines deterministic and probabilistic elements.

Keywords and Phrases: Market entry games, Coordination success, Large groups, Adaptive learning, Nash equilibrium solution.

JEL Classification Numbers: C72, C92, D83.

1 Introduction

Strategic interactions in which people or firms attempt to avoid congestion (e.g., choosing a driving route, picking a vacation spot, entering a new market) often

* This research has been supported in part by NSF Grant SBR-9512724 on market entry games, and in part by grant CA98/99.BM01 form the Hong Kong Research Grants Council. We thank Lisa Ordóñez for her help, and the Hong Kong University of Science and Technology for its hospitality.

display an incredible level of tacit coordination. However, the process by which such coordination is achieved cannot be determined by casual observation alone nor by sheer speculation. The present paper addresses the issue of the emergence of tacit coordination in large groups playing an n-person noncooperative game called a *two-market entry game* (TMEG). We present and then analyze experimental results in which individuals face a decision whether to enter one of two markets or stay out. We iterate the game, changing the capacities of the two markets from period to period, and then develop and test an adaptive learning model to account for the dynamics by which symmetric players learn to coordinate their actions in order to solve the congestion problem.

Similar to the game of pure coordination studied by Ochs (1990) and the binary allocation game investigated by Meyer, Van Huyck, Battalio, and Saving (1992), the *market entry game* (MEG) was originally devised to investigate experimentally the coordination problem which is at the heart of the inefficiencies that plague decentralized markets. In its original form, the MEG is played by a group N of n players. It begins with a public announcement of a positive integer c ($1 \leq c \leq n$) interpreted as the "commonly known capacity of the market". To enter the market, each player i ($i \in N$) is charged an entry fee h_i, which is private knowledge. Individual decisions are made simultaneously and anonymously. Once the value of c becomes publicly known, each player i must decide whether to pay his entry fee h_i and enter the market ($d_i = 1$) or stay out of it ($d_i = 0$). Communication between players is strictly forbidden. Individual payoffs are determined according to the payoff function $H_i(\mathbf{d})$:

$$H_i(\mathbf{d}) = \begin{cases} v, & \text{if } d_i = 0 \\ k + r(c - m) - h_i, & \text{if } d_i = 1 \end{cases}$$

where v, k, and r are real-valued constants, m ($0 \leq m \leq n$) is the (endogenously determined) number of entrants, and $\mathbf{d} = (d_1, d_2, \ldots, d_n)$ is the vector of individual (binary) decisions.

When $h_i = 0$ for all $i \in N$, the MEG is characterized by symmetric players, complete information, simultaneous play, zero entry fees, and an incentive to enter the market that decreases linearly in the number of entrants. If the values of h_i differ across players (Rapoport, Seale, and Winter, 1997), the effects of asymmetry between types of players (where types are determined by the entry fee) can be studied. In yet another variation (Camerer and Lovallo, 1999), asymmetry is introduced by conditioning the individual payoff of an entrant on his rank relative to other entrants, where relative ranks are determined by skill at solving puzzles or answering trivia tests. In all these cases, the individual choice on each period is between taking a costless action which determines his payoff with certainty ("staying out"), or taking another action ("entering") whose outcome becomes less favorable the larger the number of players taking this same action.

When the n players are symmetric, the MEG has multiple pure- strategy Nash equilibria in which a subset of the players enter and the remaining players stay out, as well as a symmetric equilibrium in mixed strategies (see e.g., Sundali,

Rapoport, and Seale, 1995; Rapoport, Erev, Seale, and Sundali, 1998). There are good reasons to expect coordination failure particularly if the MEG is iterated in time with the values of v, k, and r kept fixed across iterations and the value of c being randomly changed from period to period. First, because the multiple equilibria in pure strategies are not Pareto rankable, there are no cues which could have helped the players to determine the particular subset of entrants. Focal points (Schelling, 1960) are irrelevant for achieving coordination in this class of games. Secondly, the equilibria in the MEG do not maximize group (and, subsequently, individual) payoff. Consequently, like in the iterated Cournot oligopoly game, our design allows for the emergence of collusive coordination in the repeated game that may increase considerably the payoffs, if *all* the players simultaneously lower their frequency of entry. Finally, the coordination failure reported by Van Huyck, Battalio, and Beil (1990, 1991), who investigated a different class of coordination games with a relatively large number of symmetric players, provides no reasons to expect coordination success in the present MEG.

Given this expectation for coordination failure, the results reported by Rapoport (1995) were quite surprising (see Ochs, 1999, for a recent review). In his preliminary study, Rapoport reported a very high degree of competitive coordination success, with the aggregate frequencies of entry converging to the equilibrium solution, together with considerable individual differences in decision rules and total frequency of entry that showed no sign of diminishing with experience. Subsequent experiments (Rapoport et al., 1997, 1998) were designed to improve the methodology and extend the original MEG in several different directions in an attempt to delineate the conditions under which coordination success in the MEG may or may not be achieved. The major findings of remarkable coordination at the aggregate level, which strongly support the Nash equilibrium solution, coupled with considerable individual differences in frequency of entry that do not diminish with practice and individual decision rules that vary considerably from one player to another, have proved to be very robust to these systematic manipulations of the payoff function, information structure, and symmetry vs. asymmetry of the players.

The present study extends this line of research in yet a different and significant way by introducing *multiple* markets. Simultaneously faced with multiple capacities, one for each market, each player in this new class of games must decide whether or not to stay out of the market, thereby avoiding competition with the other group members and ensuring a fixed (positive or negative) payoff. Then, if she decides to enter, she must determine which market to enter. Games of this type capture a coordination problem that commuters face when determining which mode of transportation to take in order to arrive as quickly as possible at their destination. For example, when trying to reach point B from point A and knowing that the roads between these two points are expected to be congested, a commuter may face a choice between taking a train ("stay out") and arriving at point B in a fixed and known amount of time, or driving her car ("enter") in which case the length of the drive may vary considerably depending on the road she chooses to take and the decisions made by the other commuters. Firms face a similar problem when deciding whether or not to enter one of several emerging markets, which may or may not be equally attractive in terms of their profitability which, in turn, depends on their capacity. Tacit coordination in this class of games is considerably more difficult than in the MEG, which includes a single market and, therefore, only a binary choice.

When the capacities of the multiple markets are varied randomly from period to period, as in the present experiment, the players are faced with a sequence of one-play coordination games with no information to form expectations that can support any of the multiple pure strategy Nash equilibria. A major objective of the present study is to determine if, and to what extent, the symmetric mixed-strategy Nash equilibrium can account for the behavioral regularities in this sequence of one-play coordination games. If, as our previous results of the simpler MEG studies lead us to believe, individual players do not mix their strategies in a way consistent with symmetric equilibrium play, a second major objective is to determine if coordination in this class of games, or any other consistent pattern of behavior that emerges with repeated play, can be achieved by some process of adaptation.

Section 2 presents the multiple-market entry game and characterizes its pure and mixed Nash equilibria. We begin the investigation of this class of games with the simplest case of two markets; extensions to a larger number of markets are obvious. Section 3 describes the experimental procedure and reports the results from three different groups of twenty players each. These results show remarkable coordination by each group and provide strong support for the Nash equilibrium solution on the aggregate but not individual level. Section 4 presents and tests a reinforcement-based learning model, which incorporates the assumptions that the decision whether or not to enter is determined solely by the combined capacity of the two markets, whereas the decision of which market to enter is determined probabilistically. The major behavioral regularities reported in Section 3 are captured quite well by this learning model. Section 5 concludes with a discussion of the major findings.

2 A multiple-market entry game

Consider a group N of n symmetric players who participate in the following TMEG. At the beginning of each period t ($t = 1, 2, \ldots, T$), a pair of positive integers (c_A, c_B) is publicly announced. The values of c_A and c_B are interpreted as the "market capacities" of two independent markets A and B, respectively. In our experiment the capacities satisfy the constraint $1 \leq c_A + c_B \leq n$. Once the values of c_A and c_B are publicly announced, each player i $(i \in N)$ must decide whether to stay out ($d_i = 0$), enter market A ($d_i = A$), or enter market B ($d_i = B$). Decisions are made anonymously, with no communication between players. The TMEG is iterated for T periods with different capacity pairs presented on each period. On each block of ten periods these capacity pairs are chosen randomly and without replacement from a prespecified set. Individual payoffs are determined by the payoff function $H_i(\mathbf{d})$, which remains unaltered for all T periods:

$$H_i(\mathbf{d}) = \begin{cases} v, & \text{if } d_i = 0 \\ k_A + r_A(c_A - m_A), & \text{if } d_i = A \\ k_B + r_B(c_B - m_B), & \text{if } d_i = B \end{cases}$$

where $k_j, r_j (j = A, B)$, and v are real-valued commonly known constants that do not depend on the period t, m_j is the actual number of entrants in market j $(j = A, B)$, and $\mathbf{d} = (d_1, d_2, \ldots, d_n)$ is the vector of individual decisions.

The experiment reported below examines a special case of the TMEG where $k_A = k_B \equiv k$ and $r_A = r_B \equiv r$. Therefore, with the exception of the possible difference between the capacities c_A and c_B, the two markets are indistinguishable. The pair of market capacities presented on each period t are sampled without replacement from the set

$$\begin{aligned}(C_A, C_B) &= \{(2,2),(4,4),(6,6),(8,8),(10,10),(2,4),(2,8),\\ &\quad (6,8),(6,12),(4,16)\}\ .\end{aligned}$$

We shall refer to the game as *symmetric* if $c_A = c_B$, and *asymmetric*, otherwise. In the experiment reported below, five games are symmetric and five are not.

Equilibrium solutions

In characterizing the pure-strategy equilibria, we distinguish between two cases depending on whether or not $v - k$ is a non-negative integer multiple of r, i.e., whether or not $v - k = qr$ for some non- negative integer q. If there exists no such integer, then there are $n!/m_A^*!m_B^*!m_N^*!$ pure-strategy asymmetric equilibria, where m_A^* players enter market A, m_B^* enter market B, and $m_N^* = n - m_A^* - m_B^*$ players stay out. The values of m_j^* are determined from

$$m_j^* = \lfloor (rc_j + k - v)/r \rfloor, j = A, B$$

where $\lfloor x \rfloor$ is the integral part of x. If there exists such an integer q, then there exist $n!/m_A^*!m_B^*!m_N^*!$ pure-strategy equilibria with m_j^* entrants and, in addition, $n!/(m_A^* - 1)!(m_B^* - 1)!(m_N^* - 2)!$ equilibria with $m_j^* - 1$ entrants, where

$$m_j^* = (rc_j + k - v)/r, j = A, B\ . \tag{1}$$

In the first class of equilibria, players who enter are indifferent between staying in or moving out, whereas in the second class of equilibria, non- entrants are indifferent between staying out or entering one of the markets.

There also exists a symmetric mixed-strategy equilibrium, where each player either enters market A, enters market B, or stays out with respective probabilities p_A, p_B, and $1 - p_A - p_B$. To establish this equilibrium, note that the expected number of entrants to market j, conditional on player i entering this market, is $Ej = 1 + (n-1)p_j$. Player i's expected payoff from entering market j is, therefore, $k + r(c_j - Ej)$. In equilibrium this expression must equal v. This yields

$$p_j = (r(c_j - 1) + k - v)/r(n-1)\ , \quad j = A, B\ . \tag{2}$$

as the equilibrium probability of entering market j. The expected number of entrants into market j (np_j) is very close to m_j^*.

Two comments about the equilibria are in order. First, the values of m_j^* and p_j only depend on the capacity of market j, not on the capacity of the other market. This implication will be tested below. Second, the equilibria-whether pure or mixed-do not maximize total group payoff. The players can increase the total group payoff considerably by lowering the frequency of entry. For example, suppose that $k = 0$, $r = 2$, $v = 1$, and $n = 20$, as in our experiment, and that $(c_A, c_B) = (8, 8)$. Then (Eq. 1), the pure-strategy equilibrium calls for 7 players to enter market A, 7 other players to enter market B, and 6 more players to stay out, for a total group payoff of 34. If, instead, only 3 players enter each market, the total group payoff increases from 34 to 76. Similarly, the mixed-strategy equilibrium probability of entering either of the two markets in this case (Eq. 2) is 0.342, and the associated expected total group payoff is $vn = 20$. However, if each player enters with probability $p_j/2 = 0.171$, the expected total group payoff is more than doubled. If they actually mix their strategies, there is a strong incentive for all the n players to collude by lowering their probability of entry, thereby increasing their payoffs. This is a testable implication that we also examine below.

3 The experiment

Method

Subjects. Sixty subjects divided into three groups of $n = 20$ subjects each participated in the experiment. Subjects in Groups 1 and 2 consisted of undergraduate and graduate students at the University of Arizona, who volunteered to take part in a two-hour computer-controlled experiment on economic decision making with payoff contingent on performance. The subjects in Group 3 were MBA students who volunteered to participate in the experiment for the same monetary incentive. None of the subjects had participated in previous experiments on the MEG.

Procedure. The computer-controlled experiment was conducted at the Group Decision Making Laboratory at the University of Arizona. Upon arriving at the laboratory, the subjects were seated at one of twenty computer terminals located in a big conference room. Communication between the subjects was strictly forbidden. The instructions were presented on individual screens in front of the subjects. Hard copies of these instructions were made available to the subjects.

Each subject was endowed with 34 "francs" (a fictitious currency converted to US dollars at the end of the experiment at the rate of 10 francs = \$1.50). The subjects were instructed that money gained or lost during the experiment would be added or subtracted from their initial endowment. After they completed reading the instructions, the subjects played 100 iterations of the TMEG with payoff contingent on performance.

Each trial (period) consisted of two parts. In the first part, the two market capacity values c_A and c_B were displayed on the subjects' individual screens. To reduce the burden of calculation, the computer displayed on each period two tables, one for each market, listing the payoffs for all possible numbers of entrants, for the given values of c_j, ranging from 1 to 20.

In the second part of the trial, each subject was requested to type in his decision (A for entering market A, B for entering market B, and N for staying out). Once all the twenty subjects entered their decisions, each was informed of the values of m_A and m_B for the trial, his payoff for the trial, and his accumulated payoff (in francs) from the beginning of the experiment.

The ten pairs of market capacities (c_A, c_B) were sampled randomly and without replacement from the set (C_A, C_B) that included ten elements (trials 1 - 10 or block 1). The same sampling procedure was repeated for ten blocks. Thus, the subjects were presented with the same pair of market capacities (c_A, c_B) a total of ten times in a possibly different random order on each block. At the end of the experiment, the subjects were paid their accumulated payoff for the entire experiment plus a \$5.00 show-up fee. The mean payoff across the three groups, excluding the show-up fee, was \$19.80. Individual payoffs would have been exactly \$20.10 (134 "francs") if the subjects were to play the mixed equilibrium solution. The experiment lasted about 100 minutes.

Results

All of the analyses that we conducted indicated no significant differences among the three groups. In particular, we observed no differences between the participants in Groups 1 and 2, most of them undergraduate students, and the MBA students of Group 3, who had acquired some knowledge of individual decision behavior and game theory in their MBA classes. We conducted several statistical tests that compared the three groups to one another in terms of the distribution of individual total number of entries across the 100 periods, frequency distributions of number of entries for each game summed across the ten blocks, rate of learning across blocks measured in terms of the difference between the (pure strategy) equilibrium and observed frequency of entry, and frequency of switches (see below) between the two markets on two successive presentations of the same game. All of these tests do not reject of the null hypothesis of no differences among the three groups. Consequently, the results presented below are collapsed across groups.

In presenting the asymmetric games to the subjects, the low and high capacity markets were interchanged, with the labels A and B in five (randomly chosen) blocks, and B and A in the other five blocks. To facilitate presentation of the results, we shall henceforth identify A with the low capacity and B with the high capacity market.

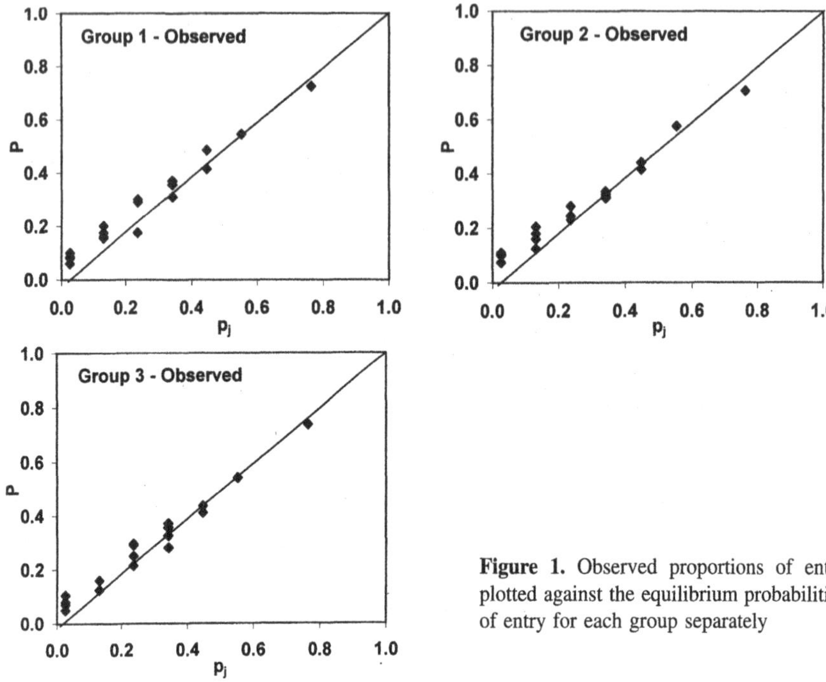

Figure 1. Observed proportions of entry plotted against the equilibrium probabilities of entry for each group separately

Frequency of entry: aggregate results. Table 1 (upper two panels) presents the mean number of entries across blocks by game, and within each game by market. The results are shown separately for the symmetric and asymmetric games. In each panel, the top row labels the market (A or B) and the second row lists the market capacity. The means and standard deviations of the observed number of entries computed across the three groups are presented in the next two rows. The fifth and sixth rows of each panel present (in boldface) the mean and standard deviation of the number of entries predicted by the mixed-strategy equilibrium solution. Thus, for example, when $c_j = 6$, $p_j = 0.237$ (Eq. 2) and the mean and standard deviation of number of entries for a group of 20 subjects are 4.74 and 1.90, respectively. The pure-strategy equilibria- which are not presented in the table–are always equal to $c_j - 1$.

The major finding is a remarkable coordination on the aggregate level. Figure 1 plots the observed proportions of entry, denoted by P, against the equilibrium probability of entry, p_j, computed from Eq. 2. Each data point is based on 200 observations. The mean observed frequencies of entries track the (pure or mixed) equilibrium values very closely across a wide spectrum of the combined market capacity values spanning the range from $c_A + c_B = 4$ to $c_A + c_B = 20$. Moreover, for the same value of c_j, there seem to be no significant differences between the two markets A and B. Also (Table 1), as predicted by the equilibrium solution, for the same value of c_j there are no significant differences between symmetric and asymmetric markets.

Table 1. Observed and simulated means and standard deviations of total number of entries

Observed

Symmetric games

	Game	(2, 2)		(4, 4)		(6, 6)		(8, 8)		(10, 10)	
	Market	A	B	A	B	A	B	A	B	A	B
	Capacity	2	2	4	4	6	6	8	8	10	10
Observed	Mean	1.80	1.57	2.83	3.50	4.97	4.37	6.90	6.60	8.27	9.07
Observed	SD	1.22	0.90	1.26	1.20	1.45	1.68	1.52	1.84	1.39	1.41
Predicted	Mean	**0.53**	**0.53**	**2.63**	**2.63**	**4.74**	**4.74**	**6.84**	**6.84**	**8.95**	**8.95**
Predicted	SD	**0.72**	**0.72**	**1.51**	**1.51**	**1.90**	**1.90**	**2.12**	**2.12**	**2.22**	**2.22**

Asymmetric games

	Game	(2, 4)		(2, 8)		(6, 8)		(6, 12)		(4, 16)	
	Market	A	B	A	B	A	B	A	B	A	B
	Capacity	2	4	2	8	6	8	6	12	4	16
Observed	Mean	1.83	3.47	1.60	6.43	5.77	6.77	5.50	11.07	3.07	14.43
Observed	SD	1.32	1.33	1.11	1.53	1.62	1.55	1.72	1.45	1.46	1.54
Predicted	Mean	**0.53**	**2.63**	**0.53**	**6.84**	**4.74**	**6.84**	**4.74**	**11.05**	**2.63**	**15.26**
Predicted	SD	**0.72**	**1.51**	**0.72**	**2.12**	**1.90**	**2.12**	**1.90**	**2.12**	**1.51**	**1.90**

Simulated

Symmetric games

		(2, 2)		(4, 4)		(6, 6)		(8, 8)		(10, 10)	
	Market	A	B	A	B	A	B	A	B	A	B
	Capacity	2	2	4	4	6	6	8	8	10	10
Simulated	Mean	1.27	1.17	3.17	2.40	4.57	4.40	7.20	7.27	8.43	9.00
Simulated	SD	0.94	0.79	1.42	1.38	1.25	1.40	2.16	2.23	1.83	1.98
Predicted	Mean	**0.53**	**0.53**	**2.63**	**2.63**	**4.74**	**4.74**	**6.84**	**6.84**	**8.95**	**8.95**
Predicted	SD	**0.72**	**0.72**	**1.51**	**1.51**	**1.90**	**1.90**	**2.12**	**2.12**	**2.22**	**2.22**

Asymmetric games

		(2, 4)		(2, 8)		(6, 8)		(6, 12)		(4, 16)	
	Market	A	B	A	B	A	B	A	B	A	B
	Capacity	2	4	2	8	6	8	6	12	4	16
Simulated	Mean	1.10	2.30	1.60	5.57	3.43	9.07	5.13	10.50	4.27	13.03
Simulated	SD	0.80	1.47	1.28	1.41	1.65	1.57	2.86	2.45	2.21	2.26
Predicted	Mean	**0.53**	**2.63**	**0.53**	**6.84**	**4.74**	**6.84**	**4.74**	**11.05**	**2.63**	**15.26**
Predicted	SD	**0.72**	**1.51**	**0.72**	**2.12**	**1.90**	**2.12**	**1.90**	**2.12**	**1.51**	**1.90**

Frequency of entry: individual results. Because the results presented in the upper two panels of Table 1 are averaged across subjects, they provide no information about individual differences. Our first attempt to assess individual differences was by counting the number of times (out of a maximum of 10) that each subject *stayed out* of the market on the ten iterations of the same TMEG. The frequency distributions of the individual total number of staying out decisions are presented in Table 2 for each of the five symmetric games separately.

Table 2. Observed and predicted frequency distributions of total number of staying out decisions by individual subjects across blocks for the five symmetric games

	Game (2,2)		Game (4,4)		Game (6,6)		Game (8,8)		Game (10,10)	
Frequency of staying out	Obs.	Pred.	Obs.	Pred.	Obs.	Pred.	Obs.	Pred.	Obs.	Pred.
10	32	34.94	22	4.06	10	0.10	3	0	2	0
9	9	19.41	7	12.54	3	0.88	1	0.01	0	0
8	3	4.85	4	17.45	8	3.57	4	0.18	0	0
7	4	0.72	4	14.39	6	8.56	4	0.72	2	0
6	3	0.07	3	7.78	3	13.48	2	2.74	1	0.01
5	3	0	3	2.89	5	14.56	6	7.12	1	0.11
4	1	0	3	0.74	6	10.92	4	12.86	0	0.79
3	2	0	4	0.13	3	5.62	6	15.92	6	3.86
2	1	0	4	0	2	1.90	5	12.93	4	12.29
1	1	0	2	0	7	0.38	10	6.23	9	23.21
0	1	0	4	0	7	0.03	15	1.35	35	19.73
Mean		8.32		6.83		5.33		3.25		1.33
Median		10		8		5.5		2.5		0

Table 2 describes the reason for the slight over-entry in markets with relative low capacities and slight under-entry in markets with relative high capacities that can be observed in Table 1. Whereas the majority of the subjects (32 out of 60) never entered either of the two markets in Game (2, 2), 12 subjects (20%) entered it at least 4 times. In equilibrium, practically no subject should enter either of the two markets in this game more than four times. With the number of entries truncated from below by zero, we observe a negatively skewed distribution of frequency of staying out with a median (10) that exceeds the mean (8.32). Similar results hold for Game (4, 4). The frequency distribution for Game (10,10) is also skewed but in the positive direction, with a similar difference between the mean (1.33) and median (0). Similar results hold for Game (8, 8).

Perhaps more important is the failure of the mixed-equilibrium solution to account for individual results. Table 2 shows the mixed- strategy equilibrium frequencies of staying out in the ten iterations of the game. In each of the five symmetric games, the difference between the observed and predicted frequency distributions-tested by the Chi Square test-is highly significant ($p < 0.01$). Whereas the aggregate results are described extremely well by the mixed-strategy equilibrium solution, individual subjects as a rule do not mix their strategies with the equilibrium probability. Rather, compared to this solution, there are many subjects who either enter too frequently or too infrequently, depending on the values of c_j. Similar results, not presented here, also obtain for the asymmetric markets. Additional tests of the mixed strategy equilibrium solution are presented below.

Individual total number of entries. Additional information about the magnitude of individual differences observed in all three groups is presented in Table 3. Columns 2–5 of this table present the observed frequency distributions of individual total number of entries across all ten games and all ten blocks (with a minimum of 0 for a player who never enters and a maximum of 100 for one who always enters). The right-hand column presents the theoretical frequency distribution of the mean number of entries computed across 2000 simulated players each who were programmed to play the mixed-strategy equilibrium. Comparison of the theoretical and observed results shows clearly that the individual differences between the real subjects are much too large to be accounted for by the symmetric mixed-strategy equilibrium. Whereas the symmetric mixed strategy equilibrium mean number of entries is for most cases between 40 and 60, the individual total number of entries varies considerably across the interval [0, 100]. However, in agreement with the aggregate results reported in Table 1, the three *means* of the individual total number of entries in Groups 1, 2, and 3 (56.30, 54.65, and 53.75) do not differ significantly from one another ($F_{93,20} < 1$). Nor does any one of them differ significantly from the equilibrium (pure or mixed) total expected number of entries.

Table 3. Frequency distribution of individual total number of entries by group

Class	Group 1	Group 2	Group 3	Total	Simulation[a]
0–9	1	0	1	2	0
10–19	0	1	0	1	0
20–29	1	0	2	3	0
30–39	2	6	2	10	0.02
40–49	3	1	4	8	5.86
50–59	5	4	3	12	13.57
60–69	3	2	2	7	0.55
70–79	2	3	3	8	0
80–89	2	2	1	5	0
90–100	1	1	2	4	0
Mean	56.30	54.65	53.75		51.58
SD	22.96	22.12	24.13		4.14
Predicted (pure strategy)	54	54	54		54

[a] Based on simulation of 2,000 players programmed to play the symmetric mixed strategy equilibrium.

Block effects in frequency of entry. Two additional analyses, each intended to explore the dynamics of play, are described next. Table 4 and 5 (upper two panels of each table) present the total number of entries for each of the two markets by block. The results are organized by symmetry (Table 4) or asymmetry (Table 5)

Table 4. Observed and simulated (by the learning model) number of entries by block and game for the symmetric games

Observed

Market A

Game	1	2	3	4	5	6	7	8	9	10	Mean	SD	Pr.
2,2	10	4	7	5	5	8	3	3	5	4	5.4	2.27	1.58
4,4	10	11	12	10	8	8	5	10	5	6	8.5	2.51	7.89
6,6	15	13	12	10	14	23	15	15	12	20	14.9	3.90	14.21
8,8	22	23	24	16	24	17	19	22	19	21	20.7	2.83	20.52
10,10	23	23	25	23	29	27	28	26	25	19	24.8	2.94	26.84
RMSE	4.30	2.75	2.69	3.72	2.37	5.12	1.71	1.40	2.48	4.57	1.98		

Market B

Game	1	2	3	4	5	6	7	8	9	10	Mean	SD	Pr.
2,2	3	5	6	3	3	5	5	7	4	6	4.7	1.42	1.58
4,4	6	11	10	10	16	14	9	11	9	9	10.5	2.80	7.89
6,6	17	14	9	22	13	5	15	11	15	10	13.1	4.70	14.21
8,8	16	18	19	20	17	24	18	26	18	22	19.8	3.22	20.52
10,10	32	23	30	29	23	26	26	27	27	29	27.2	2.90	26.84
RMSE	3.55	2.71	3.56	3.80	4.39	5.41	2.03	3.97	1.68	2.95	1.92		

Simulated

Market A

Game	1	2	3	4	5	6	7	8	9	10	Mean	SD
2,2	3	3	3	5	5	4	7	4	2	2	3.8	1.55
4,4	4	8	13	6	7	12	14	8	11	12	9.5	3.34
6,6	14	16	14	16	11	12	19	9	14	12	13.7	2.87
8,8	23	16	17	15	26	24	32	23	23	17	21.6	5.34
10,10	18	32	24	21	29	33	26	28	21	21	25.3	5.12

Market B

Game	1	2	3	4	5	6	7	8	9	10	Mean	SD
2,2	2	2	4	3	3	4	1	4	6	6	3.5	1.65
4,4	7	7	3	11	11	6	4	10	7	6	7.2	2.74
6,6	16	12	13	11	16	14	7	17	12	14	13.2	2.94
8,8	35	29	26	27	16	18	10	19	19	25	21.8	6.34
10,10	19	20	28	31	23	19	26	24	31	31	27.0	5.66

Table 5. Observed and simulated (by the learning model) number of entries by block and game for the asymmetric games

Observed

Market A

Game	Block 1	2	3	4	5	6	7	8	9	10	Mean	SD	Pr.
2,4	10	6	9	4	8	3	4	0	9	2	5.5	3.41	1.58
2,8	8	9	4	4	5	2	5	5	0	6	4.8	2.62	1.58
6,8	24	16	14	20	19	13	15	20	15	17	17.3	3.40	14.21
6,12	29	18	19	18	18	10	14	14	14	11	16.5	5.38	14.21
4,16	9	13	6	9	8	10	11	8	11	7	9.2	2.10	7.89
RMSE	9.08	4.18	3.40	4.25	2.27	2.36	3.09	3.68	2.78	2.91	1.74		

Market B

Game	Block 1	2	3	4	5	6	7	8	9	10	Mean	SD	Pr.
2,4	6	13	8	13	10	8	10	12	7	17	10.4	3.20	7.89
2,8	19	15	20	21	27	15	19	17	19	21	19.3	3.29	20.52
6,8	17	26	20	20	18	26	20	19	19	18	20.3	3.00	20.52
6,12	26	33	35	30	35	33	37	33	35	35	33.2	2.99	33.16
4,16	48	39	50	39	45	42	41	44	42	43	43.3	3.41	45.79
RMSE	3.50	4.71	2.08	4.07	3.35	3.87	2.99	2.64	2.15	4.49	1.68		

Simulated

Market A

Game	Block 1	2	3	4	5	6	7	8	9	10	Mean	SD
2,2	3	5	4	3	2	4	3	1	4	4	3.3	1.16
2,8	3	12	6	3	6	2	2	4	4	6	4.6	2.97
6,8	16	13	8	13	8	3	9	7	12	14	10.3	3.95
6,12	35	20	22	16	10	13	9	10	9	10	15.4	8.33
4,16	19	4	13	16	21	16	14	5	12	8	12.8	5.67

Market B

Game	Block 1	2	3	4	5	6	7	8	9	10	Mean	SD
2,4	2	5	7	7	9	7	8	10	7	7	6.9	2.18
2,8	17	8	15	19	16	20	20	18	18	16	16.7	3.50
6,8	26	25	30	24	27	34	28	30	25	23	27.2	3.36
6,12	18	30	26	31	39	32	36	34	35	34	31.5	5.93
4,16	34	49	39	36	31	36	38	47	40	44	39.4	5.74

of the game, and for each game by the particular market (A or B). Thus, for example, when Game (2, 2) was presented for the first time (in block 1), 10 of the 60 players entered market A and 3 entered market B (47 players stayed out). The corresponding frequencies in Game (10, 10) of block 1 are 23 and 32 (with 5 players staying out). The mean numbers of entries across blocks (third column form the right) are the same as those displayed in Table 1, once they are divided by 3.

With the exception of block 1, there is hardly any indication in the table for systematic improvement in coordination due to experience. For each of the two types of game (symmetric and asymmetric) and each of the two markets, we measured the deviation between predicted and observed frequencies by a root mean square error RMSE = $[\sum(f_o - f_p)^2/5]^{1/2}$, where f_o and f_p denote the observed and predicted frequencies, respectively. The RMSE scores are displayed in Tables 4 and 5 below the ten columns; they do not show any systematic tendency to decrease across blocks. Spearman correlations (not shown here) were computed for each market and game separately between the ten observed frequencies and the block number (1 through 10). Only 3 of the 20 correlations were significant ($p < 0.05$, two-sided t-test).

Switches in decision. On the individual level, we observe no such consistency in the decisions. Rather, across iterations of the same game individual subjects continued changing their decisions whether or not to enter and, if deciding to enter, which market to enter. To measure this block-to-block change in decision, consider any of the ten games presented on block t and code the subject's decision as N, if he stays out; A, if he enters market A; and B, if he enters market B. When the same game is presented (after ten periods on the average) on block $t + 1$, the subject is faced with the same decision task. We say that he *switches* his decision, if, for the same game, he responds k on block t and k' on block $t + 1$, where $k \in \{N, A, B\}$, $k' \in \{N, A, B\}$, and $k \neq k'$. We can construct a 3 by 3 (k by k') transition matrix separately for each subject and each pair of adjacent blocks. An analysis of the transition matrix between blocks t and $t + 1$ allows measuring the *consistency* of the subject's decision rule over iterations. If, for example, a subject uses the same deterministic decision rule for the entire duration of the game, no switches will be recorded for this subject; rather, only the cells on the major diagonal of the 3 by 3 transition matrices will include positive elements.

Table 6 (upper two panels) presents the 3 by 3 transition matrices with the observed frequencies summed across the 60 subjects. The matrices are presented by game, and for each game they are summed across the sixty subjects in the three groups and across the nine possible transitions from block t to block $t + 1$ ($t = 1, 2, \ldots, 9$). Consequently, there are $9 \times 60 = 540$ observations in each transition matrix. The frequencies of switches appear in the six off- diagonal cells of the transition matrix. For example, the subjects in Game (2, 2) switched their decisions from staying out to entering market A between successive presentations of the same game on blocks t and t+1 a total of 18 times, whereas in 21 times they

Table 6. Observed and simulated (mixed strategy equilibrium) number of switches across blocks by game

Observed: Symmetric games

	(2, 2)			(4, 4)			(6, 6)			(8, 8)			(10, 10)		
	N	A	B	N	A	B	N	A	B	N	A	B	N	A	B
N	**417**	18	14	**313**	26	26	**226**	27	37	**106**	35	37	**40**	14	14
A	21	**17**	12	16	**38**	25	27	**75**	27	27	**100**	59	16	**142**	71
B	14	9	**18**	37	11	**48**	39	32	**50**	40	50	**86**	19	69	**155**
	.163			.261			.350			.459			.376		

Observed: Asymmetric games

	(2, 4)			(2, 8)			(6, 8)			(6, 12)			(4, 16)		
	N	A	B	N	A	B	N	A	B	N	A	B	N	A	B
N	**343**	23	34	**252**	17	57	**138**	26	35	**52**	13	24	**41**	4	20
A	29	**11**	13	16	**10**	16	34	**71**	52	19	**72**	63	7	**36**	42
B	25	11	**51**	58	13	**101**	33	52	**99**	27	51	**219**	24	43	**323**
	.250			.328			.430			.365			.259		

Mixed strategy equilibrium: Symmetric games

	(2, 2)			(4, 4)			(6, 6)			(8, 8)			(10, 10)		
	N	A	B	N	A	B	N	A	B	N	A	B	N	A	B
N	**485**	13	14	**294**	49	55	**144**	66	70	**56**	58	60	**7**	25	27
A	14	**0**	0	50	**9**	9	69	**31**	30	58	**60**	64	27	**109**	106
B	13	0	**0**	54	10	**10**	68	32	**30**	60	64	**60**	24	109	**105**
	.100			.420			.620			.674			.590		

Mixed strategy equilibrium: Asymmetric games

	(2, 4)			(2, 8)			(6, 8)			(6, 12)			(4, 16)		
	N	A	B	N	A	B	N	A	B	N	A	B	N	A	B
N	**386**	12	59	**219**	8	117	**95**	56	77	**24**	26	64	**6**	14	37
A	12	**0**	2	9	**0**	4	54	**29**	45	27	**31**	67	14	**29**	84
B	59	2	**7**	116	5	**63**	79	43	**63**	64	68	**169**	37	84	**235**
	.270			.479			.654			.585			.500		

switched their decision from entering market *A* to staying out. Below each table is the proportion of switches in the game, which is simply the total frequencies in the off-diagonal cells divided by 540.

The analysis reveals several patterns. In the symmetric games (top panel), the frequency of switches between markets *A* and *B* sharply increases in the value of c_j: the combined frequencies of switches between these two markets are 21, 38,

59, 109, and 140 for Games (2, 2), (4, 4), (6, 6), (8, 8), and (10, 10), respectively. The number of switches between N and A is about the same as from N to B in all five symmetric games, showing no particular response bias in favor of one of the two markets. In the asymmetric games (second panel), we observe more switches in Game (6, 8) than in any other game. Particularly surprising is the high rate of switches in Games (6, 12) and (4, 16)), where one market is considerably more attractive than the other. These patterns are of particular importance for assessing the performance of any learning model that presumes to capture the dynamics of play.

The lower two panels of Table 6 present the theoretical frequencies of switching based on the results of 10,000 players each programmed to play the symmetric mixed-strategy equilibrium for 100 periods. The frequencies were multiplied by 540/10,000 (and rounded up to integers) so that the total of the frequencies in each 3 by 3 transition matrix would equal 540 and the observed and theoretical frequencies would be readily comparable. Comparison of the theoretical and observed results shows similar patterns in the switching data with regard to game effects. However, except of Game (2,2), the theoretical frequencies of switching (off diagonal elements) considerably exceed the corresponding observed frequencies.

Table 7 presents the same observed transition frequencies by block (rather than by game) summed across the ten games (rather than blocks) and all sixty subjects. As there are only nine transitions, the frequencies in each transition matrix sum up to 600. The major finding here is a gradual decline in the proportion of switches across blocks from a proportion of 0.42 in the transition from block 1 to block 2 to 0.267 in the final transition from block 9 to block 10. This trend is clearly inconsistent with mixed strategy equilibrium play. It is mainly due to the transitions between N and either A or B. The number of switches between the two markets, which assumes the values 73, 70, 79, 73, and 71 in the last five transition matrices, hardly changes across blocks. These results suggest that with experience, more subjects might have converged to using decision rules that allow for switches between markets but no switches between entering and staying out. We shall test this hypothesis in Section 4. The systematic decrease in the frequency of switches over time should also be accounted for by any adaptive learning model that attempts to capture the dynamics of play.

Cutoff decision rules. Each subject must first decide whether or not to stay out and then, if deciding to enter, which market to enter. Switches in decision between presentations of the same game on two successive blocks might be due to the first, the second, or both decisions. Our results suggest that most subjects slowly converge to using a deterministic decision rule as far as the first choice, but not the second, is concerned. However, if they decide to enter in the present study, then their choice of which market to enter may change as a function of their previous gains or losses in the same or different games. We tested the hypothesis that individual decision policies converge to a cutoff decision rule of the type "enter if and only if $c_A + c_B \geq c^*$", where c^* is an individual cutoff value of the

Table 7. Observed number of switches across games by block

	Block 1			Block 2			Block 3			Block 4			Block 5		
	N	A	B	N	A	B	N	A	B	N	A	B	N	A	B
N	**183**	26	41	**206**	32	29	**204**	13	44	**209**	28	37	**207**	25	23
A	43	**63**	54	16	**69**	51	28	**63**	42	14	**64**	41	29	**66**	43
B	41	47	**102**	39	31	**127**	42	43	**121**	32	46	**129**	45	30	**132**
	.420			.330			.353			.330			.325		

	Block 6			Block 7			Block 8			Block 9		
	N	A	B	N	A	B	N	A	B	N	A	B
N	**222**	22	37	**226**	24	31	**232**	12	26	**239**	21	30
A	24	**62**	35	17	**61**	41	29	**62**	32	12	**62**	41
B	35	35	**128**	27	38	**135**	29	41	**137**	26	30	**139**
	.313			.297			.282			.267		

sum of capacities of the two markets that may vary from one subject to another. This rule does not specify which market to enter, nor does it distinguish between symmetric and asymmetric games.

Even if the subject adheres to a cutoff decision rule, the ten games in the present experiment do not allow to determine c^* uniquely. If, for example, the subject always stayed out when $c_A + c_B \leq 6$ (Games (2, 2) and (2, 4)) but entered on all other games for which $c_A + c_B \geq 8$ (e.g., Games (2, 8), (4, 4), (6, 6)), then c^* may take any value in the interval (6, 8]. We estimated c^* to be the upper limit of the interval, i.e., 8 in the present example.

All the ten decisions of 35 of the 60 subjects (58.3%) in block 10 were accounted for, without a single violation, by this cutoff decision rule. The decisions of ten more subjects violated this rule only once (e.g., entering on Games (6, 12) and (10, 10), but staying out on Game (4, 16)). Table 8 (second row) presents the frequency distribution of the estimated cutoff values c^* for these 35 subjects. Thus, 7 of these 35 subjects always entered one of the two markets in all ten periods of block 10, 10 subjects entered if $c_A + c_B \geq 6$, and so on. Three of the 35 subjects never entered; the estimated cutoffs for these subjects were arbitrarily set at 21.

As if by some magical hand, the *estimated cutoffs* in block 10 are distributed uniformly on the set of values that the combined market capacity $c_A + c_B$ assumed in the present study. The null hypothesis of a uniformly distributed cutoff values for this subset of subjects could not be rejected by a Chi Square test ($p > 0.25$). If all the sixty subjects were to use cutoff rules with cutoff values sampled from this distribution, the Nash equilibrium solution would have been supported.

Table 8. Observed and simulated (by the learning model) frequency distribution of estimated cutoff values c^* in block 10

Estimated c^* value	4	6	8	10	12	14	16	18	20	21	Sum
Observed Frequency	7	3	1	5	3	4	3	4	2	3	35
Simulated Frequency	8	3	7	4	4	11	5	2	8	8	60

Taken together, the results reported above leave us with the following puzzle. On the one hand, the mixed-equilibrium solution describes the aggregate results extremely well. There is no indication for deviations from equilibrium play in the direction of fewer entry decisions that would result in considerably higher group payoffs. On the other hand, we observe considerable individual differences in the total number of entries (Table 3), frequency of switches (Tables 6 and 7), and estimated number of cutoff policies in the last block (Table 8) that cannot be accounted for by the symmetric mixed-strategy equilibrium, together with evidence for learning reflected in the decrease in number of switches across blocks.

4 Reinforcement-based adaptive learning

These behavioral regularities present a challenge, which we have tried to meet by specifying and then testing an adaptive learning model based on the principles of reinforcement learning. Our goal here is not to find the best fitting model; as there is an obvious trade-off-which many learning models opt to ignore-between goodness of fit and number of free parameters incorporated by the model, goodness of fit can always be improved by adding more parameters. Rather, our goal is to identify the minimum cognitive sophistication that is required to account simultaneously for the considerable individual differences in decision rules and coordination success on the aggregate level that we reported above. In pursuing this goal, strong emphasis is placed on parsimony of the model and psychological interpretability of its parameters. We accept the premise (e.g., Cooper and Feltovich, 1996) that the simplest model should be tried first, and that models postulating higher levels of cognitive sophistication should only be employed as the first ones fail.

We make no attempt to explicate the concept of "cognitive sophistication" (see Cooper and Feltovich, 1996, for such an attempt). Rather, we only make the intuitive claim that reinforcement-based learning models are cognitively less demanding than belief-based learning models. The latter models typically require the players to set up a probability distribution over the opponents' actions, update their beliefs, and choose a best-response strategy which maximizes their expected payoff, given their beliefs (Camerer and Ho, 1999). In contrast, reinforcement-based learning models make weaker demands on the rationality and reasoning ability of the players (Erev and Roth, 1998; Roth and Erev, 1995). They assume that players only care about the payoffs strategies yielded in the past, not about the history of play that resulted in these payoffs. Subjects are simply supposed to recall what worked well for them in the past, and then do it more frequently in the future.

Model specification

The model that we propose to account for individual learning in the TMEG combines elements from the reinforcement-based probabilistic approach to learning in games advocated by Roth and Erev (1995) and the deterministic approach to learning in the MEG (Rapoport et al., 1997). Whereas the decision whether or not to enter any of the markets in the TMEG is assumed to be made *deterministically* in terms of a cutoff point on the combined market capacity $c_A + c_B$, the decision which of the two markets to enter is assumed to be determined *probabilistically*. Specifically, the model assumes that the decision on each period is actually broken into two stages. In the first stage, the player decides whether or not to stay out: he enters if and only if the combined market capacity exceeds some individual cutoff value. This cutoff value is updated individually in a deterministic manner as explained below. In the second stage, if he decides to enter, a decision is made as to which market to enter. For asymmetric markets, the model assumes that this decision is made probabilistically in terms of two propensities that reflect the differential attractiveness of the two markets. When the two markets are symmetric, each is assumed to be entered with equal probability. These assumptions are now formalized below.

Preliminaries. We maintain the distinction between the symmetric ($c_A = c_B$) and asymmetric ($c_A \neq c_B$) TMEG. For the latter class of games, we distinguish between low (L) and high (H) markets, where $c_L < c_H, L \in \{A, B\}, H \in \{A, B\}$, and $L \neq H$.

The decision process. At the beginning of each period t, $t = 1, 2, \ldots, T$, each player $i \in N$ is assumed to be characterized by three non-negative numbers (the subject index i is suppressed): c_t, $P_t(L)$, and $P_t(H)$.

The decision whether or not to enter a market is made as follows:

If $c_t \leq c_{L,t} + c_{H,t}$, enter on period t,

If $c_t > c_{L,t} + c_{H,t}$, stay out on period t.

If condition $c_t \leq c_{L,t} + c_{H,t}$ is satisfied, then the decision *which* market to enter depends on whether or not the game is symmetric.

If the game is symmetric, each player i is assumed to enter either of the two markets with equal probability. If the market is asymmetric, he is assumed to enter either market L or market H with respective probabilities $q_t(L)$ and $q_t(H)$ that are proportional to the capacities: $q_t(L) = P_t(L)/(P_t(L) + P_t(H))$ and $q_t(H) = P_t(H)/(P_t(L) + P_t(H))$. Note that $q_t(L) + q_t(H) = 1$.

Outcome. If the game is symmetric, then set arbitrarily $A = L$ and $B = H$.

At the end of period t exactly one of four outcomes obtains:

Case 1: $c_{H,t} \leq m_{H,t} and c_{L,t} \leq m_{L,t}$ ("both markets are overcrowded").
Case 2: $c_{H,t} > m_{H,t} and c_{L,t} > m_{L,t}$ ("both markets are undercrowded").
Case 3: $c_{H,t} \leq m_{H,t} and c_{L,t} > m_{L,t}$ ("market H is overcrowded and market L is undercrowded").
Case 4: $c_{H,t} > m_{H,t} and c_{L,t} \leq m_{L,t}$ ("market H is undercrowded and market L is overcrowded").

In these four expressions, $m_{H,t}$ and $m_{L,t}$ denote the actual frequencies of entry into markets H and L on period t, respectively, where $0 \leq m_{H,t} + m_{L,t} \leq n$.

Updating of the propensities. In updating the two propensities $P_t(H)$ and $P_t(L)$ and the individual cutoff point c_t, the learning model distinguishes between two cases, depending on whether or not player i stayed out on period t.

Supposing that player i *entered* one of the two markets on period t. Denote the market he entered by $X(t)$ and the one he did not enter by $Y(t)$, where $X(t) \in \{H, L\}$, $Y(t) \in \{H, L\}$, and, of course, $X(t) \neq Y(t)$. The three parameters that characterize each player are assumed to be updated according to which of the four cases specified above obtained on period t.

If Case 1 obtains, then

$$c_{t+1} = c_t + w_t^+[v - (k + r(c_{X(t)} - m_{X(t)}))] ,$$

$$P_{t+1}(H) = P_t(H) ,$$

$$P_{t+1}(L) = P_t(L) .$$

In words, if both markets were "overcrowded" (and, consequently all subjects who entered loss money), then the individual cutoff increases in proportion to the difference in payoff between the outcome for staying out and the (negative) payoff for entering, whereas the two propensities $P_t(H)$ and $P_t(L)$ remain intact. Without specifying an upper limit on c_{t+1}, this updating rule allows for the possibility that $c_{t+1} > c_H + c_L$, resulting in no entry for any pair of capacities on period $t + 1$.

If Case 2 obtains, then

$$c_{t+1} = \max\left\{c_t + w_t^-[v - (k + r(c_{X(t)} - m_{X(t)}))], 0\right\} ,$$

$$P_{t+1}(H) = P_t(H) ,$$

$$P_{t+1}(L) = P_t(L) .$$

If both markets are "undercrowded" (and, consequently, all subjects who entered made money), then the individual cutoff value is assumed to decrease, whereas the two propensities remain unaltered. The equation for updating c_t has the same form as in Case 1, with the only exception that c_{t+1} is assumed to be bounded from below by zero.

If either Case 3 or Case 4 ("mixed" market) obtains, then

$$c_{t+1} = c_t ,$$

$$P_{t+1}(X(t)) = \max\{1, P_t(X(t)) + z_t[r(c_{X(t)} - m_{X(t)}) - r(c_{Y(t)} + m_{Y(t)})]\} ,$$

$$P_{t+1}(Y(t)) = P_t Y(t) .$$

In these two latter cases, where one market is overcrowded and the other is undercrowded, only the propensity of the market actually entered is assumed to change; the other propensity and the cutoff value remain unchanged. The term in square brackets is the difference between the payoffs associated with the two markets.

Supposing that the player *stayed out* on period t. Then,

$$c_{t+1} = \begin{cases} c_t - w_t^+ G_t, & \text{if } G_t < 0, \\ \max\{c_t - w_t^- G_t, 0\}, & \text{if } G_t \geq 0, \end{cases}$$

$$P_{t+1}(H) = P_t(H) ,$$

$$P_{t+1}(L) = P_t(L) ,$$

where G_t is the expected payoff (positive or negative) *foregone* by not entering. This expected payoff is determined from

$$G_t = q_t(H)[k + r(c_{H,t} - m_{H,t} - 1) - v] + q_t(L)[k + r(c_{L,t} - m_{L,t} - 1) - v] .$$

In words, if player i stayed out on period t, the propensities for entering either of the two markets are assumed not to change, whereas the cutoff value changes according to the sign of the expected payoff foregone by not entering on the previous period. The max operator for the case of $G_t \geq 0$ assures that the cutoff value remains non-negative.

Discounting of the weight parameters. The effects of the three weight parameters w^+, w^-, and z, are assumed to be discounted with time:

$$\begin{aligned} w_{t+1}^+ &= (1-d)w_t^+ , \\ w_{t+1}^- &= (1-d)w_t^- , \\ z_{t+1} &= (1-d)z_t , \end{aligned}$$

where $0 < d \leq 1$ is a common discount factor.

In summary, the learning model includes four parameters. Two of these (w^+ and w^-) determine the proportional change in the cutoff point for overcrowded and undercrowded markets, respectively. A third parameter (z) determines the rate of change in the propensity to reenter the same market when the markets on the previous period were "mixed". The fourth (d) parameter determines the discounting of the effects of the former three parameters. In its present form, the model assumes that the three parameters w^+, w^-, and z have the same values for all the n players and that all three of them are discounted at the same rate; these assumptions could be relaxed. The discounting assures that after considerable experience with the game the decision whether to enter one of the two markets (rather than which market to enter) will converge to a cutoff rule on the combined market capacity.

Initialization. The model allows for no individual differences in the rate of learning. The only differences between individual players are due to the initial cutoff values $c_{t=1}$. To express the fact that individual differences exist before the experiment commences, we assume that the initial cutoff values are randomly drawn from some prior distribution. We can make no statement about these heterogenous "home made" priors without having information about the history of the players before the game commences. As for the initial propensities for entering the markets, we set them up at $P_1(H) = P_1(L) = 1$. Because of the evidence reported above for slow learning, we restricted the discount factor by setting $d \leq 0.05$.

Model testing

Testing of the learning model was conducted individually. For each of the sixty subjects in Groups 1, 2, and 3, we first estimated the initial cutoff value $c_{t=1}$ from his decisions in the first ten periods of block 1. Next, for each group of subjects separately, we estimated the best fitting values of the four parameters $w^+_{t=1}$, $w^-_{t=1}$, z_1, and d in the following way. Using a high-speed computer, we conducted a systematic search of a four-dimensional grid of parameter values. For each combination of the four parameter values we simulated the results of each group member across the 100 periods using the leaning model equations for determining a decision and updating the cutoff value and the two propensities. At the end of each simulation, across the twenty players of each group, we computed the criterion measure of goodness of fit

$$Q = \sum_{t=1}^{T} \left[\left(m^{(s)}_{H,t} - m_{H,t} \right)^2 + \left(m^{(s)}_{L,t} - m_{L,t} \right)^2 \right] ,$$

where $m^{(s)}_{H,t}$ and $m^{(s)}_{L,t}$ denote the simulated number of entrants for the small and large markets on period t, respectively, $m_{H,t}$ and $m_{L,t}$ are as defined above, and summation is taken across all the 100 periods. After finding combinations of parameter values with relatively small values of Q, the search proceeded with a "fine grain" four-parameter grid. Random numbers were generated and used to determine which of the two markets to enter according to the values of the two propensities $P_t(H)$ and $P_t(L)$ when the game was asymmetric. When it was symmetric, each market was chosen with equal probability of 0.5.

Using this estimation procedure, we generated data for three groups each including twenty "simulated" players (called "stat-rats" by Bush and Mosteller, 1955), with parameter sets that vary between but no within group. To simplify the comparison between simulated and observed decisions, we generated only one set of simulated results for each group.

The comparison between observed and simulated decisions was, therefore, conducted separately for each group. Table 1 (lower two panels) presents the means and standard deviations of the simulated number of entries. The means and standard deviations are computed across all ten blocks. The results are presented in the same format as in the upper two panels of Table 1. Inspection of the lower two panels suggests that the learning model simulates the observed mean number of entries rather closely. This is clearly evident in Figure 2, which plots for each group separately and for the total across the three groups the simulated against the observed mean number of entries for each of the two markets in each of the ten games. The linear correlations between simulated and observed means for Groups 1, 2, 3, and for the total are 0.95, 0.95, 0.93, and 0.96, respectively ($p < 0.01$ in each case). The corresponding slopes of the regression line of the simulated on the observed means (0.845, 1.140, 0.917, and 0.974) do not differ significantly from unity ($p > 0.01$ in each case). These results provide strong support for the learning model. Comparison of the upper and lower parts of Table 1 together with a careful inspection of the four plots in Figure 2 shows that the only major discrepancy between simulated and observed mean number of entries is due to Game (6, 8). In this game, the learning model underestimates the frequency of entry in market A ($c_A = 6$) and overestimates it in market B ($c_B = 8$). The comparison further shows that the standard deviations of the simulated number of entries are only slightly higher than the observed values: whereas the standard deviations for the observed entry decisions range between 0.90 and 1.84, those for the simulated entry decisions range between 0.79 and 2.86.

Our next comparison concerns the effect of experience on decision. Tables 4 and 5 (lower two panels in each table) present the simulated number of entries for each group separately by game and block. Comparison of the standard deviations in the upper and lower panels of each table shows larger block-to-block variations in the simulated than the observed frequencies of entry. A line-by-line comparison shows that the simulated results follow the observed results rather closely. For both sets of data, in most of the markets there seem to be no systematic changes in frequency of entry across blocks. When the observed frequencies increase across blocks (see, in particular, market B of Games (2, 4) and (6, 12)), so do the simulated frequencies, and when they decrease across blocks (market A of Games (2, 4) and (6, 12)) again so do the simulated results. The only notable difference is in market A of Game (4, 4) where the trends in observed and simulated frequencies of entry across blocks have opposite directions.

Yet, another comparison between the two sets of data concerns the individual total number of entries across the 100 periods of the experiment. One would expect the learning model to account not only for the mean trends but also for the individual differences. Figure 3 displays side by side the frequency distributions of observed (presented in Table 3) and simulated individual total number of entry decisions. Although the simulated players seem to be slightly more heterogeneous than the observed subjects, with more simulated players having either very low (0 - 20) or very high (81 - 100) individual total number of entries, the difference between the two frequency distributions is not significant ($\chi^2(8) = 3.09, p > 0.8$).

The major difference between the observed and simulated data sets is in the frequency of switches in decision between adjacent blocks. Several analyses

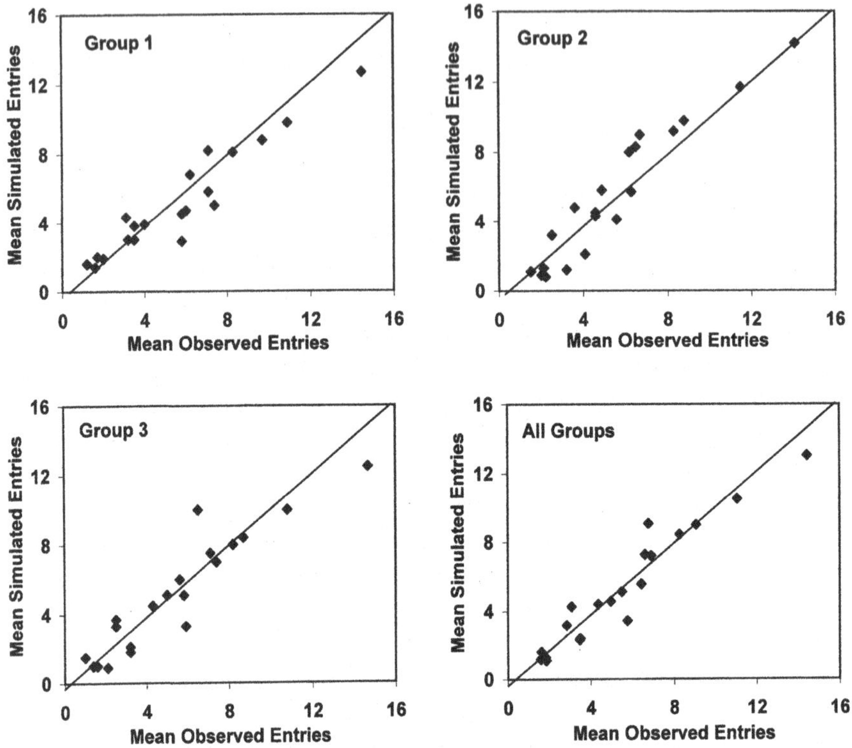

Figure 2. Scatter plots of simulated (by the learning model) against observed mean number of entries across symmetric and asymmetric games

not reported here show that the learning model underestimates the frequency of switches. (In contrast, as shown before, the mixed-strategy equilibrium solution overestimates it.) The major reason for this discrepancy is that the simulated players converge faster to cutoff decision rules (which allow for switches between markets A and B but not for switches between N and either A or B). Table 8 (bottom row) shows that all the sixty simulated players converged to cutoff rules by block 10 (with no violations) in comparison to only 35 of the real subjects. As the proportion of subjects using cutoff rules consistently gets higher, the frequency of switches between N and either A or B in the 3 by 3 transition matrices in Tables 6 and 7 approaches zero. Consequently, the discrepancy between the proportions of switches of real and simulated subjects is considerably reduced when the effect of the difference in the proportion of real and simulated players using cutoff decision rules consistently is accounted for.

Although the simulated subjects started playing on period 1 with estimated cutoff values that were not distributed uniformly and which kept being updated across trials, Table 8 shows that the learning mechanism postulated by our model caused these values to converge to a frequency distribution which is uniform across the values of the combined market capacity. This distribution constitutes a sufficient condition for equilibrium play. As noted earlier, the same results were obtained for the 35 real subjects who used cutoff rules consistently in block 10

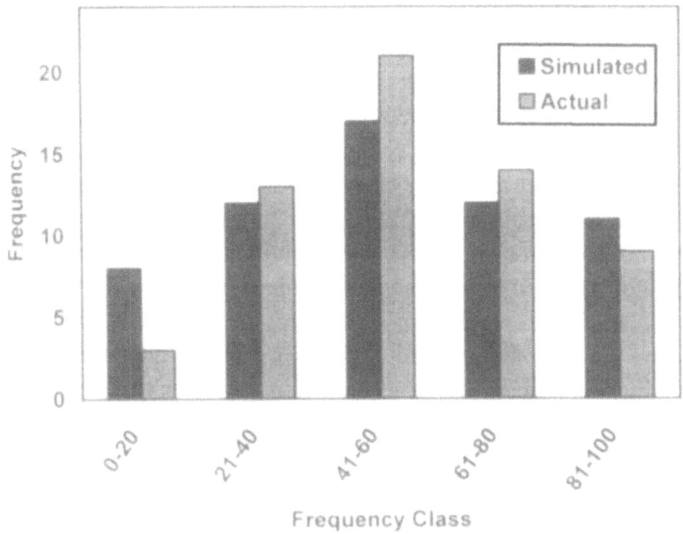

Figure 3. Frequency distribution of observed and simulated (by the learning model) individual total number of entry decisions

(see row 2 of Table 8). The null hypothesis that the frequency distribution of the sixty cutoff values of the simulated players in Table 8 (bottom row) is uniform could not be rejected ($\chi^2(9) = 12, p > 0.25$).

5 Discussion and conclusions

Three independent sets of twenty agents each repeatedly played a sequence of coordination games with two markets and capacity pairs that were varied randomly from period to period. Our results show that across all games, the symmetric Nash equilibrium hypothesis organizes the aggregate entry decisions remarkably well. The degree of coordination success achieved by our subjects exceeds the one reported by Meyer et al. (1992) and Ochs (1990), who devised different games to study coordination, and is of the same order of magnitude as reported in previous MEG studies with either symmetric or asymmetric players. We extended the original paradigm and devised this experiment with the expectation of eliciting coordination failure. Contrary to our expectation, the present results testify to the robustness and generality of the original finding of tacit coordination that is accounted for quite well by a Pareto deficient equilibrium solution.

Similarly to all of our previous experiments on tacit coordination in large groups, we find no evidence for collusion. This is somewhat surprising given the fact that, particularly in high market capacity games, collusion could double and even triple the subjects' payoff. We attribute this finding to the large size of the group that renders collusion impossible. Similar results were reported in the experimental literature on the n-person Prisoner's Dilemma game in which the equilibrium solution is also Pareto deficient. Whereas collusion is frequently observed in the repeated Prisoner's Dilemma game if $n = 2$, its frequency declines dramatically once $n \geq 3$.

We have hypothesized that each player perceives the TMEG as consisting of two rather than a single stage, where in the first stage he must decide whether or not to enter the market, and if deciding to enter, in the second stage he must determine which market to enter. Our results support this hypothesis. With experience gained in playing the TMEG, larger proportions of the subjects learn to solve the decision problem on the first stage by using cutoff decision rules in which the cutoff is stated in terms of the combined capacities of the two markets (Table 8). Although the players are symmetric, the individual cutoffs converge through some process of adaptation to different values whose distribution guarantees coordination success. These results are in line with those reported earlier by Rapoport et al. (1997) in their study of the MEG with asymmetric players in which 85 percent (compared to 58.3 percent in the present study) of the subjects used cutoff values consistently in the last block of trials. We attribute this discrepancy to the difference in the complexity of the two games, and conjecture that with more experience with the TMEG the proportion of players using cutoff decision rules will increase. In contrast, the results (Table 7) seem to suggest that the solution of the decision problem on the second stage is made probabilistically as some function of the relative capacities of the two markets. This combination of deterministic and probabilistic elements in the same adaptive learning model goes beyond the formulation of learning models in our previous market entry studies.

The adaptive learning model incorporates this two-stage decision process and organizes the basic behavioral regularities that we have observed quite well. In particular, it accounts for the coordination success on the aggregate level, the distribution of individual total number of entries, the convergence to cutoff decision rules stated in terms of the combined capacities of the two markets, and the dynamics of play captured by the analysis of switches in decision. The major difference between real and simulated players is due to a faster learning rate postulated by the model. Although we could have reduced this difference by increasing the number of parameters, introducing random error in the individual cutoff decision rules that diminishes with experience, or both, we have opted not to do so. Parsimony of the model has been judged at this stage to compensate for the possible gain in goodness of fit.

References

Bush, R.R., Mosteller, F.: Stochastic models for learning. New York: Wiley 1955

Camerer, C.F., Ho, T.H.: Experience-weighted attraction learning in games: A unifying approach. Econometrica **67**, 827–874 (1999)

Camerer, C.F., Lovallo, D.: Overconfidence and excess entry: An experimental approach. American Economic Review **89**, 306–318 (1999)

Cooper, D.J., Feltovitch, N.: Selection of learning rules: Theory and experimental evidence. University of Pittsburgh, Department of Economics, Working paper No. 305 (1996)

Erev, I., Roth, A.E.: Predicting how people play games: reinforcement learning in experimental games with unique, mixed strategy equilibria. American Economic Review **88**, 848–881 (1998)

Meyer, D.L., Van Huyck, J.B., Battalio, R.C., Saving, T.R.: History's role in coordinating decentralized allocation decisions. Journal of Political Economy **100**, 292–316 (1992)

Ochs, J.: The coordination problem in decentralized markets: An experiment. Quarterly Journal of Economics **105**, 545–558 (1990)

Ochs, J.: Coordination in market entry games. In: Budescu, D.V., Erev, I., Zwick, R. (eds.) Games and human behavior: Essays in Honor of Amnon Rapoport. Mahwah, NJ: Erlbaum 1999

Rapoport, A.: Individual strategies in a market entry game. Group Decision and Negotiation **4**, 117–133 (1995)

Rapoport, A., Seale, D.A., Erev, I., Sundali, J.A.: Equilibrium play in large group market entry games. Management Science **44**, 129–141 (1998)

Rapoport, A., Seale, D.A., Winter, E.: Coordination and learning behavior in large groups with asymmetric players. Hong Kong University of Science and Technology, Department of Marketing. Working Paper 97.098 (1997)

Roth, A.E., Erev, I.: Learning in extensive-form games: Experimental data and simple dynamic models in the intermediate term. Games and Economic Behavior **8**, 164–212 (1995)

Schelling, T.: The strategy of conflict. Cambridge, MA: Harvard University Press 1960

Sundali, J.A., Rapoport, A., Seale, D.A.: Coordination in market entry games with symmetric players. Organizational Behavior and Human Decision Processes **64**, 203–218 (1995)

Van Huyck, J.B., Battalio, R.C., Beil, R.O.: Tacit coordination games, strategic uncertainty, and coordination failure. American Economic Review **80**, 234–249 (1990)

Van Huyck, J.B., Battalio, R.C., Beil, R.O.: Strategic uncertainty, equilibrium selection, and coordination failure in average opinion games. Quarterly Journal of Economics **106**, 885–910 (1991)

Experience-weighted attraction learning in sender-receiver signaling games*

Christopher M. Anderson and Colin F. Camerer

Division of Humanities and Social Sciences, California Institute of Technology, Pasadena, CA 91125, USA (e-mail: cma@hss.caltech.edu)

Summary. We apply Camerer and Ho's experience-weighted attraction (EWA) model of learning to extensive-form signaling games. Since these games often have many equilibria, logical 'refinements' have been used to predict which equilibrium will occur. Brandts and Holt conjectured that belief formation could lead to less refined equilibria, and confirmed their conjecture experimentally. Our adaptation of EWA to signaling games includes a formalization of the Brandts-Holt belief formation idea as a special case. We find that the Brandts-Holt dynamic captures the direction of switching from one strategy to another, but does not capture the *rate* at which switching occurs. EWA does better at predicting the rate of switching (and also forecasts better than reinforcement models). Extensions of EWA which update unchosen signals by different functions of the set of unobserved foregone payoffs further improve predictive accuracy.

Keywords and Phrases: Learning, Game theory experiments, Signaling games, Equilibrium refinement.

JEL Classification Numbers: C72, C92.

1 Introduction

For a noncooperative game of any complexity, it is likely that people learn how to play the game through experience, rather than figure it out by reasoning. A

* This research was supported by NSF SBR 9511001. Thanks to Jordi Brandts and Charlie Holt for supplying their raw data. Helpful comments were received from audiences at the Universities of California (Berkeley) and Texas (Austin), Ohio State University, the Fall 1998 ESA Meetings, and from guest editor Charles Noussair and an anonymous referee.

general theory of learning is therefore crucial for understanding equilibration theoretically, and for explaining the changes in strategic behavior observed in the lab and in the field.

In this paper we apply Camerer and Ho's (1999) experience-weighted attraction (EWA) model to experimental data from sender-receiver signaling games. Signaling games are very widely used to model economic and political phenomena in which actions—perhaps apparently irrational ones leading to avoidable inefficiencies—are taken to convey asymmetric information. Applications include signaling product quality by price and advertising, strategic delay in strikes, excess capacity building by firms, choice of insurance policies, entry and pricing in monopolistic markets, incentives and personnel policies in labor markets with hidden action and hidden information, "money-burning" models of gift-giving, and many more (see, for example, Tirole, 1988, and Gibbons, 1992).

Signaling games are especially interesting because they often have many equilibria. The equilibria are theoretically distinguished by a variety of 'refinement' concepts which are routinely used to justify why some equilibria are empirically likely and others are not. However, these refinements usually assume players are reasoning particularly logically. But if players learn equilibria rather than figure them out, it is an open question whether their learning will lead to logically refined equilibria more often than unrefined equilibria. In our discussion, we present an example of how historical circumstances may lead all firms to offer dividends, a possible unintuitive equilibrium.

Experimental tests of refinement concepts have yielded somewhat pessimistic results about the ability of refinements, even fairly simple ones, to predict to which equilibria experimental subjects will converge (e.g., Banks, Camerer and Porter, 1994). Brandts and Holt (1992, 1993) suggest players in their experiments are instead using a particular learning process (essentially a form of belief learning) (see also Cooper, Kagel and Garvin, 1997a,b). They observed that players were led to an equilibrium which violated the Cho-Kreps 'intuitive criterion'. During equilibration, players left empirical 'footprints' at all information sets by choosing strategies which later turned out to be rarely chosen. When behavior eventually crystallized around the (unintuitive) equilibrium, and players thought about which types of players were likely to make out-of-equilibrium moves, they used their previous experience to form beliefs. In this game, these empirical beliefs contradicted purely logical arguments about which players would choose the out-of-equilibrium move. The learning process supports an equilibrium which is not supportable by standard game-theoretic logic.

Although Brandts and Holt's belief-based dynamic provides the intuition for 'unrefined' play, is it the best model to characterize the relationship between the history of play and eventual convergence? Camerer and Ho (1999a,b) introduced a general model of learning in games called 'experience-weighted attraction learning' (EWA). EWA hybridizes the two most popular approaches to learning in games—reinforcement and belief formation (like Brandts and Holt)—and includes these as parametric special cases. The EWA model has been estimated on 29 experimental data sets, and outperforms the familiar special cases in 25-27 of the 29 cases, correcting for extra parameters (see Camerer, Ho, and Hsia, 2000).[1] The general conclusion is that combining features of reinforcement with belief learning in a particular way is helpful for explaining observed learning.

Since the Brandts-Holt model is a special kind of belief learning model, and belief models are nested in EWA as a special case, by applying EWA to data from signaling experiments we can test the Brandts-Holt theory, and see whether adding additional EWA elements improves the fit. However, extensive-form games with incomplete information, like signaling games, demand special modifications which extend EWA's scope (see Vriend, 1997). The key problem is that players do not always know the foregone payoff to a signal which they did not choose (because its payoff depends on other players' reactions to the unsent signal, which is usually not known). Because foregone payoffs are used to update unchosen strategies, an extension is necessary when foregone payoffs are not known. Since players know the *set* of possible foregone payoffs, we extend EWA by reinforcing unchosen strategies according to some mixture of the foregone payoffs in that set.

Our paper therefore makes three contributions. We extend EWA to extensive-form games with incomplete information, in which there is imperfect information about foregone payoffs. We extend earlier experiments on signaling games, running them for 32 periods to see if sharper convergence occurs. Finally, we estimate the extended EWA model (which includes the Brandts-Holt dynamics as a special case) on the new data.

The paper is organized as follows. First we describe the games and the adaptive dynamics conjectured by Brandts and Holt. Then the EWA model is described and the modifications necessary to fit it to the signaling data are detailed. Data from new experiment are then presented and we investigate how well EWA, its various extensions, and the belief and reinforcement special cases, explain the data.

2 Adaptive dynamics and equilibrium selection

The main purpose of this paper is to apply EWA to signalling games, where it may be able explain how learning dynamics can lead players to unrefined equilibria.

This phenomenon is illustrated by two games taken from Brandts and Holt (1993), extending work by Banks, Camerer and Porter (1994) and Brandts and Holt (1992). Tables 1 and 2 show their Games 3 and 5. Nature chooses Type I or Type II (with equal probabilities) and the sender is told which half the table will be used to determine payoffs. The sender then selects m_1 or m_2, and the

[1] In-sample estimation is done in weak-link coordination games (Camerer and Ho, 1999a). Out-of-sample forecasting has been done for median-action coordination games and dominance solvable "p-beauty contests" (Camerer and Ho, 1999b), call markets (Hsia, 1998), "unprofitable games" (Morgan and Sefton, 1998), centipede games (Camerer, Ho and Wang, 1999), and bilateral call markets (Camerer, Ho, and Hsia, 2000). In some constant-sum games, EWA predicts slightly worse than belief learning (Camerer and Ho, 1999b).

receiver is notified of the sender's choice, but not the type. The receiver then chooses an action, a_1, a_2 or a_3. Payoffs are determined from the cell in the table described by the type-message-action triple; the sender's payoff is on the left and the receiver's payoff is on the right.

Table 1. Game 3

	Type I			Type II		
	a_1	a_2	a_3	a_1	a_2	a_3
m_1	45,30	15,0	30,15	30,30	0,45	30,15
m_2	30,90	0,15	45,15	45,0	15,30	30,15

Table 2. Game 5

	Type I			Type II		
	a_1	a_2	a_3	a_1	a_2	a_3
m_1	45,30	0,0	0,15	30,30	30,45	30,0
m_2	30,90	30,15	60,60	45,0	0,30	0,15

In both games, in the unintuitive sequential equilibrium both types of senders choose m_2. Receivers respond with $a_2|m_1$ and $a_1|m_2$.[2] In the sequential equilibrium which satisfies the Cho-Kreps (1987) intuitive criterion both types of senders choose m_1 and receivers respond with $a_1|m_1$ and $a_2|m_2$.[3]

2.1 Experimental data and the Brandts-Holt adjustment dynamic

In their Game 3 experiments, Brandts and Holt observe significant initial type separation—in the early periods, m_1 is three times more likely to be chosen by a t_1 sender than a t_2 sender. BH conjecture that senders start with a diffuse prior on what the likely action responses will be. With a diffuse prior, the expected payoffs for t_1s are 30 (=(45+15+30)/3) and 25 (=(30+0+45)/3) for the two messages, so t_1s tend to choose message m_1 more. The expected payoffs for t_2s are 20 and 30, so t_2s choose m_2 more often.

[2] Since both types choose m_2, receivers come to realize this, Bayesian-update and form posteriors $P(t_1|m_2) = P(t_1) = .5$ and $P(t_2|m_2) = P(t_2) = .5$. Their best response is then to choose a_1, which gives expected payoff 45 $(.5 \cdot 0 + .5 \cdot 90)$. The sequential equilibrium coheres only if receivers choose a_2 in response to message m_1, which only happens when receivers believe that m_1 choices were more likely to be made by t_2s (more specifically, $P(t_2|m_1) > 2/3$ to justify a choice of a_2 by receivers). This belief does *not* satisfy the intuitive criterion because t_2s earn 45 in equilibrium (from choosing m_2 and getting response a_1) and could not possibly benefit from switching to m_1, whereas t_1s earn 30 in equilibrium and could conceivably benefit if they choose m_1.

[3] In the intuitive equilibrium both types choose m_1 and are met with the response a_1, yielding t_1s 45 and t_2s 30. Since t_2s could conceivably earn more (45) by choosing m_2 instead, the equilibrium only sticks if defections to m_2 are met with responses of a_2. The a_2 response to m_2 can only be justified by the belief that m_2 defections are more likely to have come from t_2s (i.e., a_2 is optimal for receivers if $P(t_2|m_2) > 5/7$). This inference *does* satisfy the intuitive criterion because, indeed, t_2 types might benefit by defecting from m_1 to m_2 whereas t_1s would never benefit. Hence, the equilibrium in which both types choose m_1 satisfies the intuitive criterion.

If receivers also start with diffuse priors on which types chose a particular message, they should assign the highest expected payoffs to action a_1 in response to m_1, and a_1 in response to m_2. Empirically, however, they are more likely to choose a_2 in response to m_2. This happens quickly (in the first two periods) so it appears that receivers have anticipated the type separation, or learned it very quickly, and use it update their beliefs that a message m_2 choice came from a t_2, which makes the action response a_2 optimal. As the game continues and t_2 players continue to receive a_2 responses to m_2, they earn payoffs of 15 and begin to switch to m_1. In periods 9–12 all the t_1s pick m_1 and about 60% of the t_2s pick m_1, so equilibration goes reasonably swiftly in the direction of the intuitive sequential equilibrium in which both types pool on m_1.

That equilibrium is supported by the belief that a message m_2 would be chosen by a t_2 (who could conceivably benefit). The receiver chooses a_2 in response, yielding a lower payoff for the type t_2 than she receives from pooling on m_1, which keeps her from defecting. The intuitive criterion requires that this be deduced from the payoff table. However, in this game, this regularity is also revealed to players through the path of play: because of the initial type separation, most of the historical choices of m_1 *were* from t_2s, so there were few observations which conflicted with the intuitive criterion.

Game 5, in contrast, is designed so that observations which are likely to emerge from early disequilibrium play will conflict with the intuitive criterion. This is indeed what happens. In early periods, nearly all t_1 senders choose m_2 and most t_2s choose m_1. Receivers seem to anticipate, or learn quickly, that different types choose different messages, and they tend toward actions which are best responses given the type separation, i.e., $a_2|m_1$ and $a_1|m_2$.

Recall that in the unintuitive sequential equilibrium in which both types choose m_2, defection to m_1 is prevented if receivers think such a defection came from a t_2. Indeed, since the empirical probability of $m_1|t_2$ is high, their belief is justified by past experience (though it conflicts with the cold logic of the intuitive criterion). This highlights the need for a theory of equilibrium selection which includes a description of the convergence path and respects the way the observed convergence affects players' later beliefs. Without a story about how observations conflict with rational conjectures about beliefs, it is hard to explain this convergence to the less refined equilibrium.

In the Brandts and Holt dynamic (1993), players start with beliefs about what others will do (i.e., play each strategy with equal probability) and revise their beliefs in the light of what they observe. Cooper, Kagel and Garvin (1997a,b) give a similar explanation for results in limit pricing experiments.[4] This belief dynamic does explain the major features of the data. However, these earlier studies typically hypothesize particular parameter values, simulate paths for those values, and show that the simulated paths resemble the data. We improve on this procedure by estimating best-fitting parameters using maximum-likelihood. This technique yields standard errors for parameter estimates, and permits a formal

[4] Their analysis has an important twist: players assume others do not violate dominance.

hypothesis test of whether adding EWA features to belief learning improves accuracy significantly (as it has in 27 of 29 other data sets).

3 EWA learning

Experience-weighted attraction learning is a generalized reinforcement model which hybridizes elements of reinforcement and belief-based theories. Since it includes familiar models as special cases, it can be used as a statistical tool to compare theories, and ask whether adding more features improves their predictive accuracy. This section will highlight the important features of the model. See Camerer and Ho (1999b) or Camerer, Ho and Chong (2000) for more details.

In EWA learning, strategies have attraction levels which are updated according to either the payoffs the strategies actually provided, or some fraction of the payoffs unchosen strategies *would have* provided. These attractions are decayed or depreciated each period, and also normalized by a factor which captures the (decayed) amount of experience players have accumulated. Attractions to strategies are then related to the probability of choosing those strategies using a response function which guarantees that more attractive strategies are played more often.

3.1 EWA in the normal form

EWA was originally designed to study n-person normal form games. The players are indexed by i ($i = 1, 2, \ldots, n$), and each one has a strategy space $S_i = \{s_i^1, s_i^2, \ldots, s_i^{\ell_i - 1}, s_i^{\ell_i}\}$, where s_i denotes a pure strategy of player i. The strategy space for the game is the Cartesian products of the S_i, $S = S_1 \times S_2 \times \ldots \times S_n$. Let $s = (s_1, s_2, \ldots, s_n)$ denote a strategy combination consisting of n strategies, one for each player. Let $s_{-i} = (s_1, \ldots, s_{i-1}, s_{i+1}, \ldots, s_n)$ denote the strategies of everyone but player i. The game description is completed with specification of a payoff function $\pi_i(s_i, s_{-i}) \in \Re$, which is the payoff i receives for playing s_i when everyone else is playing the strategy specified in the strategy combination s_{-i}. Finally, let $s_i(t)$ denote i's actual strategy choice in period t, and $s_{-i}(t)$ the vector chosen by all other players. Thus, player i's actual payoff in period t is given by $\pi_i(s_i(t), s_{-i}(t))$.

The EWA model updates two variables after each round. The first variable is the experience weight $N(t)$, which is like a count of 'observation-equivalents' of past experience. The second variable is $A_i^j(t)$, the i's attraction to strategy j *after* period t has taken place. $N(t)$ and $A_i^j(t)$ begin with initial values $N(0)$ and $A_i^j(0)$, which are driven by pregame thinking due to introspection or learning transferred from similar games (e.g., Samuelson, 2000).[5]

[5] In a full model, the initial conditions would be determined by some theory of which decision rules players use (e.g., Costa-Gomes, Crawford and Broseta, 2000) or a disequilibrium theory like asymmetric response equilibrium (Weiszacker, 2000).

After a period of play, experience weights are updated according to

$$N(t) = \rho \cdot N(t-1) + 1. \tag{1}$$

Attractions are reinforced by a weighted payoff for i's j^{th} strategy, $[\delta + (1-\delta) \cdot I(s_i^j, s_i(t))] \cdot \pi_i(s_i^j, s_{-i}(t))$ (where $I(x,y)$ is the indicator function which equals one if $x = y$ and zero otherwise). The model weights hypothetical payoffs that unchosen strategies would have earned by a parameter δ, and weights the payoff actually received, from chosen strategy $s_i(t)$, by an additional $1-\delta$ (so it receives a total weight of 1). Updated attractions $A_i^j(t)$ are a depreciated, experience-weighted lagged attraction, plus an increment for the received or foregone payoff, normalized by the new experience weight. In formal terms,

$$A_i^j(t) = \frac{\phi \cdot N(t-1) \cdot A_i^j(t-1) + [\delta + (1-\delta) \cdot I(s_i^j, s_i(t))] \cdot \pi_i(s_i^j, s_{-i}(t))}{N(t)}. \tag{2}$$

The factor ϕ is a discount factor that depreciates previous attractions. Along with ρ, ϕ determines the limiting values of the attractions: if $\phi > \rho$, then attractions are not bounded by payoffs, and can grow arbitrarily far apart.

Finally, attractions determine choice probabilities using the logit form[6]

$$P_i^j(t+1) = \frac{e^{\lambda \cdot A_i^j(t)}}{\sum_{k=1}^{m_i} e^{\lambda \cdot A_i^k(t)}}. \tag{3}$$

The parameter λ measures sensitivity of players to differences among attractions: $\lambda = 0$ is equal likelihood and as λ increases, it converges to a best-response function.

3.1.1 Special cases of EWA

Simple algebra shows that parameter restrictions on $N(0), \delta$, and ρ reduce the general model to special cases of historical interest. For example, when $N(0) = 1, \rho = \delta = 0$ the model reduces to a simplified form of cumulative reinforcement in which only chosen strategies are reinforced (see Harley, 1981; Roth and Erev, 1995; Erev and Roth, 1998; Roth et al., 2000). When $\rho = \phi$ and $\delta = 1$, the model reduces to belief-based learning in which players form beliefs according to weighted fictitious play and choose strategies with high expected payoffs given those beliefs (e.g., Fudenberg and Levine, 1998, and many others).[7] When $\phi = 0$ the model corresponds to Cournot best-response dynamics, in which players

[6] Camerer and Ho (1998) show that the logit form fits slightly better than a power (or exponentiated ratio) form, and has the advantage of being usable even when attractions are negative.

[7] The key insight is that the belief updating equation can be written as a function of lagged beliefs and plugged the computation of expected payoffs. When the expected payoff computation is written as a function of its lag, the result is $E_i^j(t) = \frac{\rho \cdot N(t-1) \cdot E_i^j(t-1) + \pi(s_i^j, s_{-i}(t))}{\rho \cdot N(t-1)+1}$. The belief term magically disappears: since the only function of beliefs is to anticipate possible payoffs, and the belief updating is based on past observations, the expected payoff impact of those past observations can be mimicked by keeping track of a foregone payoff history directly. Beliefs are an unnecessary 'middleman'.

simply choose a best response to what happened in the previous period. When $\phi = 1$ the model corresponds to fictitious play, in which beliefs about what an opponent will do are an arithmetic average of what she has done in the past. This kinship between reinforcement and belief learning is surprising because most previous researchers had thought the two were fundamentally different. Instead, the EWA framework shows that belief learning is simply a kind of generalized reinforcement learning in which unchosen strategies are reinforced as strongly as chosen ones and reinforcements are weighted averages of lagged reinforcements and payoffs.

Many researchers who have studied these models have suggested that, while simple, they may be useful approximations. This begs the question of how to judge usefulness. We adopt conventional statistical criteria which enable us to judge precisely when adding parameters helps, or when omitting parameters hurts.

3.2 Extending EWA to signaling games

The first question to address in extending EWA to signaling games is what constitutes a strategy. In these games, we denote types by t_i, messages by m_j, and actions by a_k. The sender and receiver earn payoffs $\pi_S(t_i, m_j, a_k)$ and $\pi_R(t_i, m_j, a_k)$, respectively. Because senders observe their own types, it is appropriate to define their strategies conditional on observed types. There are two options.

First, one could define contingency strategies which specify a message for each type. For example, $(m_1|t_1, m_2|t_2)$ is a strategy in which the sender plays m_1 if t_1 is observed, and m_2 if t_2 is observed. This approach assumes that a sender chooses a complete strategy in each period (a strategy for each type), but only 'uses' the portion which is relevant for her observed type. In games in which complete strategies are elicited this modeling approach seems reasonable. However, in our experiments complete strategies are not elicited. The complete-strategy approach then begs the question of how to update attractions for several complete strategies which have the same 'used' portion but different 'unused' portions. For example, suppose the sender is t_1 and the chosen message is m_2. How does one update both $(m_2|t_1, m_1|t_2)$ and $(m_2|t_1, m_2|t_2)$?

We take a second approach, which is to assume that players have different strategy sets at each reachable node, which are not linked to form complete strategies. This is similar to the 'agent form' game in which each node is played by a different 'agent' for a single player, and all the agents have the same payoff.[8] In the example above, we simply update the attraction to $m_2|t_1$, and the attraction for the same message chosen by the 'unrealized type', $m_2|t_2$. In extensions of the model, we also allow updating of $m_1|t_1$, which the player could have chosen but did not, and $m_1|t_2$. Similarly, we assume receivers have strategies which are conditional on the message they observed the sender choosing, but not on the

[8] A referee wondered whether there is evidence that players act as if they use agent-normal-form reasoning. Note that our model allows the amount of cross-node dependence to vary parametrically, because δ, μ_1, and μ_2 express the degree of dependence of the choice at one node on future behavior at other nodes.

sender's type. A receiver's strategy to choose action k in response to message j will be denoted $a_k|m_j$.

Initial attractions for t_1 senders are denoted $A^{m_1t_1}(0)$ and $A^{m_2t_1}(0)$, and for t_2 senders, the initial attractions are $A^{m_1t_2}(0)$ and $A^{m_2t_2}(0)$. (In the logit form one of the attractions in each pair must be fixed for identifiability.) The initial experience counts are $N_S^{m_1}(0)$ and $N_S^{m_2}(0)$.

For receivers who observe message m_1, initial attractions are $A^{a_1m_1}(0)$, $A^{a_2m_1}(0)$, and $A^{a_3m_1}(0)$. For receivers who observe message m_2, the initial attractions are $A^{a_1m_2}(0)$, $A^{a_2m_2}(0)$, and $A^{a_3m_2}(0)$. (One of the attractions in each triple must be fixed for identifiability). The initial experience counts are $N_R^{m_1}(0)$ and $N_R^{m_2}(0)$.

3.3 The baseline model

This section discusses how the EWA model presented in Eqs. 1 and 2 can be adapted to signaling games. For receivers, this is a simple problem because they can condition only on the sender's message. Thus, the receivers know their foregone payoffs at the end of each period: they update their attraction to their chosen strategy with their realized payoff, and to other strategies with (δ times) the foregone payoff given by the actual type and message. If the receiver had chosen a_1 in response to m_1 when the sender was a t_2, for example, she would update according to:

$$N_R^{m_1}(t+1) = \rho \cdot N_R^{m_1} + 1 \tag{4}$$

$$A^{a_1m_1}(t+1) = \frac{\phi \cdot A^{a_1m_1}(t) \cdot N_R^{m_1}(t) + \pi_R(t_2, m_1, a_1)}{\rho \cdot N_R^{m_1}(t) + 1} \tag{5}$$

$$A^{a_2m_1}(t+1) = \frac{\phi \cdot A^{a_2m_1}(t) \cdot N_R^{m_1}(t) + \delta \cdot \pi_R(t_2, m_1, a_2)}{\rho \cdot N_R^{m_1}(t) + 1} \tag{6}$$

$$A^{a_3m_1}(t+1) = \frac{\phi \cdot A^{a_3m_1}(t) \cdot N_R^{m_1}(t) + \delta \cdot \pi_R(t_2, m_1, a_3)}{\rho \cdot N_R^{m_1}(t) + 1}. \tag{7}$$

Since she does not observe m_2, $N_R^{m_2}(t+1) = N_R^{m_2}(t)$ and $A^{a_km_2}(t+1) = A^{a_km_2}(t)$ for $k = 1, 2, 3$.

The sender's chosen strategies are updated according to the realized payoffs in the same way. However, with senders, it is more difficult to define foregone payoffs to unchosen strategies. There are two complications.

First, conditioning on a sender's type, the foregone payoff to the unchosen message is not known perfectly because it depends on the receiver's unobserved response. The sender knows the *set* of possible payoffs, but she does not know which payoff in the set would have resulted. Of course, this is generally the case in extensive-form games with unreached information sets. Below we consider several ways of choosing a foregone payoff in the set, or some mixture of those payoffs, to update the attraction on the unchosen message. In the baseline model, however, we simply leave attractions for unchosen sender messages unreinforced (and thus do not decay their experience counts).

The second complication is that belief models implicitly require that the attraction for the chosen message by the *unrealized type* also be updated by that type's foregone payoff. Remember that the sender knows the receiver's strategies may be message-dependent but cannot be type-dependent. Therefore, how a receiver reacts when a t_1 sender chooses a message informs the sender's belief about the receiver's reaction when a t_2 sender chooses the *same* message. For example, if a t_1 sender sends message m_1 and gets response a_1, she receives payoff $\pi_S(t_1, m_1, a_1)$ and updates $A^{m_1 t_1}$ accordingly. But she also knows that if she had been a t_2 and chosen m_1, she *would have* earned $\pi_S(t_2, m_1, a_1)$. Therefore, the sender updates according to

$$N_S^{m_1}(t+1) = \rho \cdot N_S^{m_1} + 1 \tag{8}$$

$$A^{m_1 t_1}(t+1) = \frac{\phi \cdot A^{m_1 t_1}(t) \cdot N_S^{m_1}(t) + \pi_S(t_1, m_1, a_1)}{\rho \cdot N_S^{m_1}(t) + 1} \tag{9}$$

$$A^{m_1 t_2}(t+1) = \frac{\phi \cdot A^{m_1 t_2}(t) \cdot N_S^{m_1}(t) + \delta \cdot \pi_S(t_2, m_1, a_1)}{\rho \cdot N_S^{m_1}(t) + 1} \tag{10}$$

and $N_S^{m_2}(t+1) = N_S^{m_2}(t)$ and $A^{m_2 t_k}(t+1) = A^{m_2 t_k}(t)$ for $k = 1, 2$.

In the belief learning restriction of EWA, cross-type updating occurs with $\delta = 1$. Just as EWA showed that belief learning is generalized reinforcement with 'full' reinforcement of unchosen strategies, belief learning in signaling games requires full reinforcement of unrealized types. If this seems behaviorally implausible, that implausibility should count as a strike against belief learning (as a predictive model).

The updating rules for the receivers are relatively straightforward. However, the notion of updating a foregone type, learning about a situation which did not occur (but could have), can be confusing. The models we propose in Section 3.4 are more complicated still because they suggest ways senders might update attractions for unchosen messages. Because updating rules for different combinations of realized and unrealized types and chosen and unchosen messages can get confusing, we will display update rules for senders in the game table. To illustrate the baseline update rule described above, we will use the following form:

Tabular representation of sender's baseline update rules

	Type I			Type II		
	$\underline{a_1}$	a_2	a_3	a_1	a_2	a_3
$\underline{m_1}$	$\frac{\phi A_S^{m1t1}(t) N_S^{m1}(t) + \pi_S(m_1, t_1, a_1)}{\rho N_S^{m1}(t)+1}$			$\frac{\phi A_S^{m1t2}(t) N_S^{m1}(t) + \delta \pi_S(m_1, t_2, a_1)}{\rho N_S^{m1}(t)+1}$		
m_2	$A_S^{m2t1}(t)$			$A_S^{m2t2}(t)$		

The underlined labels indicate that these rules represent an example where the realized type is t_1, the chosen message is m_1 and the chosen action is a_1. There are four cells in the table, one for each strategy-information set combination to which the sender has an attraction. In each cell is the attraction update rule for that cell given the the realized type and chosen message and response. In this case, Eq. 9 is represented in the upper left cell, where the sender increases her attraction with the full weight of the realized payoff. The upper right cell demonstrates how the foregone type is used: since the receiver's choice is message and not type dependent, the sender knows that had the type been t_2, she would have realized $\pi_S(m_1, t_2, a_1)$, but because this is only hypothetical, it is weighted by the imagination parameter, δ (Eq. 10). The cells in the lower row do not have an update rule, indicating that the attractions are just copied from one period to the next; there is no new information. To further simplify presenting the update rules, the denominator of the cells indicates how the experience counts are updated. For the baseline rule, the experience count of the chosen message is updated according to Eq. 8, and the experience count of the unchosen message is simply copied into the next period.

3.3.1 Special cases of EWA

The choice reinforcement and belief-based special cases of EWA discussed above apply, without much modification, to this adaptation of EWA for signaling games. The reinforcement model is still realized if $\delta = 0$, $\rho = 0$ and $N_R^{m_1} = N_R^{m_2} = N_S^{m_1} = N_S^{m_2} = 1$. However, the extension of the belief-based model is less obvious because it requires estimating initial belief counts rather than initial attractions.

In addition to setting $\delta = 1$ and $\phi = \rho$, we implement the belief model's implicit constraints on the $A(0)$s by estimating them indirectly: we estimate belief counts for each of the opponent's strategies and computing $A(0)$s by using these estimates to compute expected values. Thus, for the sender we estimate $N^{a_1 m_1}(0)$, $N^{a_2 m_1}(0)$, $N^{a_3 m_1}(0)$, which must sum to $N_S^{m_1}(0)$ and $N^{a_1 m_2}(0)$, $N^{a_2 m_2}(0)$, $N^{a_3 m_2}(0)$, which must sum to $N_S^{m_2}(0)$, and for the receiver we estimate $N^{m_1 t_1}(0)$, $N^{m_1 t_2}(0)$ which sum to $N_R^{m_1}(0)$ and $N^{m_2 t_1}(0)$, $N^{m_2 t_2}(0)$, which must sum to $N_R^{m_2}(0)$.

3.4 Unchosen message models

The appeal of the baseline model is that the sender is making all valid inferences: the receiver would have chosen the same action had the type been different, so the sender knows what her exact payoff would have been in the unrealized type case. However, the baseline model does not build in an answer to the sender's natural question, "Did I choose the right message, or should I have chosen the other message, given my realized type?" The alternative models presented here consider the possibility that a sender tries to force an answer to that question, using various imperfect inferences about what her payoff would have been had she chosen the other message.[9]

[9] Another model is that the sender assumes the receiver would have chosen the same (observed) action even if the sender had sent the other message. This neglects the sender's knowledge that the

3.4.1 Convex combinations of minimum and maximum payoffs

Another way to update foregone payoffs is to take a convex combination of the minimum and maximum possible payoffs. Because the weights used can vary from game to game, the model need not be sensitive to extremely high or extremely low outlier payoffs, yet it can be more robust to attractive payoffs than a median rule.[10] For example, if players frequently switch to unchosen messages, their 'grass-is-greener-on-the-other-side' switching could be captured by assuming they are optimistically putting a lot of weight on the maximum foregone payoff. Alternatively, if their message switching is slow their inertia could be modeled by assuming they are pessimistically putting a lot of weight on the minimum foregone payoff.

Let $\Pi(m_2, t_1) = \alpha MIN(\pi_S | m_2, t_1) + (1 - \alpha) MAX(\pi_S | m_2, t_1)$, where α is the parameter of the convex combination. Then the convex combination model can be written

Tabular representation of sender's convex combination model update rules

	Type I	Type II
	$\underline{a_1}$ a_2 a_3	a_1 a_2 a_3
$\underline{m_1}$	$\frac{\phi A_S^{m1t1}(t) N_S^{m1}(t) + \pi_S(m_1, t_1, a_1)}{\rho N_S^{m1}(t) + 1}$	$\frac{\phi A_S^{m1t2}(t) N_S^{m1}(t) + \delta \pi_S(m_1, t_2, a_1)}{\rho N_S^{m1}(t) + 1}$
m_2	$\frac{\phi^\tau A_S^{m2t1}(t) N_S^{m2}(t) + \mu_1 \tau \Pi(m_2, t_1)}{\rho^\tau N_S^{m2}(t) + \tau}$	$\frac{\phi^\tau A_S^{m2t2}(t) N_S^{m2}(t) + \mu_2 \tau \Pi(m_2, t_2)}{\rho^\tau N_S^{m2}(t) + \tau}$

The second row of the table gives the update rules for the unchosen message, for both the realized and unrealized type. The new parameters μ_1 and μ_2 are similar to δ; they represent the weight, or vividness of imagination, used in updating the attractions to the unchosen message for realized and unrealized types, respectively.

The other new parameter, τ, allows for the possibility that updating an unchosen message by the median foregone payoff does not have as much psychological impact as updating chosen messages, and hence is not the same as a single period of 'real' experience. In addition to being the increment to the unchosen message experience counter, τ is also an exponent of ϕ and ρ for unchosen messages and it multiplies μ_1 and μ_2. These additional appearances of τ in the updating equation allow it to be interpreted as the fraction of a period's experience in the unchosen message gained in conjecturing about and updating the unchosen message attraction. To see this, suppose that $\tau = 1$. The unchosen message rules then reduce to the chosen message rules with δ equal to μ_1 and μ_2 for the realized and unrealized types respectively. On the other hand, if $\tau = 0$, the unchosen

receiver's action choices could be message-dependent. It seems unlikely to fit the data better so we have not investigated it empirically.

[10] We also estimated a model where unchosen messages were updated with the median of the set of foregone payoffs. Results were virtually identical to the convex combination model presented here, and are available in Anderson and Camerer (1999).

message attractions are not discounted and no payoff is added to them, so they are unchanged.

3.4.2 Mirror sophistication: internal models of other players

A second model of unchosen-message foregone-payoff formulation assumes that players use all information available to them, including using their own behavior as a proxy for others'.[11] Since subjects played both roles in the course of the experiment, it is not necessary for senders to use a rule of thumb to guess about the receiver's response—a sender can appeal to the attractions of her receiver alter-ego's actions to compute the probability of each action in response to the unchosen message. Her expected payoff from the unchosen message will be the expected payoff from playing somebody like herself. We call this 'mirror sophistication' because players form a guess about what a player in another role will do by looking in a proverbial mirror at their own behavior when they were in that role.[12]

Let $p_S(a_1|m_1)$ denote the probability with which the sender's receiver alter-ego would choose a_1 given the message m_1. Let $\Pi(m_2,t_1) = \sum_{j=1}^{3} p_S(a_j|m_2) \times \pi_S(m_2,t_1,a_j)$. Then the simple sophistication model can be expressed as in the table below.

Tabular representation of sender's mirror sophistication model update rules

	Type I			Type II		
	a_1	a_2	a_3	a_1	a_2	a_3
m_1	$\frac{\phi A_S^{m1t1}(t)N_S^{m1}(t)+\pi_S(m_1,t_1,a_1)}{\rho N_S^{m1}(t)+1}$			$\frac{\phi A_S^{m1t2}(t)N_S^{m1}(t)+\delta\pi_S(m_1,t_2,a_1)}{\rho N_S^{m1}(t)+1}$		
m_2	$\frac{\phi^\tau A_S^{m2t1}(t)N_S^{m2}(t)+\mu_1\tau\Pi(m_2,t_1)}{\rho^\tau N_S^{m2}(t)+\tau}$			$\frac{\phi^\tau A_S^{m2t2}(t)N_S^{m2}(t)+\mu_2\tau\Pi(m_2,t_2)}{\rho^\tau N_S^{m2}(t)+\tau}$		

The parameters are interpreted exactly as in the previous model. However, this model departs slightly from the spirit of EWA because this implementation of mirror-sophistication implies a belief-based interpretation of attractions. In standard EWA, the learner never directly asks herself what her opponent will do in order to best respond, as she does when using a belief-based model. In

[11] Weber (2000) found this phenomenon in a repeated dominance-solvable beauty-contest game, when subjects made strategy choices repeatedly with *no feedback* after each choice. He found that players 'learned' even without feedback. Weber estimated a model in which one's own choice is taken as a proxy for what others might have done, and found a significant influence like mirror sophistication.

[12] Obviously, this rule will be sensitive to the experimental protocol and does not apply if the players do not switch roles. We regard this protocol-sensitivity as an advantage. There is a strong intuition among experimentalists that players do learn faster when they switch roles, which is supportive of such a rule. (For example, Binmore, Shaked and Sutton, 1985, found that second-moving bargainers learned to make subgame perfect offers in one trial after they reversed roles and became first-movers.) This is easily testable, by comparing experiments with different degrees of role-switching, but we know of no such experiments.

this model, however, the sophisticated learner does ask herself what she believes her opponent will do. Although the beliefs are still determined by EWA, this makes the mirror-sophistication model incompatible with a reinforcement interpretation. This is not particularly surprising because reinforcement learning uses only realized payoff streams, and has no further provision for thinking about how opponents adapt.

4 Experimental results

In order to test our baseline adaptation of EWA, its choice reinforcement and belief-based special cases, and our unchosen message updating extensions, we use Brandts and Holt's Games 3 and 5. However, while the 12 periods of data on 24 subjects they generated is sufficient to grasp the intuition behind the Brandts and Holt story, estimating a structural model as complex as EWA, and distinguishing it from special cases, requires more statistical power. Therefore, we replicated Brandts and Holt's (1993) Games 3 and 5 with 32 subjects playing 32 periods.

For our replication, we recruited Caltech undergraduates who did not necessarily have any training in economics, although many had participated in other experiments. We used a standard signaling game software which presented the game table as in Tables 1 and 2.[13] In each period, the senders were randomly selected, informed of the type and prompted for their message. When all senders had selected a message, receivers were notified of their paired sender's choice and were asked to choose a response. They knew that each type was equally likely *ex ante*. At the end of each period the realized cell of the payoff table was highlighted and subjects wrote down their payoffs. There were four cohorts of eight subjects, and we used a counterbalanced design, so two cohorts played Game 3 first and two played Game 5 first. Subjects earned an average of about $27 in about two hours, and were paid in cash as they left the laboratory. The only protocol difference between our experiment and Brandts and Holt's is that our pairings were random, with replacement; we made no attempt to ensure subjects did not play subjects they had played previously.

Figure 1 presents the data from our experiment, averaged across sessions in 4-period blocks. The results in Game 3 replicate BH closely, and confirm that with more experience, play converges reasonably sharply to the intuitive sequential pooling equilibrium at m_1, supported by action responses a_1 and a_2 to the two messages.

The Game 5 results are a little more surprising. We expected additional periods to cause t_2s to choose m_1 less and less frequently, cementing convergence to the unintuitive equilibrium at m_2. Our hunch was wrong: additional periods do not eliminate the separation between messages. The senders' strategies during the 13th through 32nd periods look much like the 9th through 12th periods of the Brandts and Holt data, suggesting the convergence to the sequential equilibrium

[13] The software we used also had a third message, which we instructed subjects never to use (they were compliant).

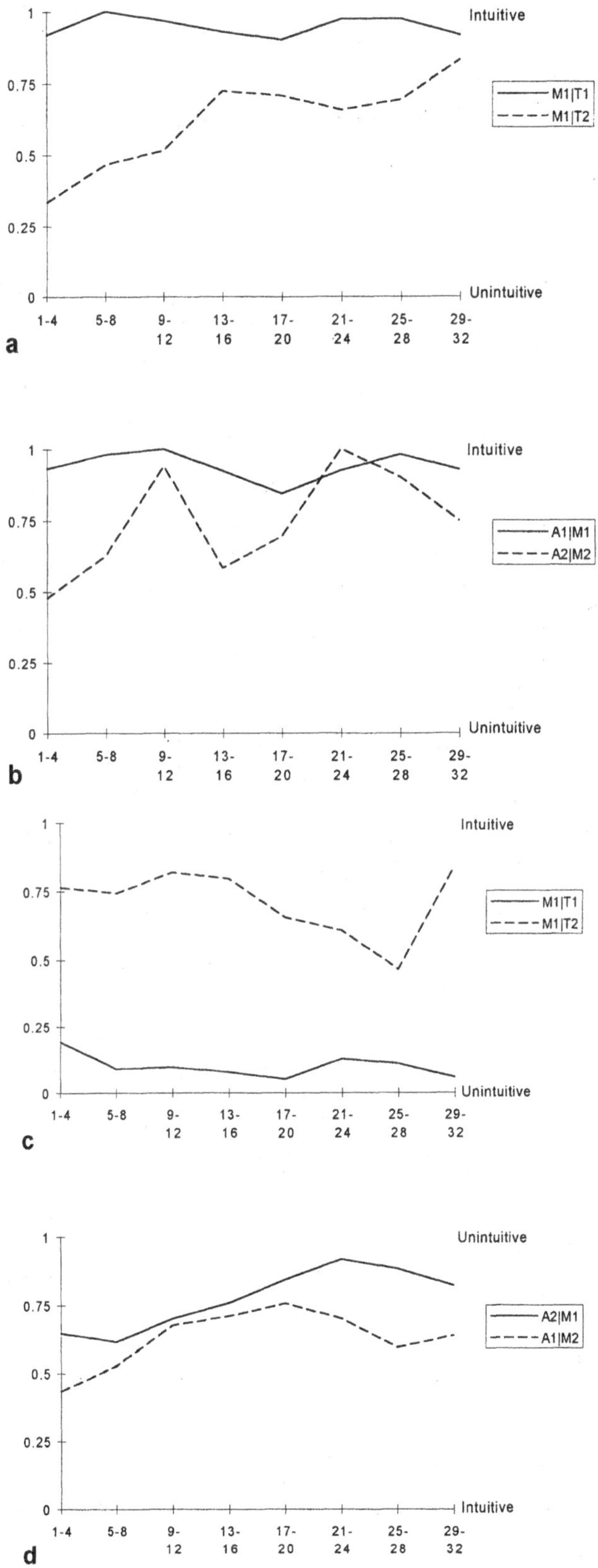

Figure 1. a Game 3 Sender. **b** Game 3 Receiver. **c** Game 5 Sender. **d** Game 5 Receiver

is not complete, even after many more periods of learning.[14] However, there is also no evidence of movement back toward the intuitive equilibrium at m_1.[15]

5 Estimation

In these models, the initial attractions, experience counts and model parameters can be estimated from our experimental data. We computed the maximum likelihood parameter estimates for each model using the constrained maximum likelihood procedure in Gauss (Aptech).[16] To simplify the estimation and make the models easily interpretable, we impose a number of restrictions on the parameter space. First, we impose bounds on the initial attractions, so that the set of possible attractions is not much larger for the EWA model than for the belief models (whose attractions are closely tied to the payoff structure).[17]

The second restriction is that message-specific experience weights should be the same for senders and receivers. That is, $N_S^{m_1}(0) = N_R^{m_1}(0)$ and $N_S^{m_2}(0) = N_R^{m_2}(0)$. While there is no *a priori* reason we think this is so, tests across a broad sample of data indicate that this is not a statistically significant restriction, and it saves two degrees of freedom.

[14] We also conducted a session with 64 periods of Game 5 only, to see if longer-run convergence was different than what we observed in only 32 periods. There was no additional movement toward either equilibrium.

[15] To test that 'zero-aversion' was preventing some subjects from switching to the unintuitive equilibrium, we ran eight subjects with a payoff table which added 15 to each payoff in Game 5 (call this Game 5'). These subjects converged to the unintuitive equilibrium in about 50 periods. However, we were concerned that the unintuitive equilibrium payoff of 105 may have been focal in that experiment, because (due to software constraint prohibiting three-digit numbers) it was indicated by a post-it note pasted on subjects' computer screens. Therefore, we ran three more sessions using a payoff table which multiplied payoffs in Game 5' by 4/5 (call this game 5*). Behavior in game 5* was indistinguishable from that in Game 5. We brought back experienced subjects from game 5* and ran them for 64 more periods, for a total of 128 periods, but did not observe further convergence.

[16] To ensure we found the peak of any local maximum we located, we used a two-step search process. From a given starting point, we used the Bernt, Hall, Hall and Hausman algorithm to search the parameter space. This algorithm estimates the Hessian, rather than calculates it exactly like Newton methods, and thus finds maxima quite quickly. However, it is not always precise. From the maximum found by the BHHH algorithm, we applied Gauss's version of the Newton gradient ascent algorithm. Using an exact (numerical) Hessian, this second algorithm often produced small improvements in fit. To ensure that the local maxima we found were global maxima, we tested a variety of starting points. We found the parameter space to be surprisingly well-behaved: in each model all of our starting points converged to the same maximum, suggesting that our estimates are in fact global maxima. The Gauss and C code used to estimate parameters is available from the first author.

[17] We look at the set of possible payoffs given the information available at the time of move and bound each initial attraction to be between the minimum and maximum attainable payoffs for each strategy. For instance, in Game 5, t_1 senders can earn payoffs $\{45, 0, 0\}$ from m_1, so $A^{m1|t1}(0)$ must be in the interval $[0, 45]$ and m_1 receivers can earn $\{45, 30\}$ from a_1, so $A^{a1|m1}(0)$ must be in $[30, 45]$. Because one of each type or message conditional strategy must be a constant in the logit form, we restrict one of the strategies in each information set to have an initial attraction equal to the minimum attainable payoff. There is no way to determine which strategy should have its attraction set to its minimum, so we estimated all possible combinations of these restrictions, and report only the one that yielded the best fit.

Our final restriction is that each of the $N(0)$s must be less than 50 and less than $\frac{1}{1-\rho}$. This prevents the model from putting so much weight on the initial attractions that there is almost no effect of the experience gained in the play of the game. Since $\frac{1}{1-\rho}$ is the asymptotic bound of $N(t)$ as t gets large, the restriction $N(0) \leq \frac{1}{1-\rho}$ forces the experience weight to increase, which means that new payoff information is getting less and less weight compared to lagged attractions; subjects do not have less perceived experience after playing the game than they brought into the game. Note that this restriction also requires $0 \leq \rho \leq 1$ (for positive N(0)).

Imposing these restrictions compromises the asymptotic normality of the maximum likelihood parameter estimates, so we construct bootstrapped confidence intervals using the percentile method.[18]

5.1 Fitting the baseline model

The objective of this paper is to test several models of how people update unchosen messages and unrealized types in signalling games. The focal point of this study is the baseline EWA model described in Section 3.3. First, we test the baseline model against models which are simpler, the choice reinforcement and belief-based special cases of EWA.

Table 3 presents the parameter estimates of EWA and and belief-based (BB) model for Game 3.[19] The predictions they generate are shown in Figure 2 (along with choice reinforcement). The estimates are performed on the first 24 periods of our data; the last eight periods are a holdout sample we try to predict. The most significant feature of the Game 3 data is that there is relatively little variance in the frequency of play of different strategies. The strategies $m_1|t_1$ and $a_1|m_1$ are played with virtually constant frequency throughout the game, $a_2|m_2$ is highly variable, but has no real trend and $m_1|t_2$ shows a steady increase. The parameter estimates show this lack of variance in the large initial experience counts and the depreciation parameters close to one. $\hat{N}^{m1}(0)$, in particular, achieves its maximum value, reflecting the relative stability of m_1 play. This stability is reinforced by $\hat{\phi} > \hat{\rho}$ which means that past attractions are amplified, so attractions are not bounded by payoffs and this convergence can be quite sharp, as it is with the m_1 data.

Table 4 presents several goodness of fit statistics which we use to compare models. The first row presents the average per period log-likelihood (summed across subjects) for the first 24 periods. This is the number that was minimized in

---

18 This nonparametric technique requires performing maximum likelihood estimates on a large number B data sets, where each data set is the sample with each subject weighted by a Poisson random number (Aptech 1995, 31). This process gives us B estimates of each parameter, and the 95% confidence interval for a parameter is given by that parameter's 2.5^{th} and 97.5^{th} order statistics. This method allows us to present the correct confidence intervals without knowing the transformation which would make the actual error distribution normal (Efron and Tibshirani 1993, 171).

19 Reinforcement model estimates are reported in Anderson and Camerer (1999). The only interesting parameter is $\hat{\phi} = .70$ in Game 3 and $\hat{\phi} = .68$ in Game 5.

Table 3. Parameter estimates for Game 3 (underlined values are fixed for identification or for model restrictions and bootstrapped 95% confidence intervals are in parentheses)

	EWA	BB		
		Value	Param	Count
δ	0.69 (0.47,1.00)	1.00		
ϕ	1.02 (0.99,1.04)	0.97 (0.97,0.99)		
ρ	1.00 (0.98,1.00)	0.97 (0.97,0.99)		
λ	0.41 (0.34,0.54)	0.32 (0.27,0.38)		
$A_S^{m_1t_1}(0)$	15.00	25.50	$N_S^{a_1m_1}(0)$	0.00 (0.00,0.00)
$A_S^{m_2t_1}(0)$	9.90 (9.04,10.72)	17.56	$N_S^{a_2m_1}(0)$	10.40 (10.39,10.41)
$N_S^{m_1}(0)$	50.00 (49.91,50.00)	34.64	$N_S^{a_3m_1}(0)$	24.24 (24.23,24.24)
$A_S^{m_1t_2}(0)$	14.72 (13.94,15.24)	20.99	$N_S^{a_1m_2}(0)$	0.00 (0.00,0.00)
$A_S^{m_2t_2}(0)$	15.00	20.85	$N_S^{a_2m_2}(0)$	15.90 (15.89,15.90)
$N_S^{m_2}(0)$	32.91 (32.81,32.94)	26.07	$N_S^{a_3m_2}(0)$	10.17 (10.17,10.18)
Receiver				
$A_R^{a_1m_1}(0)$	30.00	30.00	$N_R^{m_1t_1}(0)$	17.39 (17.39,17.39)
$A_R^{a_2m_1}(0)$	25.12 (24.61,25.74)	22.40	$N_R^{m_2t_1}(0)$	6.88 (6.86,6.88)
$A_R^{a_3m_1}(0)$	15.00 (15.00,15.00)	15.00	$N_R^{m_1t_2}(0)$	17.25 (17.24,17.25)
$N_R^{m_1}(0)$	50.00 (49.91,50.00)	34.64	$N_R^{m_2t_2}(0)$	19.19 (19.19,19.20)
$A_R^{a_1m_2}(0)$	20.08 (19.37,20.32)	23.76		
$A_R^{a_2m_2}(0)$	21.78 (21.16,21.87)	26.04		
$A_R^{a_3m_2}(0)$	15.00	15.00		
$N_R^{m_2}(0)$	32.91 (32.81,32.94)	31.57		

Table 4. Goodness of fit statistics for Game 3 (parameters estimated to minimize In Sample LL)

	EWA	BB	CR	Convex	Soph	Freq
Calibration						
In Sample LL	-12.06	-12.55	-14.15	-11.16	-11.17	-12.74
Out Sample LL	-9.02	-10.28	-9.83	-7.56	-7.55	-10.22
AIC	-12.56	-13.05	-14.48	-11.83	-11.80	-12.99
BIC	-13.71	-14.21	-15.26	-13.38	-13.25	-13.57
Fit						
In Sample Miss	0.147	0.184	0.177	0.131	0.134	0.178
Out Sample Miss	0.090	0.090	0.125	0.062	0.062	0.102
In Sample MSD	0.095	0.101	0.108	0.087	0.087	0.106
Out Sample MSD	0.063	0.076	0.082	0.051	0.052	0.083
DOF	12	12	8	16	15	6

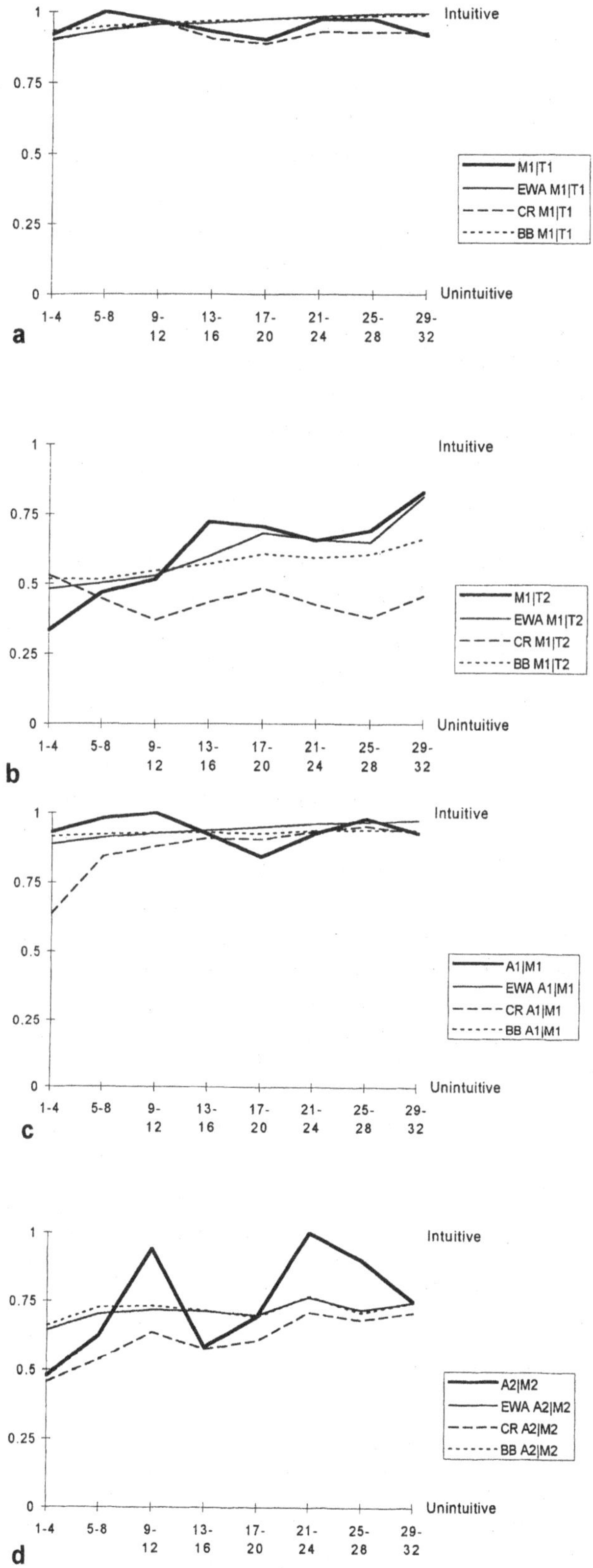

Figure 2. **a** Game 3 $m_1|t_1$. **b** Game 3 $m_1|t_2$. **c** Game 3 $a_1|m_1$. **d** Game 3 $a_2|m_2$

estimation. The second row presents the same statistic for the 25th through 32nd periods, the holdout sample into which we hope to predict. These statistics can be used to compare models, but there is no penalty for extra degrees of freedom (because a more general model will not automatically fit better out-of-sample). The third row presents the Akaike information criterion (AIC) and the fourth row presents the Bayesian information criterion (BIC).[20] These statistics can be compared directly and used for model calibration; they are designed to reach a maximum value at an optimal tradeoff between improvement of fit and additional parameters, even when models are non-nested.[21]

The second section of Table 4 presents more measures of fit. The first two rows present the in and out of sample 'miss rate'. The miss rate is the percentage of the time the subject *does not* pick the strategy that is predicted most likely to be chosen by the model. (It is one minus the hit rate.) The second two rows give the average per-period mean squared deviation.[22]

The columns of Table 4 all represent models discussed in this paper, except for the last one. The last column (Freq) is a model we propose for comparison. It is determined by taking as the model prediction the frequency of play of each strategy throughout the 24-period calibration sample; its predictions are horizontal lines determined by the data.

Choice reinforcement's out-of-sample miss rate and MSD are much worse than EWA and barely better than the frequency model. It does not fit the data well because it does not use foregone payoff information. Figure 2 shows that reinforcement mistakenly predicts t_2 senders move slightly away from m_1, when in fact they move strongly toward it. The reason is that m_2 is usually met with the response a_2, so it yields a payoff of 15 for t_2s. This payoff reinforces that choice positively and leads them to choose it again, moving them away from m_1. But in EWA and belief learning, what the t_1s learn from choosing m_1 influences the attractions for t_2 (through updating of the unrealized type attractions). Then t_2s gradually learn that message m_1 would pay 30, which is better than 15, and this indirect learning moves them toward m_1. As a result, we can see the AIC

[20] The AIC is the total in-sample log-likelihood minus the number of model parameters, divided by the number of sample periods (24). It is widely used for model comparison, but not motivated by any optimality considerations. The BIC is the total in-sample log-likelihood minus half the number of model parameters times the natural log of the number of observations, divided by the number of sample periods (24). Under certain regularity conditions (which are not satisfied if parameters are either estimated on or restricted to a boundary), the BIC can be interpreted as follows: if model i has a higher BIC than j, then $\exp\{-24 * (BIC_i - BIC_j)\}$ is an approximation to the posterior odds ratio, $\Pr(BIC_i)/\Pr(BIC_j)$, of a Bayesian observer with equal priors (Carlin and Louis, 1996).

[21] Note that the conventional way to make nested comparisons is a χ^2 test. We do not use it because the fact that some parameters are estimated and/or restricted to be on their boundaries violates the assumptions of the Central Limit Theorem necessary to show that $2(LL_i - LL_j)$ has a χ^2 distribution. However, the bootstrapped CIs from the model parameters which significantly influence fit (δ, ρ, ϕ) suggest different conclusions are very unlikely.

[22] This is calculated by creating, for each subject in each period, a vector with length equal to the number of strategies. The strategy chosen by the subject in that period is assigned a 1, and all others are zero. The MSD is the sum of squared differences between the created vector and the corresponding vector of choice probabilities predicted by the model, averaged across subjects and periods (but not across strategies).

and BIC both suggest that the additional parameters of EWA are more than justified by the improvement in fit, although the BIC also reflects the flatness of the data, suggesting that EWA does not represent a significant improvement over the frequency model.

The belief-based special case fits much better than the choice reinforcement model, but still not as well as EWA. Again, the belief model fails to adequately track the increasing frequency of play $m_1|t_2$. The problem is that the large value of $\hat{\phi}$ = .98, along with large initial experience counts, means it takes a lot of experience to alter the t_2 sender's beliefs, so the belief model does not allow learning which is fast enough. This subtle point also illustrates why we wanted to apply EWA to these type of data. The original BH story about belief formation is not precise about strength of prior, fictitious play weight, and other parameter values.[23] By estimating EWA one is forced to be very precise about the details of the model. It may not be possible to find configurations of belief model parameters which can fit the initial conditions, the basic trend, and also get the speed of convergence right. The sluggish belief learning of $m_1|t_2$ in Game 3 shows that while the belief account gets the direction right, it converges too slowly. Adding EWA parameters improves the fit considerably, producing a plot which hugs the data except at the start and in periods 13-16.

Table 5 presents the parameter estimates for Game 5. Figure 3 shows the predictions they generate. Table 6 presents goodness of fit statistics. Unlike Game 3, there is a significant trend to track in all the information sets. Look first at the central parameters, δ, ϕ and ρ. The estimated $\hat{\delta} = 0.54$, which is consistent with previous findings in games of complete information. This suggests that senders update the foregone type about half as much as they do their realized type. The depreciation parameters are also well within the range found in complete information games, and $\hat{\phi} = 0.68 > \hat{\rho} = 0.46$, which means that attractions are growing over time, and are not bounded by payoffs.

The initial experience counts are only 0.62 and 3.37. Together with depreciation parameters much less than one, low experience counts mean initial attractions are fairly quickly swamped by the experience gained in the play of the game. The initial attractions for the sender suggest the observed initial type dependence, and suggest receivers should respond to m_1 with a predominance of a_2, and to m_2 with about equal frequencies of a_1 and a_3 (which represent most of the non-a_1 responses to m_2). In this environment, where m_2 comes mostly from t_1s, a_3 has a lower expected utility for the receiver a_1, but it has the appeal of equity, which may be particularly strong in the case when subjects must switch roles from period to period.

Reinforcement does not fit well because it misses the gradual decrease in the frequency of play of m_1 given t_2 (until the sharp jump in the last block). Since m_1 gets reinforced by 30 for t_2, it is getting strongly reinforced. As in Game 3, the t_1 choices of m_2 demonstrate to players that if t_2s were to switch to m_2, they might get 45; this unrealized type reinforcement helps explain why

[23] In their entry games, Cooper et al. (1997a,b) do specify parameter values, as do Brandts and Holt (1994).

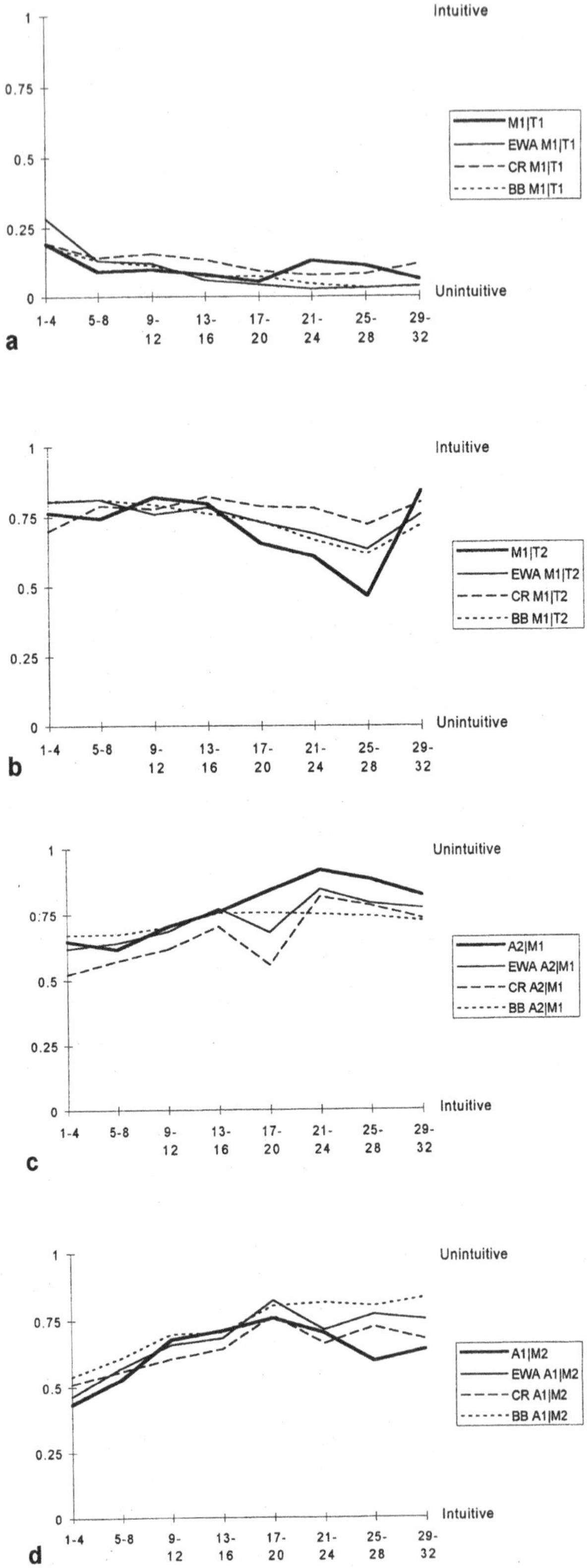

Figure 3. **a** Game 5 $m_1|t_1$. **b** Game 5 $m_1|t_2$. **c** Game 3 $a_2|m_1$. **d** Game 5 $a_1|m_2$

Table 5. Parameter estimates for Game 5 (underlined values are fixed for identification or for model restrictions and botstrapped 95% confidence intervals are in parentheses)

	EWA	BB		
		Value	Param	Count
δ	0.54 (0.45,0.63)	1.00		
ϕ	0.65 (0.59,0.71)	0.88 (0.88,0.94)		
ρ	0.46 (0.39,0.54)	0.88 (0.88,0.94)		
λ	0.09 (0.07,0.11)	0.15 (0.12,0.18)		
Sender				
$A_S^{m_1t_1}(0)$	18.25 (18.25,18.26)	34.70	$N_S^{a_1m_1}(0)$	2.18 (2.17,2.20)
$A_S^{m_2t_1}(0)$	30.00	44.62	$N_S^{a_2m_1}(0)$	0.00 (0.00,0.01)
$N_S^{m_1}(0)$	0.62 (0.59,0.66)	2.84	$N_S^{a_3m_1}(0)$	0.65 (0.62,0.67)
$A_S^{m_1t_2}(0)$	30.00	30.00	$N_S^{a_1m_2}(0)$	3.60 (3.58,3.61)
$A_S^{m_2t_2}(0)$	11.34 (11.34,11.34)	19.95	$N_S^{a_2m_2}(0)$	0.56 (0.54,0.59)
$N_S^{m_2}(0)$	3.37 (3.37,3.38)	8.12	$N_S^{a_3m_2}(0)$	3.96 (3.95,3.97)
Receiver				
$A_R^{a_1m_1}(0)$	30.00	30.00	$N_R^{m_1t_1}(0)$	0.63 (0.58,0.65)
$A_R^{a_2m_1}(0)$	37.26 (37.26,37.26)	35.04	$N_R^{m_2t_1}(0)$	3.06 (3.04,3.07)
$A_R^{a_3m_1}(0)$	0.00 (0.00,0.01)	3.32	$N_R^{m_1t_2}(0)$	2.21 (2.20,2.23)
$N_R^{m_1}(0)$	0.62 (0.59,0.66)	2.84	$N_R^{m_2t_2}(0)$	5.06 (5.05,5.08)
$A_R^{a_1m_2}(0)$	41.88 (41.88,41.88)	33.93		
$A_R^{a_2m_2}(0)$	15.00	24.35		
$A_R^{a_3m_2}(0)$	43.26 (43.26,43.26)	31.96		
$N_R^{m_2}(0)$	3.37 (3.37,3.38)	8.12		

they switch. By leaving out unrealized type reinforcement, choice reinforcement cannot account for the basic trend in $m_1|t_2$. Similarly, it is too slow to adjust to the initial decrease and subsequent increase in the frequency of a_2 given m_1, so it must fit the initial periods with an essentially smooth function. Table 6 shows that adding EWA parameters to reinforcement improve in- and out-of-sample fit by every statistic.

The belief-based model does a better job of capturing the gradual decrease and sudden increase in the frequency of play of m_1 given t_2; it does almost as well as EWA. However, it predicts essentially constant play for a_2 in response to m_1, and an essentially constant rate of increase for a_1 in response to m_2. This mirrors our findings for Game 3: the belief-based model, which formalizes the BH

Table 6. Goodness of fit statistics for Game 5 (parameters estimated to minimize In Sample LL)

	EWA	BB	CR	Convex	Soph	Freq
Calibration						
In Sample LL	-13.49	-16.41	-15.26	-13.13	-13.15	-18.00
Out Sample LL	-17.44	-20.40	-19.20	-17.69	-17.29	-18.60
AIC	-13.99	-16.91	-15.60	-13.80	-13.78	-18.25
BIC	-15.15	-18.07	-16.37	-15.35	-15.23	-18.83
Fit						
In Sample Miss	0.181	0.221	0.190	0.171	0.172	0.253
Out Sample Miss	0.184	0.227	0.199	0.191	0.191	0.254
In Sample MSD	0.105	0.127	0.116	0.101	0.101	0.146
Out Sample MSD	0.113	0.138	0.127	0.116	0.113	0.145
DOF	12	12	8	16	15	6

dynamic, gets the direction right, but adding the flexibility of EWA substantially improves tracking of convergence.

5.2 Fitting the unchosen message models

The estimates show that EWA is not too complicated (and that reinforcement and belief models are 'too simple'), because the baseline EWA model offers significant improvement over the special cases. Now we look at the more complicated alternative models of unchosen message reinforcement.

Table 7 presents the parameter estimates for Game 3 for the two alternative models.[24] While they both offer significant improvement over EWA, they have similar parameter estimates and offer similar fits. The fit statistics in Table 4 show that there is a significant feature of senders' behavior EWA is not capturing, and both unchosen message models capture it equally well.

The unchosen message models make slight improvements over EWA in a number of places. We highlight one such improvement to convey the subtle nature of the dynamic the unchosen message models predict.

The $m_1|t_1$ time series (Figure 2a) shows that t_1s almost always choose m_1, but occasionally choose m_2 (0-10% of the time, depending on the block). Starting with period 13, the m_2 rate fluctuates from about 3-10% and increases over time. The EWA model underpredicts this rate, starting about 4% and decreasing over time, to 1% in the last few periods. The unchosen message models do much better because they predict that the $m_2|t_2$ rate is around 4%, and slightly increasing over time.

How do the unchosen message models keep $m_2|t_2$ around while EWA all but extinguishes it? Consider how EWA updates when t_1s choose m_1, which they

[24] We do not include figures for the unchosen message models because their predictions are very similar to EWA and to each other in both games. Figures are available in Anderson and Camerer (1999).

Table 7. Parameter estimates for Game 3 (underlined values are fixed for identification or for model restrictions and bootstrapped 95% confidence intervals are in parentheses)

	EWA	Convex	Soph
δ	0.69 (0.47,1.00)	0.59 (0.56,0.64)	0.48 (0.33,0.66)
ϕ	1.02 (0.99,1.04)	0.77 (0.76,0.81)	0.76 (0.71,0.81)
ρ	1.00 (0.98,1.00)	0.98 (0.98,1.00)	0.98 (0.98,1.00)
λ	0.41 (0.34,0.54)	1.11 (1.10,1.15)	1.20 (1.13,1.36)
τ		1.07 (1.04,1.11)	1.05 (0.95,1.15)
μ_1		0.00 (0.00,0.00)	0.00 (0.00,0.01)
μ_2		0.46 (0.41,0.50)	0.57 (0.48,0.72)
α		0.00 (0.00,0.00)	
Sender			
$A_S^{m_1t_1}(0)$	15.00	15.00	15.00
$A_S^{m_2t_1}(0)$	9.90 (9.04,10.72)	12.52 (12.51,12.52)	12.79 (12.77,12.81)
$N_S^{m_1}(0)$	50.00 (49.91,50.00)	29.32 (29.32,29.32)	32.35 (32.35,32.35)
$A_S^{m_1t_2}(0)$	14.72 (13.94,15.24)	14.59 (14.58,14.59)	14.57 (14.48,14.57)
$A_S^{m_2t_2}(0)$	15.00	15.00	15.00
$N_S^{m_2}(0)$	32.91 (32.81,32.94)	50.00 (50.00,50.00)	50.00 (50.00,50.00)
Receiver			
$A_R^{a_1m_1}(0)$	30.00	30.00	30.00
$A_R^{a_2m_1}(0)$	25.12 (24.61,25.74)	27.38 (27.38,27.38)	27.55 (27.52,27.57)
$A_R^{a_3m_1}(0)$	15.00 (15.00,15.00)	15.00 (15.00,15.00)	15.00 (15.00,15.00)
$N_R^{m_1}(0)$	50.00 (49.91,50.00)	29.32 (29.32,29.32)	32.35 (32.35,32.35)
$A_R^{a_1m_2}(0)$	20.08 (19.38,20.32)	17.20 (17.19,17.21)	16.97 (16.93,17.03)
$A_R^{a_2m_2}(0)$	21.78 (21.16,21.87)	17.81 (17.80,17.82)	17.52 (17.46,17.57)
$A_R^{a_3m_2}(0)$	15.00	15.00	15.00
$N_R^{m_2}(0)$	32.91 (32.81,32.94)	50.00 (50.00,50.00)	50.00 (50.00,50.00)

almost always do. Because of the receivers' actions, $m_1|t_1$ is almost always updated with its highest payoff, 45. What happens to the attraction of $m_2|t_1$? Since m_2 is rarely chosen by t_1s, most of the updating of that attraction comes from unrealized type updating when t_2s choose m_2. Because the unrealized type reinforcement is estimated to be strong ($\hat{\delta} = 0.69$), and given how receivers respond to m_2, the attraction for $m_2|t_1$ usually gets reinforcement of 0, and sometimes $\delta \cdot 30$ (about 21). Because $m_1|t_1$ is typically getting reinforced by 45, and $m_2|t_1$ is typically getting reinforced by 0 or 21, EWA predicts $m_2|t_1$ gets more and more infrequently chosen over time.

The unchosen message models correct this subtle 'overlearning' predicted by EWA by using choices of a different message by a different type, $m_1|t_2$, as a cognitive opportunity to think again about the possible payoffs from m_2 for t_1s. In the convex model, for example, μ_2 is estimated to be 0.57 (and all the weight is on an unchosen message's highest payoff), so that when $m_1|t_2$ is chosen, $m_2|t_1$ is reinforced by 0.57 (45), around 26. Updating of 'distant' choices is like a reminder that a message which is rarely chosen by a particular type may yield a good payoff after all. Since that message-type combination is not directly reinforced very often, this indirect reinforcement is necessary to maintain a substantial probability that it may occur in the future. The baseline model does not allow this kind of reinforcement, and hence, overpredicts how quickly $m_1|t_1$ is distinguished by direct experience. The difference is small in percentage terms, but is important in statistical estimation and gives a substantial predictive advantage (especially out-of-sample) to the unchosen message models.[25]

The unchosen message model estimates for Game 5 are presented in Table 8. As with Game 3, the behavior of both of Game 5's alternative models is similar to EWA, and very similar to each other. This again suggests that if there is some significant pattern in the data not captured by EWA, these models capture it in the same way. Table 6 shows, however, that the improvement in fit from modeling the unchosen messages is barely worth the extra degrees of freedom. AIC favors the unchosen message models, but BIC does not, and the out-of-sample statistics are similar for EWA and the unchosen message models.

6 Discussion

Our first objective in this paper was to replicate Brandts and Holt's results. We closely replicated their results. However, the additional periods we ran demonstrated that the convergence in Game 5 is slower than expected, and even 64 (or 128) periods is not enough to converge to equilibrium.

Using these data, we tested our adaptation of EWA to signalling games. The baseline model updates the attractions to a sender's unrealized *type*. This allows the sender to make all valid inferences given that receivers are playing message-contingent strategies. This model performed significantly better than its choice reinforcement and belief-based special cases. The belief-based case is of particular interest because it formalizes the BH dynamic. Our results indicate that while the BH dynamic captures the direction of the frequency trends, the formal belief-based restrictions underestimate the speed of learning.

Although EWA performs better than its special cases, it may also be that EWA itself is too simple. Looking at the results from both games, updating unchosen

[25] While the frequencies of $m_2|t_2$ are small, they can have a large impact on estimation (particularly when log likelihood is the fit measure). Because the logarithms of small positive numbers can be hugely negative, it makes a big difference whether a model predicts that a rare event is impossible, or just very unlikely.

Table 8. Parameter estimates for Game 5 (underlined values are fixed for identification or for model restrictions and bootstrapped 95% confidence intervals are in parentheses)

	EWA	Convex	Soph
δ	0.54 (0.45,0.63)	0.51 (0.49,0.52)	0.50 (0.48,0.54)
ϕ	0.65 (0.59,0.71)	0.70 (0.69,0.72)	0.70 (0.68,0.71)
ρ	0.46 (0.39,0.54)	0.86 (0.86,0.87)	0.84 (0.84,0.85)
λ	0.09 (0.07,0.11)	0.28 (0.26,0.32)	0.25 (0.23,0.30)
τ		0.57 (0.56,0.58)	0.48 (0.45,0.49)
μ_1		0.05 (0.04,0.05)	0.00 (0.00,0.01)
μ_2		0.22 (0.20,0.23)	0.14 (0.09,0.14)
α		0.00 (0.00,0.00)	
Sender			
$A_S^{m_1 t_1}(0)$	18.25 (18.25,18.26)	23.83 (23.83,23.83)	23.30 (23.30,23.30)
$A_S^{m_2 t_1}(0)$	<u>30.00</u>	<u>30.00</u>	<u>30.00</u>
$N_S^{m_1}(0)$	0.62 (0.59,0.66)	5.01 (5.01,5.01)	4.36 (4.36,4.36)
$A_S^{m_1 t_2}(0)$	<u>30.00</u>	<u>30.00</u>	<u>30.00</u>
$A_S^{m_2 t_2}(0)$	11.34 (11.34,11.34)	24.41 (24.41,24.41)	23.79 (23.79,23.79)
$N_S^{m_2}(0)$	3.37 (3.37,3.37)	7.09 (7.09,7.09)	6.32 (6.32,6.32)
Receiver			
$A_R^{a_1 m_1}(0)$	<u>30.00</u>	<u>30.00</u>	<u>30.00</u>
$A_R^{a_2 m_1}(0)$	37.26 (37.26,37.26)	31.87 (31.87,31.87)	32.09 (32.09,32.09)
$A_R^{a_3 m_1}(0)$	0.00 (0.00,0.01)	0.00 (0.00,0.00)	0.00 (0.00,0.01)
$N_R^{m_1}(0)$	0.62 (0.59,0.66)	5.01 (5.01,5.01)	4.36 (4.36,4.36)
$A_R^{a_1 m_2}(0)$	41.88 (41.88,41.88)	23.77 (23.77,23.77)	24.88 (24.88,24.88)
$A_R^{a_2 m_2}(0)$	<u>15.00</u>	<u>15.00</u>	<u>15.00</u>
$A_R^{a_3 m_2}(0)$	43.26 (43.26,43.26)	24.06 (24.06,24.06)	25.17 (25.16,25.17)
$N_R^{m_2}(0)$	3.37 (3.37,3.37)	7.09 (7.09,7.09)	6.32 (6.32,6.32)

messages does improve upon EWA's ability to fit the data and to predict out of sample. In developing the alternative models, we expected to capture a few specific features of the subjects' learning process. One such feature is the relative size of imagined experience, represented by τ. Since the unchosen message is updated, it is necessary to update its experience count as well. Because this experience is a result of the learner's conjecture, we hypothesized it is less valuable than actual experience. This was weakly supported, as τ is about one in Game 3 and about 0.5 in Game 5.

A second feature we hoped to capture was the imagination coefficient on the realized-type, unchosen-message payoff and unrealized-type, unchosen-message payoff. We expected them to have a multiplicative effect: μ_1 requires only one level of counterfactual reasoning, but μ_2 requires two, suggesting μ_2 would be on the order of $\mu_1 \cdot \delta$. This expectation is not realized in our estimates, however, as μ_1 is zero in both games and μ_2 is greater than zero. This result is surprising because it implies that imagination is not necessarily nested: senders will go through two counterfactuals without learning from one. It may also suggest that updating same-type, different-message attractions does not allow quick-enough convergence to type-conditional messages.

Finally, we hoped to gain some insight into how subjects reinforce unchosen messages. The two unchosen message models we examine produce essentially similar fits on the two games we have examined. Because of its extra parameter, we conclude the convex combination payoff model is inferior to the sophisticated payoff model (indeed its AIC and BIC are higher for both games). However, the models are so similar that it is difficult to conclude the intuition behind the mirror sophistication model is more compelling than that of the convex combination model. Thus, while we have been able to determine that senders do update the unchosen message attractions, we have not been successful in explaining what determines the value added to the attractions of the unchosen message. Based on these results, one might use either of the unchosen message models and expect to do adequately.

Using formal learning models can provide insight into how unintuitive equilibria might arise in natural markets. One example of historical convergence to an unintuitive equilibrium is dividend policy of firms (see Bhattacharya, 1979). From a tax point of view, firms should not pay dividends because they are taxed as regular income of investors; if the firms' cash were instead reinvested, the result would be higher investor capital gains, which are taxed at a lower rate. So why do firms pay dividends? Suppose there are two types of firms: low-quality ones, which do not always have enough cash to meet a regular dividend payment (and cannot borrow to finance it), and high-quality ones which have plenty of cash and good business prospects. Decades ago, security analysts were less able to learn about a firm's financial health from accounting data and company sources than they are today; and struggling firms were less able to borrow. In this era, regular dividend payments signal a firm's financial health: low-quality firms cannot afford to commit to dividends (and often miss regular payments), but high-quality firms can. This part of corporate history corresponds to a temporary separating-equilibrium phase in which low- and high-quality firms are distinguished by their dividend policies.

But as firms realized how important dividends are, and credit markets developed, low-quality firms soon realized they had to pay dividends (or else reveal their type) and could borrow to do so. So a pooling equilibrium emerged in which all firms paid dividends. However, this equilibrium is unintuitive if high-quality firms have good investment opportunities and would prefer to plow dividend payments into those investments. If security analysts have the (newly-developed) capacity to guess a firm's investment prospects, these high-quality firms could conceivably benefit from cutting the dividend (if the capital markets interpreted this as a signal of having good opportunities). However, low-quality firms cannot benefit as much (if security analysts can see they have few good opportunities). Thus, firms may be 'forced' to continue to pay dividends if they think capital markets will interpret a cut as a signal of low-quality—since only low-quality firms did not pay dividends in the past. In this analytical narrative, dividend policy is an unintuitive equilibrium which emerged because the trace of the past, in which only low-quality firms did not pay dividends, inhibits high-quality firms from breaking the pooling equilibrium (even though their perception of the capital markets' likely reaction does not obey the intuitive criterion).

Although we cannot identify the exact form of unchosen message updating, we have replicated earlier results that empirical histories which conflict with logical refinements can be generated, and that these histories interfere with convergence to more refined equilibria. The Brandts-Holt dynamic provides the intuition for how this conflict might arise. Adding the flexibility of EWA improves our understanding (and predictive accuracy) considerably. Therefore, carefully specified formal learning models provide insight into how agents combine information about the history of play with the payoff table to make strategic decisions and help us to understand when play might be inconsistent with logical refinements.

References

Anderson, C.M., Camerer, C.F.: Experience-weighted attraction learning in sender-receiver signaling games. California Institute of Technology Social Science working paper number 1058 (1999)

Aptech: Constrained maximum likelihood (1995)

Banks, J., Camerer, C.F., Porter, D.: An experimental analysis of Nash refinements in signaling games. Games and Economic Behavior **6**(1), 1–31 (1994)

Bhattacharya, S.: Imperfect information, divided policy and 'the bird in the hand' fallacy. Bell Journal of Economics **10**, 259–270 (1979)

Binmore, K., Shaked, A., Sutton, J.: Testing noncooperative bargaining theory: A preliminary study. American Economic Review **75**, 1178–1180 (1985)

Brandts, J., Holt, C.: An experimental test of equilibrium dominance in signaling games. American Economic Review **82**(5), 1350–1365 (1992)

Brandts, J., Holt, C.: Adjustment patterns and equilibrium selection in experimental signaling games. International Journal of Game Theory **22**(3), 279–302 (1993)

Brandts, J., Holt, C.: Naive bayesian learning and adjustment to equilibrium in signaling games. University of Virginia Department of Economics working paper (1994)

Brown, G.: Iterative solution of games by fictitious play. In: Activity, analysis of production and allocation. New York: Wiley 1951

Camerer, C.F., Ho, T., Chong, K.: Sophisticated learning and strategic teaching in repeated games. Caltech working paper (2000)

Camerer, C.F., Hsia, D., Ho, T.: Ewa learning in coordination games and bilateral call markets. Caltech working paper (2000)

Camerer, C.F., Ho, T.: EWA learning in games: Heterogeneity, time-variation, and probability form. Journal of Mathematical Psychology **42**, 305–326 (1998)

Camerer, C.F., Ho, T.: Experience-weighted attraction learning in games: Estimates from weak-link games. In: Budescu, D., Erev, I., Zwick, R. (eds.) Games and human behavior, pp. 31–52. Mahwah, NJ: Lawrence Erlbaum 1999a

Camerer, C.F., Ho, T.: Experience-weighted attraction learning in normal form games. Econometrica **67**(3), 827–874 (1999b)

Camerer, C.F., Ho, T., Wang, X.: Heterogeneous EWA learning with payoff learning in centipede games. California Institute of Technology Social Science working paper (1999)

Carlin, B., Louis, T.: Bayes and empirical Bayes methods for data analysis. New York: Chapman and Hall 1996

Cho, K., Kreps, D.: Signaling games and stable equilibria. Quarterly Journal of Economics **102**(2), 179–221 (1987)

Cooper, D., Garvin, S., Kagel, J.: Signalling and adaptive learning in an entry limit pricing game. RAND Journal of Economics **28**(4), 662–683 (1997a)

Cooper, D., Garvin, S., Kagel, J.: Adaptive learning vs. equilibrium refinements in an entry limit pricing game. Economic Journal **107**(442), 553–575 (1997b)

Costa-Gomes, M., Crawford, V., Broseta, B.: Cognition and behavior in normal-form games. Econometrica (forthcomming)

Cournot, A.: Researches in the mathematical principles of the theory of wealth (translated by Bacon, N.). London: Haffner 1960

Efron, B., Tibshirani, R.: An introduction to the bootstrap. New York: Chapman and Hall 1993

Erev, I., Roth, A.: Predicting how people play games: Reinforcement learning in experimental games with unique, mixed strategy equilibria. American Economic Review **88**, 848–881 (1998)

Fudenberg, D., Levine, D.: The theory of learning in games. Cambridge, MA: MIT Press 1998

Gibbons, R.: Applied game theory for economists. Princeton, NJ: Princeton University Press 1992

Harley, C.: Learning the evolutionarily stable strategy. Journal of Theoretical Biology **89**, 611–633 (1981)

Hsia, D.: Learning in call markets. USC Department Economics (1998)

Morgan, J., Sefton, M.: An experimental investigation of unprofitable games. Princeton University Woodrow Wilson School manuscript (1998)

Roth, A., Erev, I.: Learning in extensive-form games: Experimental data and simple dynamic models in the intermediate term. Games and Economic Behavior **8**, 164–212 (1995)

Roth, A., Erev, I., Slonim, R., Barron, G.: Equilibrium and learning in economic environments. The predictive value of approximations. Harvard University working paper (2000)

Samuelson, L.: Analogies, adaptations, and anomalies. University of Wisconsin working paper (2000)

Tirole, J.: The theory of industrial organization. Cambridge, MA: MIT Press 1988

Vriend, N.: Will reasoning improve learning? Economic Letters **55**, 9–18 (1997)

Weber, R.: Learning without feedback in beauty contests. Carnegie Melon University working paper (2000)

Weiszacker, G.: Ignoring the rationality of others: Evidence from experimental normal-form games. Harvard Business School working paper (2000)

Theoretically robust but empirically invalid? An experimental investigation into tax equivalence*

Rudolf Kerschbamer and Georg Kirchsteiger

Department of Economics, University of Vienna, Hohenstaufeng.9, 1010 Vienna, AUSTRIA
(e-mails:rudolf.kerschbamer@univie.ac.at; georg.kirchsteiger@univie.ac.at)

Summary. The idea that the final distribution of the tax burden (economic incidence) does not depend on the initial distribution of tax liabilities (statutory incidence) is referred to as the Liability Side Equivalence principle. This paper tests this principle in the laboratory and finds that subjects who actually have to pay the tax carry a higher tax burden. It is argued that this violation of Liability Side Equivalence is due to the fact that a change in the distribution of tax liabilities induces a shift in behaviorally relevant social norms. This shift, in turn, affects the impact of the tax. Our results explain some striking empirical observations and have important theoretical and practical implications.

Keywords and Phrases: Tax equivalence, Tax incidence, Social norms.

JEL Classification Numbers: H21, H22, H30, C91, C92.

1 Introduction

"...it is a matter of indifference whether a general tax on transactions is assessed on the seller's or on the buyer's side of the market" (Musgrave 1959, p. 351). This principle, which we refer to as Liability Side Equivalence (LSE), underlies virtually any theoretical and empirical investigation on the effects of taxes. LSE is theoretically very robust. It holds independently of the nature of the transaction upon which the tax is imposed and independently of the framework in which the

* We are indebted to Gerhard Clemenz, Uwe Dulleck, Simon Gächter, Stefan Krasa and Clemens Puppe for helpful comments. Special thanks for useful suggestions are due to an anonymous referee and to the editor of this special issue. We also gratefully acknowledge financial support from the Austrian Chamber of Labor. Of course, the usual disclaimer applies.

transaction is conducted (unilateral or bilateral monopoly, oligopoly, or perfect competition), provided that prices can adjust and that the behavior of the parties in the transaction depends on net payoffs only.[1,2]

Despite its theoretical robustness, policy makers and other economic agents do not act in accordance with the normative implications of LSE. For example, if LSE holds, the policy decision about the legal distribution of tax liabilities should be guided exclusively by the objective to minimize administration and compliance costs. Minimization of these costs would, in general, require imposing the tax liability on one side of the transaction only. Nevertheless, most tax codes distinguish between an employer's and an employee's contribution to the social security/payroll tax. Similarly, reductions of labor taxes imposed on producers' side are frequently discussed as remedies against unemployment. At the same time taxes on the same factor inserted on the household side are rarely mentioned.[3] Apparently, public policy discussions proceed on the assumption that the former are mainly borne by producers while the latter impose few if any burden on them – a belief that is obviously inconsistent with LSE.

The discrepancy between economic theory on the one side and actual tax policy and public discussion of policy issues on the other raises the question of LSE's empirical validity. An empirical test of LSE with field data seems difficult since in the real world economic variables typically depend on a variety of continuously changing factors and not only on taxes.[4] An alternative way of testing LSE are experiments in the laboratory. This method has the advantage that the observations are made under circumstances controlled by the researcher.

In perfectly competitive markets with price taking agents, there is little freedom in price determination. It is therefore not surprising that the policy discussions inconsistent with LSE concentrate on markets characterized by frictions, heterogeneous goods and bargaining. One possible reason for a violation of LSE in such markets is price rigidity. For example, if after a change in tax code the entire social security tax has to be paid by the employers, it is unlikely that gross wages would fall immediately. The labor market would be out of equilibrium in the short run and firms would have to bear a larger share of the tax burden as before. However, even in labor markets, markets with pronounced institutional impediments to price flexibility, wages are adjusted quite regularly. Hence, wage

[1] Cf., for example, Musgrave (1959), Kotlikoff and Summers (1987), and Pisauro (1991).

[2] Several other equivalencies between taxes (e.g. the equivalence between a uniform tax on wages and a uniform tax on consumption, or the equivalence between a uniform tax on income and a uniform tax on output) have been recognized in the public finance literature. These equivalencies hold, however, only under fairly restrictive conditions and are not the focus of our paper.

[3] See e.g. the " Programm für mehr Wachstum und Beschäftigung" (program for more growth and employment) of the German Government (1996).

[4] Cf., however, Lockwood and Manning (1993). These authors analyze the effects of a non-linear tax system for wage bargaining in unionized economies and then test their theoretical predictions empirically. In the empirical investigation they find that income taxes (incident on households) and payroll taxes (incident on firms) have very different long run effects. They argue that "...this presents something of a problem, as any theory based on maximization behavior...would suggest that the formal incidence of a tax should be irrelevant in the long run" (p. 20), and attribute this result to either data problems or model misspecification.

rigidity alone can cause only short run violations of LSE. In addition, changes in thc tax system can be foreseen very often by both, trade unions and employers. Thus, wages could be adjusted accordingly even in advance.

An alternative explanation for violations of LSE in not perfectly competitive markets starts from the assumption that the behavior of economic agents is not entirely based on their net gains. For example, it seems quite plausible that the legal obligation to pay a tax is regarded as a moral obligation to bear it (to a certain degree) also economically. Whenever this is the case, individual behavior is affected by gross gains and a shift in the legal liability to pay a tax may influence its economic outcome.

The present paper tests this second explanation in the laboratory. As mentioned earlier, violations of LSE seem to be observed in markets that are not perfectly competitive but rather characterized by bargaining. A very simple bargaining situation, which is also easy to implement in an experiment, is the one stage ultimatum game. In this game one person, the proposer, makes an offer about the division of a given amount of money (the "cake") between her and her partner, the responder. The responder may accept or reject this offer. If he accepts, the proposer receives the cake minus the offer as gross earnings. The responder's gross earnings are the offer. If the offer is rejected both players receive nothing.

Assuming rational subjects whose decisions are based on their own net earnings only, the subgame perfect equilibrium of this game is straightforward. Since in case of rejection the responder earns nothing, he is prepared to accept any offer that gives him a nonnegative gain. Knowing this, the proposer will offer zero, and this offer will be accepted by the responder. Thus, in subgame perfect equilibrium the proposer receives the whole cake and the responder earns nothing.[5] It is well known that in one stage ultimatum experiments subjects do not behave in accordance with the subgame perfect equilibrium [see Güth et al. (1982), and Roth et al. (1991), among many others; for an overview see Camerer and Thaler (1995), or Güth (1995)]. The net offers differ substantially from zero, and low but positive offers are frequently rejected. Most authors explain these results by the impact of social norms (or manners) on individual behavior. Subjects have a propensity to punish those who offer them an amount that is, according to their fairness norm, too low, even if punishment is costly for them [see e.g. Bolton (1991), Kirchsteiger (1994), or Camerer and Thaler (1995)].[6] This implies that there exists a threshold such that the offer is rejected if it is below this threshold.

[5] When the proposer can choose his offer continuously then this is the unique sugame perfect Nash equilibrium. Whit a discrete stategy space there exists a second subgame perfect equilibrium in which the responder accepts any strictly positive offer and rejects zero, and in which the proposer offers the smallest monetary unit allowed.

[6] Other explanations are adaptive learning (see Roth and Erev, 1995; Gale et al., 1995) and the "lack of anonymity" hypothesis, i.e. the assertion that the very act of observing behavior by the experimenter influences play away from the subgame perfect equilibrium toward a "fair" allocation (see Hoffman et al., 1994). Abbink et al. (1996) designed an experiment to distinguish between adaptive learning and punishment hypothesis. They found much more evidence for the latter than for the former. Bolton and Zwick (1995) conducted experiments that discriminate between the anonymity and the punishment explanation. They present evidence supporting the conclusion that the punishment

In order to test for LSE, we modify the simple ultimatum game such that in case of acceptance one of the partners has to pay a tax from her/his gross earnings. In half of the sessions the proposer has to pay it, in the other half the responder. In each session subjects play the simple ultimatum game several times (several "rounds") with different partners. We conduct our experiment in two versions. In the "Non Insurance" ("N") version the subjects are divided equally into the groups P (for proposer) and R (for responder), and each subject gets a partner each round. In this version we give subjects no justification for the collection of taxes. In the "Insurance" ("I") version the R group consists of two subjects more than the P group. Hence, two responders have no partner in each round. These subjects are not allowed to participate in the respective round and they get a monetary compensation for that. Now the collection of taxes is motivated with the argument that funds are needed to finance the compensations to inactive players – in the Insurance version the responders were "insured" against not being matched.

Assuming rational individuals whose decisions depend only on their own net earnings the subgame perfect equilibrium prediction of all designs of our experiment is that the responder earns nothing and the proposer receives the cake minus the tax, regardless of upon whom the tax is levied.[7] Hence, LSE should hold if subjects are fully rational and egoistic. On the other hand, if there exists a threshold such that offers below this threshold are rejected, the validity of LSE depends crucially on the nature of the social norm underlying the threshold. As long as this threshold (measured in net earnings) remains unaffected by a change in the statutory distribution of tax liabilities, LSE should hold. If, however, the legal obligation to pay a tax is regarded as a moral obligation to bear it, the threshold should change with a change in the statutory distribution of tax liabilities and LSE should be violated.

We found that LSE is violated in both versions of our experiment. Subjects have, on average, strictly higher after-tax earnings if their partners have to pay the tax than if they have to pay it themselves. Thus, the final distribution of the tax burden depends on the initial distribution of tax liabilities. The reason for conducting the Insurance version was the following. If LSE is violated and if this is due to gross gain based social norms, we thought that the violation of LSE should be more pronounced when people think that there is a plausible reason for taxation. It turned out that the increase in the responders' offered after-tax earnings when the tax liability was shifted from them to the proposers was on average higher in the I than in the N version of the experiment. A statistical test revealed, however, that the difference between the two versions is not significant.

hypothesis explains much more of the deviation from subgame perfect equilibrium than does the anonymity hypothesis.

[7] Since the subjects in the experiment have a discrete strategy space there is actually a second subgame perfect Nash equilibrium outcome in which the responder earns the smallest monetary unit allowed and the proposer the rest (see Footnote 5 above).

The rest of the paper is organized as follows. The next section describes the experimental design. In Section 3 we present the results, in Section 4 we interpret them. Section 5 contains some concluding remarks.

2 Experimental procedures

On the whole we conducted 20 experimental sessions. For each session we recruited – depending on the experimental design (for details see below) – either 10 or 12 prospective subjects. Participants obtained a show up fee of 100 Austrian Shillings (ATS).[8] In addition they received all earnings resulting from their decisions during the experiment. They were paid in cash at the end of the respective session. To rule out interdependencies between different sessions subjects were allowed to participate in a single session only.

At the beginning of each session subjects were randomly allocated to one of two groups. One group was designated to take the role of the proposer, the other group took the role of the responder. Each group was placed in a separate room with one of the experimenters being the only other person present. A total of four to six rounds of the following simple game was played: At the outset of each round the experimenters formed pairs consisting of one proposer and one responder each. To rule out the possibility of reputation formation and of rewarding or punishing a subject's previous behavior, each player obtained a new partner in each round and the identity of the partners was kept secret. Each proposer was asked to make a written proposal on how to divide an amount of 70 ATS between her and her partner, i. e., the responder in the respective pair.[9] The amount of 70 ATS remained the same in all sessions, in all rounds and for all pairs and was common knowledge among subjects. The division proposals ("offers") of all proposers were collected by the experimenter and transmitted to the other experimenter via phone. To rule out any kind of group pressure or herd behavior the transmission was conducted in codified form so that other proposers did not know a given proposer's offer. This codification changed in each round. After having decodified the transmitted offers the experimenter in the responders' room forwarded them to the respective responder. The identity of the person who made the offer was not revealed. Then each responder was asked to decide whether to accept this offer or not. The experimenter in the responders' room collected the acceptance/rejection decisions and transmitted them in codified form via phone to the other experimenter. The latter decodified and forwarded this information in written form to the respective proposer. This concluded the respective round and a new round began or the session was over.

The basic game just described was played in four different designs (see Table 1). Each design was implemented in (all rounds of) 5 of the 20 sessions.

[8] When the experiments were conducted 10 ATS were about 1 US$.

[9] In conducting our experiment we tried to avoid the use of suggestive language that could have an influence on subjects' behavior. For example, we called the partners "Person X" and "Person Y" (and not, for instance, "employer" and "employee"), the offers "division proposals" (and not, e. g., "wage offers"), the payoffs "earnings" (and not, e. g., "profits" and "incomes").....

Table 1. Different designs

	Non insurance version	Insurance version
Tax paid by P	NP	IP
Tax paid by R	NR	IR

Non insurance version: The subjects are divided equally into the groups P and R and each subjet gets a partner each round. Insurance version: The group R consists of two more subjects than the group P: two subjects R remain unmatched; unmatched Rs get a lump sum payment ("insurance" against not being matched).

With respect to the number of participating responders the four designs can be divided in two design-pairs ("versions"). For the **Non Insurance (N)** version of the experiment we recruited 12 students per session. The subjects who showed up on time were divided evenly into the groups P and R, i. e., in this version there were as many proposers as responders.[10] The earnings of the players were determined as follows: If the offer was accepted the gross earnings of the proposer were given by the 70 ATS minus the offer. The gross earnings of the responder were just the offer. If the offer was rejected, both partners received nothing and the 70 ATS expired. In case of acceptance (and only in this case) one of the partners had to pay a tax of 20 ATS from his gross earnings. In 5 of the 10 N sessions the proposer had to pay the tax (we denote this design in what follows by **NP**), in the other 5 N sessions the tax was imposed on the responder (design **NR**).[11] The players got neither a justification for the collection of the tax nor for the choice of the party on which the tax was imposed.

For the **"Insurance" (I)** version of the experiment we recruited 10 students per session. The subjects who showed up on time were again divided into the groups P and R. In contrast to the N version of the experiment the participants were not divided evenly but in such a way that group R had two members more than group P. Hence, in each round two responders were ignored in the process of forming pairs. These two players remained without partner and could not participate in the respective round of the session. The pair formation process was designed in such a way that (i) each active player received a new partner in each round and (ii) each responder remained unmatched for exactly two rounds. Unmatched responders obtained a compensation of 20 ATS in the respective round. The gross earnings of the active players were determined exactly as in the N version of the experiment. In the case of acceptance (and only in this

[10] Unfotunately some subjects who had signed up for the expiriment did not show up on time. If an odd number of subjects came to a session one person was chosen randomly. This person got the show-up fee of 100 ATS but was not allowed to participate.

[11] To rule out uncontrollable side effects caused by queer subjects the range of possible offers in each design was constrained in such a way that each of the two partners in a pair could potentially get net earnings between 0 and 50 ATS. That is, each proposer was asked to choose an offer between 0 and 50 ATS in design NP and an offer between 20 and 70 ATS in design NR. In the experiment no proposer ever tried to make a forbidden offer.

case) one of the partners in each pair had to pay a tax of 20 ATS from his gross earnings. In (all rounds of) 5 of the 10 I sessions the proposer had to pay the tax (we denote this design by **IP**), in the other 5 I sessions the tax was imposed on the responder (design **IR**). The collection of taxes was now motivated with the argument that funds to finance the compensations to the inactive players were needed. Notice, however, that the payments to unmatched responders were pure lump-sum, i.e. did not depend on any decision made by any subject.[12] As in the N version of the experiment the subjects got no motivation for the choice of the "side of the market" on which the tax was imposed.

To make sure that all subjects had effectively understood the rules of the respective session they were asked to compute their own gains and the gains of their partners in four hypothetical examples. After all subjects had solved these examples correctly, and after the experimenters had answered all remaining questions the experiment started with a trial round (i.e., a round without monetary consequences for the players). Then the first round of the "real" game in the respective session began. The number of rounds was such that each person played once with each member of the other group.

3 Experimental results

As described above each design was implemented in 5 of the 20 sessions. Each session lasted about 45 minutes. Each set of students participated in a single session only. Most of the subjects were students of law, some were students of computer sciences and a few were students of psychology. The subjects were comparable in age and education to undergraduates. None of them was a student of the economics department.

In total 48 subjects participated in the NP, 52 subjects in the NR, 46 subjects in the IP and 44 subjects in the IR sessions.[13] Subjects made 116 offers in the NP, 138 offers in the NR, 102 offers in the IP and 93 offers in IR sessions. In the NP sessions 24 offers, in the NR sessions 33 offers, in the IP sessions 22 offers and in the IR sessions 15 offers were rejected. Subjects earned on average 189 ATS (about US$ 19) per session.[14]

[12] We decided against making the transfers of unmatched responders dependent on the decisions of matched subjects for the following reason: Since players in ultimatum games seem to be motivated not only by their own monetary (net) payoff but also by their relative payoff standing, this would have introduced additional effects that are absent in the N version of the experiment. (For instance, the desire to help unmatched subjects could have induced a matched responder to accept an offer that otherwise had been rejected.) This would have rendered the interpretation of any differences in the results between the two versions rather difficult. For example, if the violation of LSE turned out to be more pronounced in the I than in the N version of the experiment, it would not be clear whether the reason is the justification of the tax or the additional distributional concerns.

[13] As mentioned before, in some sessions not all subjects enrolled for the experiment actually showed up. For example, in none of the NP sessions more than 10 subjects arrived at the agreed upon date. Hence, in this design a maximum of 5 rounds was played per session.

[14] Note that this amount exceeds by far the hourly net wage an average Austrian student could earn in his best alternative job (about 80 ATS).

The mean offered net-income for the responder (i.e., offer net of taxes) over all sessions was 19.3, i.e. a little less than 40% of the net cake. This replicates roughly the results of previous studies (see Camerer and Thaler, 1995). However, the offers differed in the different designs as our first result shows. In this result reference is made to the term *net offer.* By net offer we mean the after-tax earnings of the responder implied by a given offer. That is, the net offer equals the actual offer in the NP and the IP design, while it is the offer minus the tax in the NR and the IR design.

Result 1. *In both versions of our experiment the net offers were, on average, higher if the tax was imposed on the proposer rather than on the responder.*

Figure 1 presents evidence for this finding. The figure compares the distributions of all net offers in the two designs belonging to the same (i.e., N or I) version of the experiment.

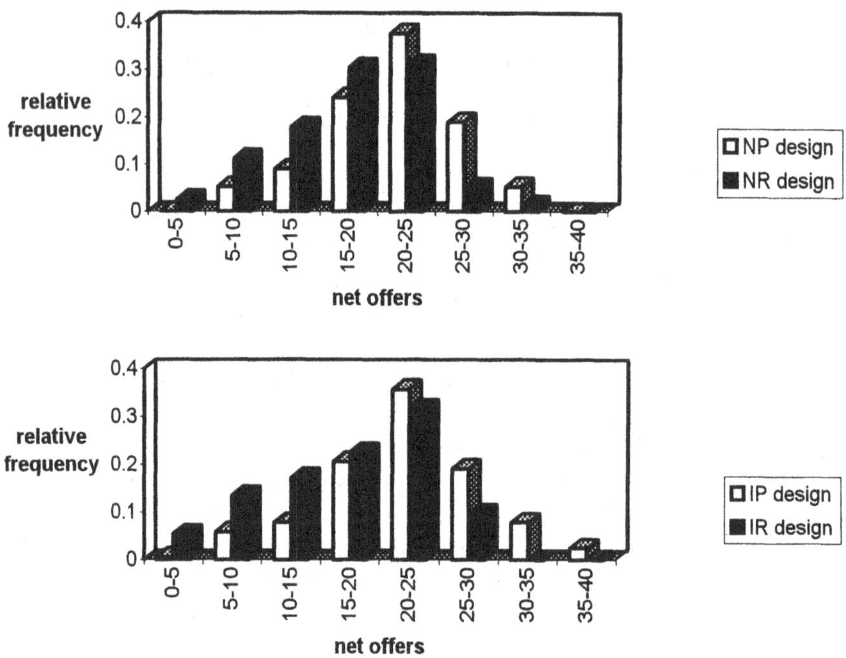

Figure 1. Distribution of net offers

It is obvious from Figure 1 that the relative frequency of low net offers (offers less than 20) was higher in the NR design (IR design) than in the NP design (IP design), whereas high net offers were made more often in the NP (IP) than in the NR design (IR design). Hence it was, on average, always better for a responder if the tax was imposed on the proposer rather than on the responder.

The same conclusion can be drawn from Table 2. In the N version of our experiment the average net offer was 21.1 ATS if the proposer had to pay the tax, whereas the net offer was only 17.6 ATS if the responder had to pay the tax, implying a difference of 3.5 ATS, i.e. 17.5% of the tax. For net earnings this implies that the responder's offered net income increased about 20 percent when the tax liability was shifted from him to the proposer. The violation of LSE was even more pronounced in the I version of the experiment. Here the average difference in net offers was 4.8 ATS, i.e. 24% of the tax. This implies that in the I version the responder's offered net income increased by about 25% if the tax liability was shifted from him to his partner.[15]

Table 2. Average net offers

Design	All net offers	Last round net offers
NP	21.1	21.4
NR	17.6	17.4
NP minus NR	3.5	4.0
IP	21.8	21.9
IR	17.0	13.9
IP minus IR	4.8	8.0

Next we tested the equivalence statistically. Since the same subjects participated in the different rounds of a given session individual observations are not independent from each other. Hence, only individual first round observations and mean offers of the different sessions were used for statistical tests. We first conducted a Wilcoxon-Mann-Whitney test with individual first round offers. For both versions the differences in net offers were statistically significant even at a 1% level.[16,17] Furthermore, we conducted a permutation test with average offers of each session as well as with the average last round offers.[18]

Table 3 shows the significance levels for the hypothesis that the session means of all net offers (or the session means of last round net offers, respectively) in the NP sessions (IP sessions) are equal to the session means of all net offers (session means of last round net offers) in the NR sessions (IR sessions). The alternative hypothesis is that the mean offers are higher in the NP sessions (IP sessions) than in the NR sessions (IR sessions). As can be seen from Table 3 the hypothesis that the session averages in the NP sessions are equal to those in the NR sessions can be rejected at the 2% confidence level. The same holds for comparisons within the I version of our experiment.

[15] We also conducted statistical tests on whether the violation of LSE is larger in the I than in the N version of the experiment. The difference turned out not to be significant, however.

[16] We also conducted a Kolmorogov- Smirnov test with the individual first round offers. The results are qualitatively the same as those of the Wilcoxon-Mann-Whitney test.

[17] We tested the equivalence also by using all individual observations. The results of these Wilcoxon-Mann-Whitney tests are qualitatively the same as those for the individual first round offers. However, as already mentioned, the individual observations were not independent from each other. Hence, tests using individual observations are not statistically valid.

[18] Since we have numerical data, the most powerful nonparametric test of location is the permutation test. However, this test is only feasible when the sample size is small. If we use all individual first round offers the sample size is too large to make a permutation test. Hence, in this case we have to conduct a Wilcoxon-Mann-Whitney test. On the other hand, the use of session means allows for the

Table 3. Results of the permutation tests

	p (NP vs. NR)	p (IP vs. IR)
Session mean of all net offers	0.012	0.012
Session mean of last round net offers	0.031	0.004

p (A vs. B): Significance level of the permutation test statistics with data of the A and the B sessions

The differences between the P and R designs within each version of the experiment did not vanish when the subjects got more experienced. On the contrary, the average differences in the net offers were even more pronounced in the last rounds (see Table 2). These last round differences were highly significant (see Table 3). Figure 2 shows the evolution of the differences in net offers. It is not clear whether these differences converged to any specific level, but obviously there was no tendency for the differences to vanish.

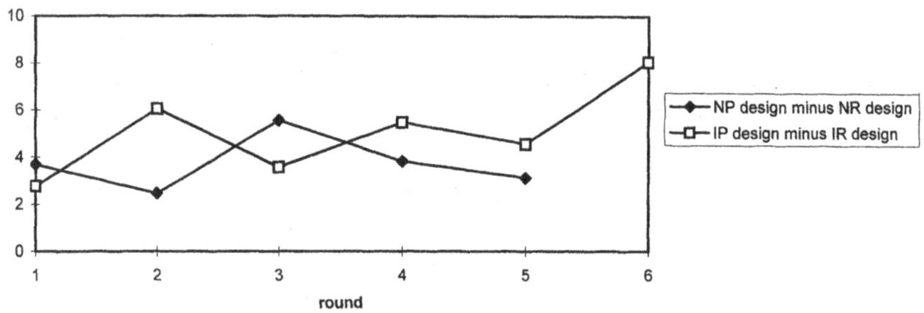

Figure 2. Differences between the mean net offers in different rounds

The failure of LSE is also reflected in the acceptance decisions of the responders, as our second result shows:

Result 2. *When controlling for the net amount offered the rejection rates were higher if the tax was imposed on the proposer rather than on the responder. This result holds for both, the Insurance and the Non Insurance version of the experiment.*

Figure 3 depicts the rejection rates following net offers of different levels. As can be seen from this figure, net offers of a given level were more often rejected if the tax was imposed on the proposer. In other words, if the responder had to pay the tax he was more willing to accept a given after-tax income than if the proposer had to pay it.[19] This holds for all levels of net offers and for both versions of the experiment. A similar conclusion can be also drawn from Table 4. This table shows the rejection rates of net offers below 20.

---

permutation test. For a discussion of the applicability of the differnt tests see Siegel and Castellan (1988, cap. 6).

[19] In the NP design no offer below 5 was ever made and in the NP as well as in the NR design all offers above 25 were accepted. In the Insurance version no offer above 30 was rejected.

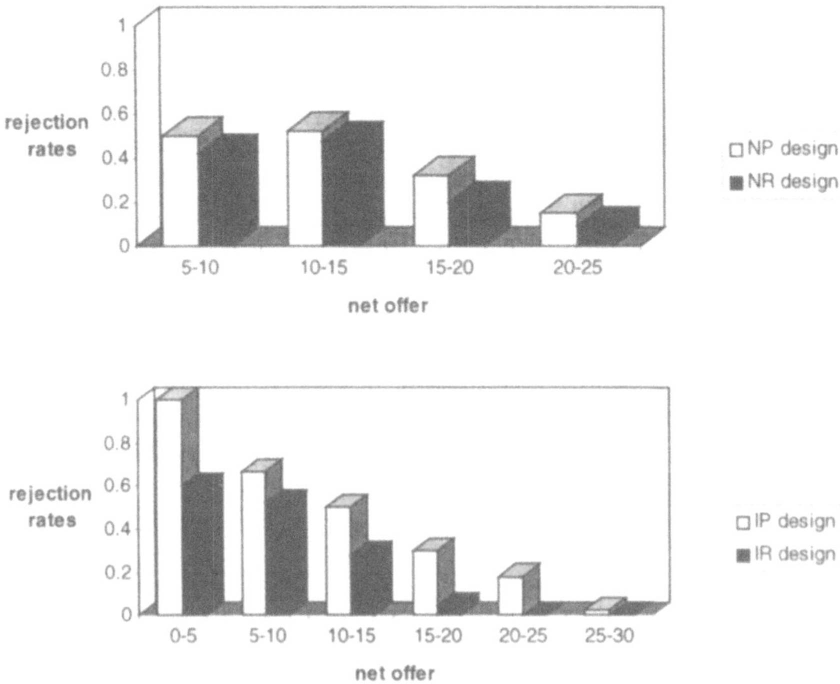

Figure 3. Rejection rates

Again these rates were higher when the tax was imposed on the proposer. To see whether these differences were significant we conducted a permutation test with the rejection rates of offers below 20 in each individual session. It turns out that the differences were significant at a 10% level in the Non Insurance version and at a 5% level in the Insurance version. Hence, we can conclude that the acceptance behavior of the responders – like the offers made by the proposers – contradicted LSE. We can therefore summarize our results to:

Statement: *Liability Side Equivalence held neither in the Non Insurance nor in the Insurance version of our experiment. In both versions subjects had, on average, strictly higher net earnings if the tax was imposed on their partners than if they had to pay the tax themselves.*

Table 4. Rejection rates of net offers below 20

NP	NR	IP	IR
0.55	0.43	0.57	0.36

4 Interpretation of the results

Having established the fact that LSE is violated in the laboratory, one has to ask for the reasons for this result. One possible explanation is, of course, money illusion. If subjects suffer from money illusion they accept lower net offers in settings in which they are asked to pay a tax because they do not really realize that taxes reduce their net earnings. However, we do not believe that money illusion explains our results, since in our experiment (i) there was no time lag between disbursement of gross earnings and payment of taxes, and since (ii) subjects repeatedly revealed their ability to calculate net earnings of both parties correctly. We believe that the violation of LSE can better be explained by the influence of social norms on the behavior of players. Recall the punishment explanation for behavioral anomalies in ultimatum experiments without taxes. In this explanation people are willing to reject strictly positive offers below a certain threshold because they consider such offers as unfair. We think that a change in the distribution of statutory liabilities to pay a tax induces a shift in subjects' fairness perceptions. This shifts subjects' acceptance thresholds and thus leads to violations of LSE.[20,21]

To facilitate the understanding of this explanation for our results it might be instructive to construct a simple model that incorporates social norms and that is consistent with the violation of LSE. Consider a game with two players indexed by $i \in \{P, R\}$ and let x^i denote the monetary net payoff of, and t^i the tax paid by, player i. Assume that each player acts to maximize the expected value of his/her utility function

$$u^i = u^i\left(x^i, s^i\right), \tag{1}$$

where $s^i \equiv x^j - x^i + \lambda\left(t^j - t^i\right)$ and $\lambda \in [0, 1]$. We assume that

(i) u^i is continuously differentiable on the domain of $\left(x^i, s^i\right)$;
(ii) $u_1^i > 0$;
(iii) $u_2^i \leq 0$ if $s^i \geq 0$, and $u_2^i < u_1^i/2$ if $s^1 < 0$.

The function specified in (1) is a statement about the objectives that motivate behavior of economic agents. It tells us that the behavior of subjects might be driven not only by their own material wellbeing, but also by the extend to which their social norm concerning the distribution of payoffs is fulfilled. To evaluate the implications of (1) for the outcome of our ultimatum game let us start with the special case where $\lambda = 0$. In this case the social norm depends on net payoffs

[20] This explanation is, in most respects, empirically indistinguishable from money illusion. However, for most policy questions it is a matter of indifference which of these two explanations is the more appropriate one.

[21] As mentioned earlier there exist other explanations for the ultimatum game results, an important one being the adaptive learning hypothesis. These alternative explanations are incompatible with our results as long as they are based on the assumption that subjects' behavior depends only on net earnings. We regard it as implausible that learning dynamics depend on gross earnings. If, however, ultimatum experiment results are really caused by learning, our results show that learning behavior is (at least partly) based on gross gains and that LSE does not hold as long as the learning dynamics do not converge to the subgame perfect equilibrium.

only. If the net payoff of player j exceeds that of player i $\left(s^i > 0\right)$ then $u_2^i \leq 0$ measures the utility loss from being worse off: For a given own material net payoff x^i player i prefers an outcome with more equal material payoffs to one in which the inequality in his opponent's favor is more pronounced. For the case in which player i has a higher monetary payoff than player j $\left(s^i < 0\right)$ we allow a wide variety of possible preferences. Player i might be spiteful and prefer a further increase in inequality $\left(u_2^i < 0\right)$ as in Kirchsteiger (1994), he may be driven by fairness/equity considerations and prefer a more equal distribution $\left(u_2^i > 0\right)$ as in Bolton and Ockenfels (1999), Fehr, Kirchsteiger and Riedl (1998), and Fehr and Schmidt (1999), or he may not care about inequality in his own favor at all $\left(u_2^i = 0\right)$. All we require is that fairness considerations do not overcome the direct impact of the own material net payoff $\left(u_2^i < u_1^i/2\right)$.

Obviously, if $\lambda = 0$, LSE holds. Also notice that, if $u_2^i = 0$ for all s^i, we are back to the standard model in which players are only interested in their own material net payoffs.

Now suppose that the legal obligation to pay a tax is regarded as a moral obligation to bear it (to a certain degree) also economically ($\lambda > 0$). Also suppose that subjects exhibit a strict aversion against being worse off $\left(u_2^i < 0 \text{ if } s^i > 0\right)$. Then a player's wellbeing does not only depend on the distribution of net- but also on that of gross- earnings. (In the limit case of $\lambda = 1, s^i$ gives weight only to the distribution of gross-earnings). What does this imply for the outcome of our ultimatum game, where, in case of an agreement, the proposer has to pay a tax of t^P and the responder one of t^R? To answer this question let the proposer's preferences be represented by $u^P = u^P\left(x^P, s^P\right)$ and assume that the proposer does not exactly know the preferences $u^R = u^R\left(x^R, s^R\right)$ of a randomly matched responder, but only the distribution of responders' preferences. Denote by c the net cake to be divided, and assume that $t^i (i = P, R)$ is lower than c, as it is the case in our experiment. First notice that, if the players knew each others preferences, then we would never observe rejected offers. To see this, look at the allocation where, for given taxes, the difference in the weighted average of gross and net earnings is zero $\left(s^i = 0\right)$. For this allocation the utilities of proposer and responder are given by $u^P\left(\left(c + \lambda\left(t^R - t^P\right)\right)/2, 0\right)$ and $u^R\left(\left(c - \lambda\left(t^R - t^P\right)\right)/2, 0\right)$, respectively. Since the impact of the own monetary net payoff on utility is strictly positive, both players strictly prefer this allocation to the disagreement allocation which yields utility $u^i = u^i(0,0)$ for $i = P, R$.

Next notice that, again if players' preferences were common knowledge, a proposer would always offer the lowest amount accepted by the responder. This follows from $u_2^i < u_1^i/2$. By the same assumption, for every responder there exists a unique threshold such that he accepts any offer higher or equal, and rejects any offer lower, than this critical one. For the case where preferences are private information these two facts together imply that a proposer's behavior does not directly depend on the distribution of responders' preferences but only on the distribution of acceptance thresholds. Now consider a given responder R. Denote his acceptance threshold by $\underline{x}^R$. Since the responder accepts an offer only if the utility from acceptance is not lower than the disagreement utility, $\underline{x}^R$

solves

$$u^R(0,0) = u^R\left(x^R, c - 2x^R + \lambda\left(t^P - t^R\right)\right). \tag{2}$$

For $x^R = 0$ the RHS of (2) is smaller than $u^R(0,0)$. Also, for $x^R = \frac{c+\lambda(t^P - t^R)}{2}$ the RHS of (2) exceeds $u^R(0,0)$ as has been shown above. Therefore, and because u^R is continuous and strictly monotone in x^R, a unique solution of (2) exists, and this solution $\underline{x}^R$ is smaller than $\left(c + \lambda\left(t^P - t^R\right)\right)/2$. Thus, $s^R > 0$ and $u_2^R < 0$ at $\underline{x}^R$. Furthermore, $\underline{x}^R$ depends on λ, t^P and t^R. Applying the implicit function theorem reveals that

$$\begin{aligned} \partial \underline{x}^R / \partial t^P &= -\lambda u_2^R / \left(u_1^R - 2u_2^R\right) > 0; \\ \partial \underline{x}^R / \partial t^R &= \lambda u_2^R / \left(u_1^R - 2u_2^R\right) < 0. \end{aligned}$$

In other words, a shift of the formal obligation to pay a tax from the proposer to the responder decreases the responder's acceptance threshold. Since this is true for all responders in the population, the distribution of acceptance thresholds for the case where $t^P = \bar{t} > 0$ and $t^R = 0$ dominates that for the case where $t^P = 0$ and $t^R = \bar{t}$ in the sense of first order stochastic dominance. The rest is trivial: Higher acceptance thresholds in the population of responders if the tax is imposed on proposers induce proposers to make more generous net offers (our Result 1), and responders to reject net offers of a given level more frequently (our Result 2). This violates LSE. Furthermore, the violation of LSE increases in λ. This might explain why the departure from LSE seems to be more pronounced in the I than in the N version of our experiment: Subjects' respect for the statuary distribution of the tax liabilities may simply be intensified if they see a reason for collecting taxes.[22]

5 Concluding remarks

The idea that the final distribution of the tax burden (economic incidence) is independent from the initial distribution of tax liabilities (statutory incidence) is referred to as the Liability Side Equivalence (LSE) principle. This paper tested this principle in the laboratory and found that the equivalence was violated. In both versions of our experiment the after-tax earnings offered by the proposers to the responders (net offers) were, on average, higher if the tax was imposed on the proposer than if the responder had to pay it. Furthermore, the responder was more inclined to accept a net offer of a certain level if he had the legal obligation to pay the tax.

[22] This explanation is in line with the finding of Hoffman and Spitzer (1985) and Hoffman et al. (1994) that the results of bargaining depend crucially on the way the bargaining positions are allocated. If there is a justification for a certain allocation of the positions, e.g. if the winners of a quiz "earn" the position of proposers as in Hoffman et al. (1994), unequal divisions of the cake are much more often demanded and accepted. On the importance of entitlements see also Kahneman et al. (1986).

We have demonstrated that a simple model, based on the premise that people are motivated by both their pecuniary after-tax payoffs and social norms concerning the distribution of payoffs, can explain the violation of LSE, provided the legal obligation to pay a tax is regarded as a moral obligation to bear it. LSE relies on two crucial assumptions: (i) People are fully rational. (ii) People are only interested in after-tax earnings.[23] In principle one can give up either of these assumptions in an attempt to explain the evidence. In our model we have chosen to relax assumption (ii) because we belief that the ultimatum game is a relatively simple game and because subjects repeatedly revealed their ability to calculate the after-tax earnings of both parties in the game correctly.

Whether a shift of the legal liability to pay a tax can influence its economic outcome depends, of course, on the environment considered. If trade takes place on competitive markets and if the characteristics of the good traded are completely specified, prices and quantities converge to the market clearing level rather quickly.[24] It seems that in such environments market forces are very strong and overcome any kind of idiosyncrasies of individual behavior.[25] If, on the other hand, prices are determined at least partially by bargaining, the outcome can be shaped by social norms which, in turn, may be influenced by statutory tax incidence. Hence, in such markets there exists the possibility that the outcome is affected by behavior that is influenced by gross gains. An important example for such a market is the labor market. This might explain why there is so much discussion on the distribution of the statutory incidence of the payroll tax but at the same time scarcely a discussion on the statutory incidence of consumption taxes.

The most immediate normative implication of LSE is that the decision over the division of statutory liabilities to pay a tax on the two sides of the market should only be guided by the goal of minimizing administration and compliance costs. This paper's findings suggest that adhering unconditionally to that implication may lead to suboptimal results in markets in which social norms have an influence on equilibrium behavior of transactors: If in such markets a shift in statutory incidence changes transactors' behaviorally relevant norms then the usual trade-off between different basic objectives of tax policy – as e.g., the minimization of the tax induced dead weight loss, the achievement of an equitable distribution of the tax burden, high employment, etc. – must be made in designing the pattern of legal liabilities.

---

[23] LSE *does not* rest on the assumption that subjects are only interested in their *own* after-tax earnings (selfish behaviour), however.

[24] For an overview of market experiments see Davis and Holt (1993, chapters 3 and 4).

[25] Notice that the kind of motivation or utility function we introduced in Section 4 to explain the violation of LSE in our ultimatum game is consistent with the experimental fact that market games tend to converge to the competitive equilibrium. This has independently been shown by Bolton and Ockenfels (1999) and by Fehr and Schmidt (1999).

References

Abbink, K., Bolton, G., Sadrieh, A., Tang, F.: Adaptive learning versus punishment in ultimatum bargaining, Discussion Paper No. B-381, University of Bonn (1996)

Bolton, G.: A comparative model of bargaining: theory and evidence. American Economic Review **81**, 1096–1136 (1991)

Bolton, G., Ockenfels, A: A theory of equity, reciprocity and competition. American Economic Review (forthcoming) (1999)

Bolton, G., Zwick, R.: Anonymity versus punishment in ultimatum bargaining. Games and Economic Behavior **10**, 95-121 (1995)

Camerer, C., Thaler, R.: Anomalies: ultimatums, dictators and manners. Journal of Economic Perspectives **9**, 209–219 (1995)

Davis, D., Holt, C.: Experimental economics. Princeton, NJ: Princeton University Press 1993

Fehr, E., Schmidt, K.: A theory of fairness, competition and cooperation. Quarterly Journal of Economics (forthcoming) (1999)

Fehr, E., Kirchsteiger, G., Riedl, A.: Gift exchange and reciprocity in competitive experimental markets. European Economic Review **42**, 1–34 (1998)

Gale, J., Binmore, K., Samuelson, L.: Learning to be imperfect: the ultimatum game. Games and Economic Behavior **8**, 56–90 (1995)

German Government: Programm für mehr Wachstum und Beschäftigung. http://www.bundesregierung.de/inland/bpa/themen/t hx059602.htm. (1996)

Güth, W.: On ultimatum bargaining experiments – a personal review. Journal of Economic Behavior and Organisation **27**, 329-344 (1995)

Güth, W., Schmittberger, R., Schwarze, B.: An experimental analysis of ultimatum bargaining. Journal of Economic Behavior and Organisation **3**, 367–388 (1982)

Hoffman, E., Spitzer, M.: Entitlements, rights and fairness: an experimental examination of subjects' concepts of distributive justice. Journal of Legal Studies **15**, 254–297 (1985)

Hoffman, E., McCabe, K., Shachat, K., Smith, V.: Preferences, property rights and anonymity in bargaining games. Games and Economic Behavior **7**, 346–380 (1994)

Kahneman, D., Knetsch, J., Thaler, R.: Fairness as a constraint on profit seeking: entitlements in the market. American Economic Review **76**, 728–741 (1986)

Kirchsteiger, G.: The role of envy in ultimatum games. Journal of Economic Behavior and Organization **25**, 373–390 (1994)

Kotlikoff, L., Summers, L.: Tax incidence. In: Auerbach, A., Feldstein, M., (eds.) Handbook of public economics, Vol.**1**, pp. 1043–1092. Amsterdam: North Holland 1987

Lockwood, B., Manning, A.: Wage setting and the tax system: theory and evidence for the United Kingdom. Journal of Public Economics **52**, 1–29 (1993)

Musgrave, R.: The theory of public finance. New York: McGraw-Hill 1959

Pisauro, G.: The effect of taxes on labor in efficiency wage models. Journal of Public Economics **46**, 329–345 (1991)

Roth, A., Prasnikar, V., Okuno-Fujiwara, M., Zamir, S.: Bargaining and market behavior in Jerusalem, Ljubljana, Pittsburgh, and Tokyo: an experimental study. American Economic Review **81**, 1068–1095 (1991)

Roth, A., Erev, I.: Learning in extensive form games: experimental data and simple dynamic models in the intermediate term. Games and Economic Behavior **8**, 164–212 (1995)

Siegel, S., Castellan, J.: Nonparametric statistics for the behavioral sciences. Singapore: McGraw-Hill 1988

The manufacturer's authorised representative in the EU is Springer Nature Customer Service Centre GmbH, Europaplatz 3, 69115 Heidelberg, Germany. If you have any concerns regarding our products, please contact ProductSafety@springernature.com

Printed and bound by CPI Group (UK) Ltd, Croydon, CR0 4YY
25/09/2026
02222896-0004